T5-BAM-050

UNITARY SYMMETRY AND ELEMENTARY PARTICLES

UNITARY SYMMETRY AND ELEMENTARY PARTICLES

D. B. LICHTENBERG

Physics Department
Indiana University
Bloomington, Indiana

1970

ACADEMIC PRESS New York and London

ACADEMIC PRESS, INC.
111 Fifth Avenue, New York, New York 10003

United Kingdom Edition published by
ACADEMIC PRESS, INC. (LONDON) LTD.
Berkeley Square House, London W1X 6BA

LIBRARY OF CONGRESS CATALOG CARD NUMBER: 71-117114

PRINTED IN THE UNITED STATES OF AMERICA

CONTENTS

4. The Symmetric Group and Identical Particles

5. Lie Groups and Lie Algebras

6. Multiplets

7. Young Tableaux and Unitary Symmetry

8. Clebsch–Gordan Coefficients

9. The Eightfold Way

10. Approximate $SU(6)$

11. The Quark Model

12. Variants of the Quark Model

PREFACE

The role of symmetry has come to be emphasized in the study of the physics of the elementary particles. One reason for this, as pointed out by Wigner, is that we do not have a satisfactory dynamical theory to describe the interactions of the elementary particles. Despite this lack of detailed knowledge of dynamics, much information can be obtained by studying the symmetry properties of elementary particle interactions. However, there is a further reason to study the symmetry properties of interactions: The beauty and simplicity of symmetry properties make their use aesthetically rewarding and valuable, even in cases for which a dynamical theory exists. Many results which can be calculated only with difficulty by a dynamical theory follow quite simply from the exploitation of the symmetry properties of a physical system.

In the years since 1936, when Breit and his collaborators postulated that nuclear forces were charge independent, symmetry under rotations in the space of isospin has been well established as an approximate symmetry of the strongly interacting particles. However, after the discovery of hyperons and heavy mesons, physicists began to look for a larger approximate symmetry.

The first important step in this direction was taken by Gell-Mann and independently by Nishijima in 1952, who postulated the existence of the quantum number of strangeness. Another important contribution was made in 1956 by Sakata and his collaborators, with a model based on primacy of the proton, neutron, and Λ hyperon. Sakata's followers in 1959 incorporated his model into the framework of the group $SU(3)$, or, as it is commonly called, unitary symmetry. Then, in 1961, came the basic work of Gell-Mann and Ne'eman who independently proposed a classification of the strongly interacting particles based on the octet version of unitary symmetry. Subsequently a number of still larger symmetries have been proposed, such as $SU(6)$, but with less success. In 1964 Gell-Mann and, independently, Zweig proposed the quark model. This model has since begun to form the basis of a simple dynamical approach to the interactions of elementary particles.

The origin of the present book was a special series of lectures on unitary symmetry and elementary particles which I gave at the University of Nebraska in 1965. Two of my colleagues there, D. W. Joseph and T. A. Morgan, kindly tape-recorded the lectures and edited them into a set of notes. Subsequently, the notes were considerably expanded into a one-semester course for graduate students at Indiana University. This book is based on the notes of that course, revised and brought up to date. The book as now constituted bears little resemblance to the notes of the original lectures.

In writing the book, I have assumed that the reader has had a course in quantum mechanics and some exposure to the theory of abstract groups and group representations. However, for convenience, I have summarized a number of the most useful concepts in the early chapters. Thus, I have quoted a number of theorems without proof in order to remind the reader of them. Often I have given references to the places where the proofs can be found. Occasionally, I have sketched a proof, usually without rigor, to bring out the flavor of the subject.

Considerable space is devoted to a description of techniques using Young tableaux. Most of these methods I discovered for myself, and therefore, my treatment may not be completely standard. I make no claim that my results are new, but I hope that I have at least given methods which are easy to understand and apply.

In my discussion of experimental matters, I have stressed a few examples of $SU(3)$ multiplets which are reasonably well established. Any attempt to include all elementary particles would become rapidly outdated. Periodically, up-to-date lists of elementary particles are published, and anyone can play the game of trying to classify them into $SU(3)$ multiplets.

Of the many topics on the quark model, I have arbitrarily selected a few of interest to me, especially those I have worked on myself. I have also omitted a wealth of other material for various reasons. I have not discussed attempts to generalize $SU(6)$ to the relativistic case because of the conceptual difficulties and contradictions this program has encountered. I also did not treat in detail the role of $SU(3)$ in the weak interactions, partly because I did not have time to discuss this in one semester, and partly because I did not wish to devote much time to the concept of currents in a course devoted primarily to symmetry.

Neither in my lectures nor in this book have I aimed for completeness, either in content or in references. But I have given references to many of the important works which either provide additional background information for the reader or enable him to pursue the subject more deeply. I particularly hope that the reader is encouraged to undertake the quest for a greater understanding of the nature and scope of unitary symmetry in physics.

Acknowledgments

I am grateful to many physicists for their valuable discussions, especially to Jerrold Franklin, Gordon Frazer, Archibald Hendry, David Joseph, Tom A. Morgan, Yuval Ne'eman, Enrico Predazzi, and Larry Schulman. I thank Frank Chilton and Johan de Swart for permission to reproduce tables photographically from their papers which appeared in *Reviews of Modern Physics*. I thank Arthur Rosenfeld, not only for permission to reproduce a table from the publication of the Particle Data Group, which appeared in the January 1969 issue of *Reviews of Modern Physics*, but also for permission to use information from the January 1970 issue, prior to publication. Those portions of my own work which are described in the book were supported in part by the National Science Foundation.

UNITARY SYMMETRY AND ELEMENTARY PARTICLES

CHAPTER 1

INTRODUCTION

"... *if we knew all the laws of nature, or the ultimate Law of nature, the invariance properties of these laws would not furnish us new information.*"

—Wigner

1.1 Uses of Symmetry

We are fortunate that there exist today the theories of electromagnetism and gravitation that can be used to describe the behavior of certain physical systems. It is useful to classify the predictions of such theories into two main types: first, the predictions which follow from the detailed dynamics of the theory, and second, those which follow from the invariance properties or symmetries which the theory satisfies.

To illustrate these two kinds of predictions, we consider an example from the domain of quantum electrodynamics. This theory enables us to calculate, for example, the differential cross section for electron–electron scattering with both beam and target particles polarized in the same direction. If we calculate this cross section both for electrons polarized parallel and antiparallel to the beam direction, we shall find we get the same answer—the two cross-sections are equal. But we can obtain this result immediately by noting that the theory is invariant under rotations and under permutations of the two electrons. Thus, by making use of the symmetry properties of the theory, we can make certain predictions more simply than by going through detailed calculations which incorporate the symmetry at every step.

In this example, the use of symmetry to obtain the equality of the two cross sections is convenient, but not necessary. However, in some other cases, the use of symmetry is essential, either because the system is so complicated that we cannot perform the calculations or because a dynamical theory does not exist. For example, in the case of proton–proton scattering, we do not have a theory which enables us to make an accurate calculation of the differential cross section. However, we can again make use of symmetry to predict that the cross sections will be equal for protons polarized parallel and antiparallel to the beam axis.

Thus, at the present time, when we do not have a good dynamical theory of some of the interactions of elementary particles, it is especially important to understand the consequences of symmetry.

1.2 Symmetries and Conservation Laws

It is a striking fact of nature that among the many properties of physical systems that continuously change with time, a few properties remain constant. These constant properties appear in so many different physical systems that they are among the most fundamental laws of physics, and are known as conservation laws. In addition to those properties which, so far as is known from experiment, are exactly conserved, there are other properties which are only approximately conserved. The oldest known conservation laws are those of linear momentum, angular momentum, and energy. A familiar conservation law which holds only approximately is the conservation of parity.

Another remarkable fact about physical systems is the symmetries or invariances they possess under certain transformations. Again, some of these symmetries are exact,[1] and some are only approximate. We have already mentioned the invariance of physical systems under rotations and permutations of identical particles. Other familiar examples are the translations in space and time. An example of an approximate symmetry is symmetry under space inversion. We shall be concerned primarily with a symmetry which holds only approximately—unitary symmetry. However, for clarity, we shall treat the concept of symmetry more generally, and discuss other symmetries in brief.

[1] Whenever we refer to a symmetry or conservation law as exact, we mean only that no violation has been observed. It is possible that future experiments will show that a symmetry now thought to be exact is really only approximate.

The intimate connection between symmetries and conservation laws, two apparently unrelated properties of physical systems, was first noticed in classical mechanics by Jacobi in 1842.[2] In his paper, Jacobi (1884) showed that for systems describable by a classical Lagrangian, invariance of the Lagrangian under translations implies that linear momentum is conserved, and invariance under rotations implies the same of angular momentum. Still later, Schütz (1897) derived the principle of conservation of energy from the invariance of the Lagrangian under time translations. Herglotz (1911) was the first to give a complete discussion of the ten constants of the motion associated with the invariance of the Lagrangian under the group of inhomogeneous Lorentz transformations. For systems not describable by a Lagrangian, the situation is more complicated (Van Dam and Wigner, 1965), and we shall not discuss it here.

Now let us consider the connection between the symmetry of a system and a conservation law in the Hamiltonian formalism, which is more convenient in quantum mechanics. In both classical and quantum mechanics, the conservation of linear momentum, angular momentum, and energy follow from the symmetries of the Hamiltonian under translations, rotations, and time translations. More generally, in quantum mechanics, whenever a conservation law holds for a physical system, the Hamiltonian of the system is invariant under a corresponding group of transformations. The converse of this statement is not true, as even if the system has a Hamiltonian which is invariant under a group of transformations, there may not be a corresponding conservation law. An example is invariance under time reversal.

Wigner (1959) showed that all symmetry transformations of a quantum mechanical state can be chosen so as to correspond to either unitary or antiunitary operators. It is the unitary transformations which are associated with the conservation laws. However, if a Hamiltonian is invariant under an antiunitary transformation, such as time reversal, this has other consequences which can be tested against experiment.

If a Hamiltonian describing a system of interacting elementary particles is invariant under a unitary transformation, then it can be shown that the scattering matrix or S matrix of a system is also invariant under the transformation. The importance of the S matrix in elementary particle physics is well recognized. In fact, some physicists have speculated that perhaps a Hamiltonian describing elementary particle interactions does not exist, and that the laws of physics are contained in the properties of the S matrix.

[2] The history given in this paragraph is based on a footnote in a paper of Van Dam and Wigner (1965) who attribute the historical study to Eugene Guth.

1.3 Symmetries and Groups

A set of symmetry transformations on a physical system has the mathematical properties that are associated with a group. For example, let us consider some properties of rotations.

(1) If we rotate a physical system and follow this rotation by a second rotation, the resulting transformation is also a rotation.

(2) Suppose we perform three successive rotations. It does not matter whether we perform the first rotation and then a rotation equal to the resultant of the other two or perform the resultant of the first two and then the third. In other words, rotating a system is an associative operation.

(3) We may rotate a system through zero degrees so as to leave its position unchanged. This is an identity operation.

(4) If we rotate a system through a certain angle, we may then rotate it back to its original position. The second operation is an inverse operation to the first.

These four properties of rotations are just the properties of a group. The translations in space and time, and indeed most symmetry transformations on physical systems, also have the group properties. It is for this reason that the study of group theory is so important for understanding the consequences of the symmetries of physical systems.

Let us now look at still another property of rotations. If we perform first one rotation on a system and then a second, the system does not necessarily end up in the same orientation as if we perform the rotations in the reverse order. In other words, the operation of rotating a system is not commutative. On the other hand, rotations about a single axis lead to the same result when performed in any order; similarly, order does not count when performing successive translations. We must therefore study groups whose members commute and those whose members do not.

The rotations and translations in space and time may be made through any angle or any distance. For this reason, there are a continuous infinity of transformations which leave certain physical systems invariant. These transformations correspond to continuous groups. In fact, the continuous transformations in which we are chiefly interested are even more restricted; they form sets of transformations corresponding to Lie groups.

One difference between the rotations and translations is that the rotations vary over a finite angular domain, while translations in space and time (in a flat space) vary over an infinite domain. It turns out that the corresponding groups have many different mathematical properties.

To the rotations and the translations in space and time may be added Lorentz transformations, which are also presently believed to be exact symmetries of elementary particle systems. Together, the translations in space and time, the rotations, and the Lorentz transformations, correspond to the inhomogeneous Lorentz group, or Poincaré group.

In addition to the symmetries which physical systems exhibit under certain continuous transformations, physical systems are invariant (or at least approximately so) under certain discrete transformations. These transformations usually constitute a finite set, and appropriate to the study of these symmetries is the theory of finite groups.

1.4 Eigenstates, Quantum Numbers, and Selection Rules

An invariance or symmetry of an interaction under a transformation does not always lead to useful restrictions on the final states that can be reached from a given initial state. However, the consequences of the symmetry of an interaction are particularly striking for certain special quantum mechanical systems, which are represented by the eigenstates of observable conserved operators.

If a physical system is represented by an eigenstate of several such observable operators, then the system can be characterized by a corresponding set of quantum numbers. In any transition of the system from an initial state to other states, only final states with the same values of the quantum numbers are reached. Therefore, there are *selection rules* against transitions to states with different values of the quantum numbers. These selection rules enable us to make predictions about which reactions are possible and which are not. For example, we know that in any reaction the total charge Q of an isolated system remains a constant; or in other words, we have the selection rule $\Delta Q = 0$ in the reaction. This is only one of a number of selection rules which are known to be satisfied in the interactions of elementary particles.

As we have already mentioned, although some symmetries are exact so far as is known, others are only approximate. As an example of an approximate symmetry we may consider the invariance under rotations in isospin space or conservation of isospin. Isospin is believed to be conserved in strong interactions, but it is known that it is not conserved in electromagnetic interactions. Therefore, if we have a reaction involving strongly interacting particles we have the approximate selection rule $\Delta I = 0$. The fact that the selection rule holds to a good approximation is a reflection of the fact we believe that the symmetry holds to a good approximation: That is, that the extent of the violation is of order of the fine structure constant $\alpha = 1/137$.

However, we can also have a selection rule which holds to a good approximation for a dynamical reason. That is, the fact that a particular selection rule holds well may not reflect an underlying symmetry at all, but may be a dynamical accident. For example, in atomic physics there is the well-known approximate electric dipole selection rule for radiative transitions of excited states of atoms. The selection rule states that the orbital angular momentum L of the atom changes by one unit during the transition. This of course is consistent with conservation of angular momentum because the photon carries off one unit of angular momentum. However, the selection rule does not reflect any basic symmetry but the accident that, at low energy, the photon tends to carry off only one unit of angular momentum. If the energy of the transition were much larger we would have quadrupole and higher multipole transitions. In fact, such transitions are observed in nuclear physics, where the gamma rays are much more energetic. So whenever we are confronted with an approximate selection rule we have the additional question of whether it is the result of true approximate symmetry or of a dynamical accident which holds in a limited energy range. If the latter is true, we call the symmetry an approximate dynamical symmetry.

1.5 A Listing of Symmetries

We now briefly discuss the symmetries of interactions, some of which we shall treat in more detail later. We also discuss their associated conservation laws if they exist.

As in classical mechanics, the symmetry under inhomogeneous Lorentz transformations is connected to ten conserved quantities. These are the momentum-energy four-vector p_μ, the three components of the angular momentum $\mathbf{J}$, and three conserved quantities which are connected with pure Lorentz transformations or *boosts*. By a boost we mean a transformation to a coordinate frame moving with constant velocity with respect to the original system, but which is unrotated and which coincides with the original system at some arbitrary time equal to zero.

The conservation laws connected with boosts do not appear to have as much use in physical problems as conservation of angular momentum. The reason is that most physical systems that we shall consider are in eigenstates of the total angular momentum and its z component. Since the operators corresponding to the components of angular momentum do not commute with the operators connected to boosts, a physical system cannot be in an eigenstate of two of these operators simultaneously. Thus we usually do not get useful selection rules from the conservation laws connected to boosts.

The symmetry under the group of inhomogeneous Lorentz transformations is usually assumed to be an exact symmetry, at least so far as elementary particle physics is concerned. If this symmetry is exact, then space is flat. But according to the theory of general relativity, it is possible for space to be curved. Perhaps then, rather than considering the Poincaré group, we should consider a larger group, of which the Poincaré group is only an asymptotic limit. However, we shall not consider this possibility.

Other important symmetries are the invariances under gauge transformations. These symmetries are associated with conservation of charge Q, baryon number B, and two lepton numbers, the electron number L_e and the muon number L_μ. (The electron number is defined to be the number of electrons plus electron neutrinos; the muon number is the number of muons plus muon neutrinos or neutrettos.) There is extremely good evidence to support exact conservation of charge and baryon number at least on a local, as opposed to a cosmological, scale. Also, no reaction in which lepton number is not conserved has been seen, but the weight of evidence is not nearly so great as for charge and baryon number conservation.

Next, consider symmetries connected with the permutation of identical particles. All observable quantities of physics must be symmetric (that is, must not change) under the permutation of identical particles. This appears to be obvious, for if it were not so, we could distinguish between two particles by the observable which was not symmetric under their interchange. Although observables must be symmetric under the interchange of identical particles, quantum mechanical states (which are not observables) need not have this symmetry. However, since observables are operators which act on quantum mechanical states, the states themselves must have definite symmetry properties under interchange of identical particles.

So far as is known from experiment, identical particles of integral spin are always in states which are symmetric under the interchange of any two of the particles, while identical particles of half-odd-integral spin are in states which are antisymmetric under this interchange. This property of states of identical particles is the content of a theorem, called the spin-statistics theorem, which can be proved from the postulates of certain field theories. See, for example, Streater and Wightman (1964). Despite the fact that no two identical elementary particles have ever been seen in states with other than the usual symmetry properties, we shall discuss later hypothetical states which possess more complicated symmetry properties under permutations. Of course, whether or not these more complicated states exist, all observables must be symmetric under the interchange of two identical particles. Even if states with complicated permutation symmetry do not exist, consideration of such states is very important. This is because the states may have complicated symmetry properties under the interchange of only some of the coordinates of identical particles.

Most of the invariances under discrete transformations, other than permutation symmetries, seem to be approximate. Invariance under inversion of space coordinates, denoted by the symbol **P**, is an approximate symmetry.

Invariance under space inversion is related to the law of conservation of parity. Also approximate is the symmetry under the transformation of replacing particles by antiparticles, called charge conjugation and given the symbol C. The conservation law related to this symmetry is called charge conjugation parity or C parity.

The symmetry corresponding to inverting space coordinates and performing charge conjugation, denoted by CP, is observed to hold experimentally, in some weak interactions at least, to a greater approximation than either C or P separately. However, the experimental evidence on K-meson decay by Christenson *et al.* (1964) is most simply interpreted as evidence that CP is also an approximate symmetry.

Next consider symmetry under inversion of the time coordinate. This symmetry is called time inversion or time reversal, and is denoted by T. Since, in quantum mechanics, time inversion is an antiunitary transformation, there is no corresponding conservation law. In a Lagrangian or Hamiltonian theory, the effects of time inversion make themselves felt as reality conditions on the coupling constants entering the theory.

The experiments on K-meson decay have been interpreted by Casella (1969) as implying that T is not an exact symmetry in weak interactions. Independently of how good this evidence is for T violation, many physicists expect this symmetry to be approximate. The reason is that in certain field theories, a theorem can be proved in which the product CPT is an exact symmetry. See, for example, Streater and Wightman (1964). If the theorem applies to physical systems, then CP noninvariance implies T noninvariance. Whether or not the postulates of these field theories are true, thus far there is no experimental evidence for any violation of CPT. Since C and P correspond to unitary transformations and T to an antiunitary transformation, CPT symmetry is antiunitary, and, like T, it does not correspond to a conservation law. However, invariance under CPT does have some very important consequences; for example, one consequence is that the mass of a particle is equal to the mass of its antiparticle.

The remaining symmetries we discuss are approximate, and seem to have most to do with the strong interactions. However, some of these symmetries appear to be conserved in electromagnetic interactions as well. The charge Q (in units of the proton charge) of every strongly interacting particle discovered thus far can be written as a sum of two numbers, I_3 called the third component of the isospin (for reasons to be discussed shortly), and $\frac{1}{2}Y$, where Y is called hypercharge. The conservation of charge, as we have said before, is exact so far as is known, and corresponds to invariance of interactions under a gauge transformation. However, the strong and electromagnetic interactions, but not the weak interactions, seem to be invariant under two gauge transformations corresponding to the conservation of I_3 and Y separately.

Although conservation of I_3 and Y can be considered to be conservation laws associated with gauge transformations, they also can be considered as certain operations in a larger group. There are two such larger groups which seem to be of particular importance, the special unitary groups in two and three dimensions, denoted by $SU(2)$ and $SU(3)$, respectively. The unitary group $U(n)$ is defined as the group of $n \times n$ unitary matrices; and the special unitary group $SU(n)$ is the subgroup of unitary matrices which have determinant unity.

The group $SU(2)$ is locally isomorphic to the rotation group in three-dimensions $R(3)$. Because of this, we can define a total isospin[3] I in an abstract three-dimensional space, analogously to the total angular momentum J in ordinary space. (In fact, the group $SU(2)$ is also appropriate to describe ordinary angular momentum.) The vector $\mathbf{I}$, which is conserved if the interactions are invariant under rotations in this space, has components I_1, I_2, and I_3, the last of which we have mentioned previously. Since I_3 is related to the charge of a strongly interacting particle, a rotation about the first or second axis in isospin space is a transformation which changes the charge of a state.

Since two particles which have different values of I_3, but are otherwise identical, have different values of the charge, their electromagnetic interactions are different. Thus, isospin invariance is an approximate symmetry of strongly interacting particles which is broken by the electromagnetic interaction. In the weak interactions, not only I, but I_3 is also an approximate or broken symmetry.

By incorporating Y with I into a still larger group $SU(3)$, we have what is often called the unitary symmetry theory of strongly interacting particles. This symmetry is very approximate, since it is broken even in strong interactions. This fact has led physicists to speculate that perhaps the category of the strong interactions should be subdivided into the very strong, which are invariant under $SU(3)$, and the moderately strong, which break the symmetry. However, not enough is known of the dynamics of strong interactions for us to say whether this picture is correct.

In addition to the symmetry $SU(3)$, a number of still higher symmetries have been postulated as being approximate symmetries of the strong interactions. Among these are the symmetries $SU(6)$ and $SU(12)$. It is not yet clear how fruitful a consideration of higher symmetries will be, although $SU(6)$ appears to have some relevance as an approximate dynamical symmetry which holds best at low energies. Since several of the special unitary groups seem to have relevance for elementary particle physics, we shall later study a number of the properties of $SU(n)$, where n is any positive integer.

[3] In addition to the word isospin, the words isotopic spin, isobaric spin, and I spin are often used to describe this space.

CHAPTER 2

SOME PROPERTIES OF GROUPS

2.1 Elementary Notions

We have seen that the symmetry transformations on physical systems have the properties of groups. Therefore, in order to exploit the consequences of symmetries, it is useful to study some of the mathematical properties of groups. These fall into two main categories: the properties of abstract groups and the properties of representations of groups. For the most part, we shall discuss representations in the next chapter.

Here we shall consider only a few of the properties of groups, and we shall usually not prove our assertions. For a fuller treatment, see Wigner (1959) or Hamermesh (1962).

A *group* is a set of elements which obey the following rules:

(1) A law of combination, usually called multiplication, of group elements is defined such that if a and b belong to a group G, then the product element ab is unique and belongs to G. We can write this property of a group as follows:

$$(a \in G \quad \text{and} \quad b \in G) \Rightarrow ab \in G, \tag{2.1}$$

where the symbol $\in$ means "belongs to" or "is a member of" and the symbol $\Rightarrow$ means "implies."

(2) The law of combination is associative, that is, given the elements a, b, and c, we have

$$(ab)c = a(bc). \tag{2.2}$$

The elements do not necessarily commute, however.

(3) An identity element e (or 1 or a_0) exists satisfying the property that for all a in G,

$$ea = ae = a. \tag{2.3}$$

(4) To every element a in G, an inverse element a^{-1} in G exists such that

$$a^{-1}a = aa^{-1} = e. \tag{2.4}$$

Sometimes postulates (3) and (4) are replaced by the weaker postulates that a left identity and a left inverse exist. From these weaker postulates, postulates (3) and (4) can be derived.

A group is said to be *finite* if it has a finite number of elements. Otherwise the group is said to be *infinite*. The number of elements of a finite group is called the *order* of the group.

The simplest group is the trivial group consisting of only one element, the identity e. The identity is always its own inverse. The next simplest group consists of two elements, e and a, with a law of multiplication

$$ee = e, \qquad aa = e, \qquad ea = ae = a. \tag{2.5}$$

The element a is its own inverse. To give a familiar example of this group, we can identify the elements e and a with the numbers 1 and -1, respectively. It is obvious that the multiplication table (2.5) is satisfied by the numbers ± 1, with the identification $1 = e$, $-1 = a$. The group consisting of the elements e and a is important in quantum mechanics because it can be associated with such discrete transformations as space reflection and charge conjugation.

If all the elements of a group commute, the group is said to be *commutative* or *Abelian*. The groups of one and two elements are clearly Abelian. The law of combination for Abelian groups is often denoted by the symbol $+$. If so, the identity element is called zero and the inverse to the element a is called $-a$.

An example of an infinite Abelian group under addition is the set of all integers. For any two integers m and n, $m + n$ is also an integer; the zero element exists such that $0 + m = m$; and for any integer m there exists the integer $-m$ such that $m + (-m) = 0$. The associative law holds, of course, for addition of integers. Thus, the integers satisfy the group properties. However, the set of positive integers is clearly not a group, since an inverse and identity do not exist. It is easy to see that the set of real numbers is an Abelian group under addition.

The smallest non-Abelian group is the group of permutations of three objects. We write the symbol

$$\begin{pmatrix} 123 \\ ijk \end{pmatrix}$$

for the permutation in which the object in position 1 is put into position i, the object in position 2 is put into position j, and the object in position 3 is put into position k. In general, if we use this notation, a permutation can be written in more than one way. For example, the symbols

$$\begin{pmatrix} 123 \\ ijk \end{pmatrix} \quad \text{and} \quad \begin{pmatrix} 312 \\ kij \end{pmatrix}$$

denote the same permutation. The permutation group of three objects has the following six elements

$$\begin{aligned} e &= \begin{pmatrix} 123 \\ 123 \end{pmatrix}, & a &= \begin{pmatrix} 123 \\ 213 \end{pmatrix}, & b &= \begin{pmatrix} 123 \\ 132 \end{pmatrix}, \\ c &= \begin{pmatrix} 123 \\ 321 \end{pmatrix}, & d &= \begin{pmatrix} 123 \\ 312 \end{pmatrix}, & f &= \begin{pmatrix} 123 \\ 231 \end{pmatrix}. \end{aligned} \tag{2.6}$$

That the permutations on n objects form a group can be readily seen as follows: Two successive permutations (which we call the product) are again a permutation, and the operation of making permutations is associative. The identity is

$$\begin{pmatrix} 123 \cdots n \\ 123 \cdots n \end{pmatrix},$$

and the inverse to the permutation

$$\begin{pmatrix} 12 \cdots n \\ ij \cdots k \end{pmatrix} \quad \text{is} \quad \begin{pmatrix} ij \cdots k \\ 12 \cdots n \end{pmatrix}.$$

The permutation group of n objects or symbols is often called the *symmetric group* of *degree n*, and is written S_n.

Another notation for the symbol

$$\begin{pmatrix} 12 \cdots n \\ ij \cdots k \end{pmatrix} \quad \text{is just} \quad (ij \cdots k).$$

With this shorter notation, it is assumed that this is a permutation on the numbers $123 \cdots n$. In this notation, there is only one symbol for a given permutation. The longer, two-rowed notation is more useful for multiplying permutations. Such a multiplication is done as follows: We write the second permutation using a symbol such that its top row is the same as the bottom row of the first permutation. Then the product permutation has the same top row as the first permutation and the same bottom row as the second. As an example, suppose we wish first to make the permutation

$$\begin{pmatrix}123\\213\end{pmatrix},$$

and then the permutation

$$\begin{pmatrix}123\\312\end{pmatrix}.$$

We proceed by writing the second permutation as

$$\begin{pmatrix}213\\132\end{pmatrix}.$$

Then the product permutation is

$$\begin{pmatrix}213\\132\end{pmatrix}\begin{pmatrix}123\\213\end{pmatrix}=\begin{pmatrix}123\\132\end{pmatrix}.$$

We have seen that S_3 is a group of order six by writing down all possible permutations of three symbols. We can count the number of elements more simply in the following way: The first object can go in any one of three places. Once it is fixed, the second object can go in either of the two remaining places. Once the second object is in its place, the third object must go in the remaining place. Thus, there are a total of $3 \cdot 2 \cdot 1 = 3!$ possible permutations. By similar reasoning, the symmetric group S_n is of order $n!$ Using the two-row notation, there are also $n!$ ways of writing a given permutation of S_n.

We have written down in Eq. (2.5) the complete multiplication table for the group with two elements. This is just the multiplication table for the symmetric group S_2, with elements

$$\begin{pmatrix}12\\12\end{pmatrix} \quad \text{and} \quad \begin{pmatrix}12\\21\end{pmatrix}.$$

Likewise, we can form a multiplication table of the group S_3 by actually carrying out the multiplication of each permutation of S_3 with every other permutation of S_3. This multiplication table has been given in so many places (e.g., Wigner, 1959) that we shall not reproduce it here.

2.2 Homomorphism, Isomorphism, and Subgroups

A group H is said to be *homomorphic* to a group G if there is a mapping of G onto H such that multiplication is preserved. In this mapping, more than one element of G may correspond one element of H. If a and b are two arbitrary elements of G and $c = ab$, then in H the corresponding elements

a', b', and c' must satisfy $c' = a'b'$. In any homomorphic mapping the identity of G is mapped into the identity of H.

In a homomorphic mapping of G onto H, if the correspondence between the elements of G and those of H is one-to-one, then H is said to be *isomorphic* to G. Clearly, if H is isomorphic to G, then G is isomorphic to H. Therefore, we usually say simply that the two groups are isomorphic. An isomorphic mapping of a group onto itself is called an *automorphism*. All isomorphic groups have the same multiplication table. In fact, all the properties of an abstract group are contained in its multiplication table.

Another important concept is that of a *subgroup* of a given group G. This is simply a subset of the group which itself has the group properties under the same multiplication law as G. Every group has at least two subgroups, the group itself and the identity or unit element. These are called improper subgroups. All other subgroups are called proper subgroups. The two-element group has only improper subgroups. The symmetric group S_3 has three proper subgroups of order two and one of order three. We denote the subgroups of order two by H_1, H_2, and H_3 with elements

$$(123),\ (132) \in H_1,$$
$$(123),\ (321) \in H_2,$$
$$(123),\ (213) \in H_3.$$

These groups are isomorphic to S_2. In fact, all groups of order two are isomorphic to S_2. This is because they all have the same multiplication table, as can be seen by the following argument.

Let the two distinct elements of the group of order two be e and a. We must have $ee = e$, $ea = ae = a$ by definition. Then the only other product is aa, but this must be $aa = e$, since if $aa = a$, $a = e$. Thus the multiplication table is unique.

2.3 Infinite Groups

If we have a group with a finite number of elements, we can label the group elements g_a where the subscript is a parameter which takes on a finite number of values. We can generalize this idea to a countably infinite group with an infinite number of discrete values of a or to a group with a continuum of elements in which the parameter a can vary continuously. In the latter case a group element is usually denoted by $g(a)$. We can generalize still further to the case where a stands for more than one continuously varying parameter.

The set of values of the parameter or parameters which characterize a group element can be considered to be points in some kind of space. The number of parameters characterizes the dimension of the space. We shall restrict ourselves to a *topological* space, that is, to a space in which *distance* between any two points is defined. Under certain conditions, we can make a correspondence between elements of a group and the points given by the values of the parameters in a topological space. Alternatively, under certain conditions we can consider the group elements themselves to be points in the space.

Suppose we have a group whose elements are labeled by a letter which can stand for either a single parameter or for several parameters which can take on a continuum of values. Furthermore, suppose we demand that if we form the product of any two group elements to give a third:

$$g(a)g(b) = g(c),$$

where a, b, and c are particular values of the parameter, then c should be a continuous function of the parameters a and b. In other words, we assume that a small change in either a or b will produce only a small change in c. If so, the group is said to be a *continuous* group, and the points in parameter space can be identified with the group elements. A mixed continuous group is a continuous group which also contains discrete parameters. Even more restrictive than a continuous group is a *Lie* group, in which, if $g(a)g(b) = g(c)$, then c is an analytic function of a and b.

In the detailed study of continuous and Lie groups, one can bring to bear the methods of functional analysis. We shall not do so in any generality but shall consider principally the concepts which we shall need later. Further information can be found in Pontriagin (1966). See also Hamermesh (1962); Wigner (1959); and Racah (1965) for discussions of continuous groups.

Let us consider groups of coordinate transformations as examples of Lie groups. Many such transformations have applicability in quantum mechanics. First consider the group of translations in one dimension, namely

$$x' = x + a. \tag{2.7}$$

This is a one-parameter Lie group and is an Abelian group under addition. The group elements can be denoted by $g(a)$ or simply by the parameter a itself. The values of the parameter are the points on a line. The line is an example of a topological space with which we associate or identify the group elements. As another example, consider the transformation

$$x' = ax + b. \tag{2.8}$$

This is a two-parameter Lie group. A continuous group which is characterized by a finite number of continuous parameters is sometimes called a *finite continuous* group.

If a Lie group is defined by a number of parameters, say r parameters, and we cannot find a smaller number of parameters to characterize the group, then the parameters are called *essential* parameters, and the group is called an r-parameter Lie group.

As an example of a group which is characterized by two parameters, only one of which is essential, we give the following:

$$x' = x + a + b.$$

In this case, an infinite number of different combinations of a and b correspond to the same group element. The group characterized by one essential parameter would be

$$x' = x + c, \qquad c = a + b.$$

Then to each c, there is only one group element.

Another example of a Lie group is the general linear group in two dimensions. This is the group of general linear transformations on a two-dimensional space and is denoted by $GL(2)$. If the parameters are real, the group is often denoted by $GL(2R)$, whereas if the parameters are complex, the group is denoted either by $GL(2C)$ or simply by $GL(2)$. The transformation is given by

$$\begin{aligned} x_1' &= a_{11}x_1 + a_{12}x_2, \\ x_1' &= a_{21}x_1 + a_{22}x_2. \end{aligned} \tag{2.9}$$

The group $GL(2R)$ is characterized by four essential parameters and $GL(2C)$ is characterized by eight essential parameters. We can write the transformation in matrix form as follows:

$$x' = Ax, \qquad A = \begin{pmatrix} a_{11} & a_{12} \\ a_{21} & a_{22} \end{pmatrix}, \qquad x = \begin{pmatrix} x_1 \\ x_2 \end{pmatrix}. \tag{2.10}$$

The linear group in n dimensions $GL(n)$

$$x' = Ax, \qquad A = \begin{pmatrix} a_{11} & \cdots & a_{1n} \\ \vdots & & \vdots \\ a_{n1} & \cdots & a_{nn} \end{pmatrix}, \qquad x = \begin{pmatrix} x_1 \\ \vdots \\ x_n \end{pmatrix} \tag{2.11}$$

is characterized by $2n^2$ essential parameters. The matrices A are the group elements. If the matrices are unitary, the group is called $U(n)$. The equalities $A_{ji}^* = (A^{-1})_{ij}$ put n^2 independent conditions on the $2n^2$ variables and reduce the number of essential parameters to n^2. A linear group satisfying the one restriction that its determinant is unity is called a *special* linear group and is denoted by $SL(n)$. The group $SU(n)$ is the special unitary group in n dimensions, i.e., the group of unitary matrices with determinant unity. It is characterized by $n^2 - 1$ parameters. We shall be dealing very much with this group.

Other groups of importance are the orthogonal group in 2, 3, or n dimensions. The group $O(n)$ is composed of those elements of $GL(n, R)$ which leave the product

$$x^2 = \sum_i x_i^2 \tag{2.12}$$

invariant. This group has $n(n-1)/2$ essential parameters. The rotation group $SO(n)$ or $R(n)$ consists of the orthogonal matrices with determinant unity. It has the same number of parameters as $O(n)$. However, $R(n)$ in an odd number of dimensions does not include the reflections $x' = -x$. The rotation group in two dimensions is Abelian, and is characterized by one essential parameter θ: The transformation is

$$\begin{aligned} x_1' &= x_1 \cos\theta - x_2 \sin\theta, \\ x_2' &= x_1 \sin\theta + x_2 \cos\theta. \end{aligned} \tag{2.13}$$

The transformation can be written in matrix form as

$$x' = R(\theta)x, \tag{2.14}$$

where

$$x = \begin{pmatrix} x_1 \\ x_2 \end{pmatrix}, \qquad R(\theta) = \begin{pmatrix} \cos\theta & -\sin\theta \\ \sin\theta & \cos\theta \end{pmatrix}. \tag{2.15}$$

Here, the rotation matrix $R(\theta)$ is the group element.

Still another example of a Lie group is the homogeneous Lorentz group (or simply Lorentz group), which is the group which leaves invariant the quantity

$$x^2 + y^2 + z^2 - c^2 t^2.$$

This is a six-parameter group in four dimensions. The group $O(3)$ is a subgroup of the Lorentz group. If we add to the Lorentz group the group of translations, we obtain the ten-parameter inhomogeneous Lorentz group, or Poincaré group.

Of importance in group theory is the idea of a compact group. We shall not give the general definition of a compact group, but just one that is useful for our purposes. To do this, we need the concepts of a bounded and closed set. A set of numbers is said to be *bounded* if none of the numbers in the set exceeds a given positive number M in absolute value. A set is said to be *closed* if the limit of every convergent sequence of points in the set also lies in the set. As an example, the real line between zero and one is bounded. It is open if either of the end points zero and one is not included in the set, but is closed if zero and one are included.

An r-parameter group is compact if the domain of variation of all the parameters is bounded and closed. The group of rotations is compact, because the angle of rotation about any axis must lie in the closed interval

$0 \le \theta \le 2\pi$. However, the translation group is not compact, as the parameter a in the transformation $x' = x + a$ can vary between plus and minus infinity. Here the domain of variation of a is unbounded. (If a varies over a bounded region, the translations do not form a group.)

The group of pure Lorentz transformations is characterized by a parameter, the velocity v, which is bounded. However, the limit point $v = c$ is not a member of the group, as a Lorentz transformation yields infinity when $v = c$. Therefore, the Lorentz group is not compact.

A continuous group is *locally compact* if in the neighborhood of any point in parameter space, a closed domain or interval exists. The point in question can be a boundary point of the domain. A compact group is, of course, locally compact. In addition, some noncompact groups such as the translation group and the Lorentz group are locally compact.

2.4 Cosets, Conjugate Classes, and Invariant Subgroups

Let H be a proper subgroup of the group G, and let g belong to G, but not to H. Then the product elements gh_i, where the h_i are all the elements of H, are said to form a *left coset* of H. The coset is denoted by gH. Similarly, we can define a *right coset* Hg with members $h_i g$. (Some authors define cosets to include sets gH and Hg where g is any member of G, including members of H.)

With our definition of a coset, it can be seen that none of the elements of a coset of H belongs to H. We show this by assuming that gh_i belongs to H and obtaining a contradiction. Since $h_i \in H$, $h_i^{-1} \in H$. If $gh_i \in H$, then $gh_i h_i^{-1} \in H$, since the product of any two elements of a group is itself an element. But

$$gh_i h_i^{-1} = ge = g. \tag{2.16}$$

Thus, if $gh_i h_i^{-1} \in H$, then $g \in H$. But this is in contradiction to the assumption contained in our definition of a coset that g does not belong to H. Therefore, none of the elements of gH belongs to H. From this theorem it can be seen that a coset is not a group. The proof follows from the fact that the identity element e, since it belongs to H, is not in any coset of H.

We state without proof two other properties of cosets. The first is that two cosets have either no elements in common or else have identical elements. The second is that the set S of the elements of all the (left or right) cosets of a subgroup H of G, plus the elements of H, contains all the elements of G. We can obtain all the cosets of H by multiplying the elements of H by all the g_i in G which are not in H. However, using this procedure, we may obtain the cosets more than once.

These properties of cosets can be used to prove Lagrange's theorem, which states that the order of a subgroup of a finite group is a factor of the order of the group. By this we mean that if g is the number of elements of G, and h is the number of elements of a subgroup H, then $g = hn$, where n is an integer. We shall not prove the theorem, which is to be found, for example, in Wigner (1959).

We now consider the concept of conjugate elements of a group. An element a belonging to G is said to be *conjugate* to an element b belonging to G if there exists an element u in G such that $a = ubu^{-1}$. The set of all elements of G conjugate to a given element is said to be the *conjugate class* of the element. The *relation* of conjugation between elements is an equivalence relation, since it can be shown that (1) a is conjugate to a, (2) if a is conjugate to b, then b is conjugate to a, and (3) if a is conjugate to b, and b is conjugate to c, then a is conjugate to c.

It is clear that the identity element of any group is in a class by itself. This follows because the identity commutes with all members of the group. Therefore

$$ueu^{-1} = euu^{-1} = e. \tag{2.17}$$

Similarly, every element of an Abelian group is in a class by itself. It should also be obvious that, except for the identity, no class is a group.

By actual multiplication of the elements of S_3, we can verify that this group contains three classes. They are the class with the identity (123), the class with the elements (132), (213), and (312), and the class with the elements (312) and (231).

We now state without proof an important theorem about conjugate elements. Let H be a subgroup of G, and let $h \in H$ and $g \in G$. Form the product elements

$$h' = ghg^{-1}, \tag{2.18}$$

for all h in H. Then the h' form a group H' which is isomorphic to H. The group H' is called the *conjugate subgroup* to H in G. Different choices of the element g, in general, give different groups H'. If, for all g in G, the elements of H and gHg^{-1} are identical, then H is said to be an *invariant* or *self-conjugate* subgroup of G. If H is an invariant subgroup of G and if $h \in H$, then $ghg^{-1} \in H$ for any g in G.

A group is called *simple* if it is not Abelian and contains no proper invariant subgroups. It is called *semisimple* if it contains no Abelian invariant subgroups except the unit element. In applying these definitions to Lie groups, we shall exclude invariant subgroups with only a finite number of elements.

We shall not prove the following theorem about invariant subgroups:

THEOREM. *A subgroup of order two in any group is invariant.*

The group $SU(2)$ contains the elements

$$e = \begin{pmatrix} 1 & 0 \\ 0 & 1 \end{pmatrix} \quad \text{and} \quad a = \begin{pmatrix} -1 & 0 \\ 0 & -1 \end{pmatrix}. \tag{2.19}$$

These elements comprise an invariant subgroup of order two, sometimes called Z_2. However, $SU(2)$ has no infinite invariant subgroups, and so is simple.

We quote another theorem without proof:

THEOREM. *The left and right cosets of an invariant subgroup H of a group G, formed with any element of G are identical.*

This theorem enables us to prove an important theorem about an invariant subgroup and its cosets.

Suppose we have a group G containing an invariant subgroup H. Consider a set S whose elements are all the distinct cosets of H plus H itself. This set is a group, as can be seen by the following argument: The elements of S are $b_i H$ and H, and b_i are all the elements of G which are not in H. Since H is invariant we have $b_i H = H b_i$. We can now perform *coset multiplication*

$$b_i H b_j H = b_i b_j H H = b_i b_j H, \tag{2.20}$$

because, since H is a group, the elements of HH are the elements of H itself. Since $b_i b_j$ belongs to G, $b_i b_j H$ is either H itself or a coset of H, depending on whether or not $b_i b_j$ is an element of H. The identity element of S is just H itself, and the inverse to the element $b_i H$ is $b_i^{-1} H$. Thus S is a group whose members are sets: namely H and the cosets of H. It is called the *factor* or quotient group and is denoted by G/H.

As an example, we note that from $SU(2)$ and Z_2 we can construct the factor group $SU(2)/Z_2$. This group is isomorphic to the three-dimensional rotation group $R(3)$.

We now introduce the notion of the direct product of two groups H and H'. The group G is called the *direct product* of H and H' if the following two conditions are satisfied:

(1) Every h in H commutes with every h' in H'.

(2) Every g in G can be written uniquely as a product of an element of H and an element of H', that is

$$(g \in G) \Rightarrow (g = hh'). \tag{2.21}$$

The direct product is written in the form $G = H \times H'$.

Another notation for an element of a direct product is to denote the element g by means of an ordered pair (h, h'), i.e.,

$$g = (h, h'), \tag{2.22}$$

with the following rule of multiplication

$$(h_1, h_1')(h_2, h_2') = (h_1 h_2, h_1' h_2'). \tag{2.23}$$

It follows from the definition of the direct product G of two groups H and H' that H and H' are isomorphic to two invariant subgroups of G having only the identity element in common. Suppose H has m elements h_i, where $h_1 \equiv e$ is the identity of H, and H' has n elements h_j' with identity $h_1' \equiv e'$. Then H is isomorphic to the invariant subgroup of G with elements $h_i e'$, and H' is isomorphic to the invariant subgroup with elements eh_j'. The common identity element of both invariant subgroups is ee'. It is often said for brevity that H and H' themselves are invariant subgroups of G with only the identity in common.

CHAPTER 3

SYMMETRY, GROUP REPRESENTATIONS, AND PARTICLE MULTIPLETS

3.1 Linear and Unitary Vector Spaces

We have seen that the transformations which can be made on physical systems and which leave them invariant have the mathematical properties of groups. Thus, the use of group theory is a valuable tool in studying symmetry. However, in order to apply the ideas of group theory to physical problems, we must study the properties of the quantum mechanical states of a system and how these states transform under symmetry operations. To do this, we need the concepts of unitary vector spaces and of unitary and antiunitary operations on these spaces. We need these concepts because the physical states are represented in quantum mechanics by vectors in a unitary vector space, and the symmetry transformations are represented by unitary and antiunitary operators which act on the space.

We begin by defining a *linear vector space* as a set of elements called vectors with the following properties:

(1) If ϕ, ψ, and χ are vectors in a linear vector space V, then a law of combination, called addition, is defined such that

$$\phi + \psi \in V. \tag{3.1}$$

(2) This operation is commutative and associative, i.e.,

$$\phi + \psi = \psi + \phi, \tag{3.2}$$

$$\phi + (\psi + \chi) = (\phi + \psi) + \chi. \tag{3.3}$$

(3) There exists a null vector 0 such that

$$\phi + 0 = \phi. \tag{3.4}$$

(4) For every vector ϕ, there exists a negative $-\phi$ such that

$$\phi + (-\phi) = 0. \tag{3.5}$$

The foregoing properties show that a vector space is an Abelian group under addition.

(5) There are further properties, namely that there exist scalars $a, b, \ldots$ (in general, complex numbers) which satisfy the requirements

$$\phi \in V \Rightarrow a\phi \in V, \tag{3.6}$$

$$a(\phi + \psi) = a\phi + a\psi, \tag{3.7}$$

$$(a + b)\phi = a\phi + b\phi, \tag{3.8}$$

$$(ab)\phi = a(b\phi). \tag{3.9}$$

(6) There exist an identity scalar 1 and a null scalar 0 such that

$$1\phi = \phi, \tag{3.10}$$

$$0\phi = 0. \tag{3.11}$$

We use the same notation for the scalar zero and the null vector. This should not lead to any confusion.

If the scalars are complex numbers, the space is often called a *complex* vector space; if the scalars are real, the space is called a *real* vector space.

A set of vectors $\phi_1, \phi_2, \ldots, \phi_n$ belonging to a vector space is called *linearly dependent* if there exist scalars $a_1, a_2, \ldots, a_n$, not all zero, such that

$$a_1\phi_1 + a_2\phi_2 + \cdots + a_n\phi_n = 0. \tag{3.12}$$

Otherwise, the vectors are said to be *linearly independent*. The *dimension* of a vector space is the maximum number of linearly independent vectors in the space. If there are a finite number of these, the space is called a finite-dimensional vector space; otherwise, the space is infinite-dimensional.

A *linear manifold* is a subspace M of a vector space V, such that M itself is a vector space. If $M = V$ or M contains only the null vector, M is said to be an improper subspace of V; otherwise M is a proper subspace. A vector space or linear manifold is *spanned* by a set of vectors $\phi_1, \ldots, \phi_n$ if every vector in the space can be written as a linear combination of the $\phi_1, \ldots, \phi_n$ with scalar coefficients. A linearly independent set of vectors which spans a vector space is called a *basis*. Any vector in the set is called a *basis vector*.

Of special importance in physics is the concept of a unitary vector space in which a scalar product is defined. A *unitary vector space* is a linear vector space in which the scalar product of any two vectors ϕ and ψ, denoted by (ϕ, ψ), has the following properties:

$$(\phi, \psi) \quad \text{is a scalar (a complex number)}, \tag{3.13}$$

$$(\phi, \psi)^* = (\psi, \phi). \tag{3.14}$$

where the asterisk denotes the complex conjugate of a scalar. It is clear from this postulate that (ϕ, ϕ) is a real number. Also

$$(\phi, a\psi) = a(\phi, \psi) \tag{3.15}$$

$$(\phi, \psi + \chi) = (\phi, \psi) + (\phi, \chi) \tag{3.16}$$

$$(\phi, \phi) \geq 0, \qquad (\phi, \phi) = 0 \Rightarrow \phi = 0. \tag{3.17}$$

Any linear vector space in which (3.13) through (3.16) hold is said to be a space with metric. If (3.17) also holds, the metric is said to be positive definite.[1] The quantity

$$(\phi, \phi)^{1/2} = \|\phi\| \tag{3.18}$$

is called the norm[2] of the vector ϕ. If $(\phi, \phi) = 1$, then ϕ is said to be normalized to unity, or more simply, normalized. The *distance* between two vectors ϕ and ψ is defined to be $\|\phi - \psi\|$. Two vectors ϕ and ψ are called *orthogonal* if

$$(\phi, \psi) = 0.$$

Clearly, the null vector is the only vector orthogonal to every vector in the space. That 0 is orthogonal to any vector ϕ is seen by writing $0 = \phi - \phi$. Then $(0, \phi) = (\phi - \phi, \phi) = (\phi, \phi) - (\phi, \phi) = 0$. That 0 is the only such vector follows immediately from the postulate $(\phi, \phi) = 0 \Rightarrow \phi = 0$.

We next come to the important concept of an orthonormal set of vectors. The vectors $\phi_1, \phi_2, \ldots, \phi_n$ form an orthonormal set if

$$(\phi_i, \phi_j) = \delta_{ij}, \tag{3.19}$$

where δ_{ij} is the Kronecker-δ defined by $\delta_{ij} = 1$, $i = j$; $\delta_{ij} = 0$, $i \neq j$. If an orthonormal set spans a vector space V, it is called a complete orthonormal

[1] Some people define such a metric as nonnegative.

[2] Some authors call (ϕ, ϕ) the norm of the vector ϕ.

set or an *orthonormal basis*. Then any vector ψ in V can be written

$$\psi = \sum_i a_i \phi_i, \tag{3.20}$$

where the a_i are scalars. By taking the scalar product of ϕ_j with (3.20) and using (3.19), we see that

$$(\phi_j, \psi) = \sum_i a_i(\phi_j, \phi_i) = \sum_i a_i \delta_{ji} = a_j. \tag{3.21}$$

An important kind of linear vector space is a *Hilbert space*. This is a unitary vector space which is also *complete*. By completeness we mean that any convergent sequence of vectors converges to a vector within the space. With this definition, any unitary space of finite dimension is a Hilbert space. An example of a Hilbert space of infinite dimension is the space of square integrable functions. The scalar product of two functions in this case is defined to be

$$(\phi, \psi) = \int \phi^*(x)\psi(x)\,dx. \tag{3.22}$$

In quantum mechanics, it is postulated that a normalized vector in Hilbert space stands for a possible state of a physical system. The quantity (ψ, ϕ) stands for a transition amplitude from the state ϕ to the state ψ. The physical significance of the amplitude (ψ, ϕ) is that the square of the absolute value of a transition amplitude represents the probability of a transition from the state ϕ to the state ψ. Thus, (ψ, ϕ) is not physically observable, but $|(\psi, \phi)|$ is.

It follows that if ψ denotes a physical state, then

$$\psi' = e^{i\alpha}\psi, \qquad \alpha \quad \text{real},$$

equally well denotes the same physical state. The set of all vectors which differ only by a phase factor (a constant of magnitude unity) is said to be a *ray* in Hilbert space. Thus, a physical state can be denoted by a ray in Hilbert space or by any vector of the ray.

3.2 Operators

An operator on a linear vector space is a mapping of the space into itself. In other words, an operator L acting on a vector ψ in a space V yields a new vector ψ' in the space. We use the notation $\psi' = L\psi$ and write this definition as follows:

$$\psi \in V \Rightarrow L\psi \in V. \tag{3.23}$$

The concept of an operator without further restrictions is too general for our purposes. It is convenient at times to restrict ourselves to operators which satisfy in addition to (3.23) the postulate

$$L(\phi + \psi) = L\phi + L\psi. \tag{3.24}$$

Especially important is a *linear operator* which, *in addition* to (3.23) and (3.24), satisfies

$$L(a\phi) = aL\phi. \tag{3.25}$$

An *antilinear* operator K satisfies the first two of these postulates, but instead of (3.25), it satisfies the following postulate:

$$K(a\phi) = a^* K\phi. \tag{3.26}$$

Suppose we have a set of operators L_i which map a vector space into itself. Consider two operators L_1 and L_2 which map any vector ϕ into corresponding vectors ψ_1 and ψ_2; i.e.,

$$L_1\phi = \psi_1, \qquad L_2\phi = \psi_2.$$

Then we can define the *sum* of L_1 and L_2 as the operator L which maps ϕ into the vector $\psi = \psi_1 + \psi_2$. With this definition, we have

$$(L_1 + L_2)\phi = L_1\phi + L_2\phi.$$

Also, the sum is commutative and associative.

Now if the vector ψ belongs to V, then so does $a\psi$ where a is a scalar. Then, given an operator L, such that $L\phi = \psi$, we can define an operator $L' = aL$ such that

$$L'\phi = aL\phi = a\psi.$$

Thus, we have defined addition of operators and multiplication of operators by scalars. We can further define a null operator which maps every vector onto the unit vector. Also, if $L\phi = \psi$, we can define $-L$ as the operator which maps ϕ into $-\psi$. With these definitions, it is easy to show that the set of operators and scalars satisfy Eqs. (3.1) through (3.11) and so, the operators *themselves* constitute a linear vector space.

Let us now define a *product* of two operators belonging to a vector space as a mapping that associates the given two operators with another operator in the space. Furthermore, suppose the products satisfy the following equations:

$$(aL_1 + bL_2)L_3 = aL_1L_3 + bL_2L_3,$$
$$L_1(aL_2 + bL_3) = aL_1L_2 + bL_1L_3.$$

Then we say that the operators form an *algebra*. (Some authors also require associativity.)

Suppose $L_1\phi = \psi$ and $L_2\psi = \chi$, where L_1 and L_2 are operators. Then we can define the product of the two operators as the operator L which satisfies

$$L\phi = \chi = L_2L_1\phi.$$

If the operators are linear, it is easy to see that they constitute an algebra. Furthermore, this algebra is *associative*, but not necessarily commutative. In Chapter 5, we shall consider a *nonassociative* algebra, in which the product of two operators L_1 and L_2 is defined as the *commutator* of the operators:

$$L = L_1 L_2 - L_2 L_1 = [L_1, L_2].$$

Such an algebra is called a *Lie algebra.* In the future, when we write the product simply as $L_1 L_2$ we shall mean an associative algebra.

We can define the *hermitian adjoint* $L^\dagger$ of a linear operator L acting on a unitary vector space as follows:

$$(\psi, L\phi) = (L^\dagger\psi, \phi). \tag{3.27}$$

A linear operator L is *hermitian* if $L^\dagger = L$. If $L^\dagger = -L$, the operator is said to be *antihermitian.*

If we operate with a linear operator L on a vector ψ, it may happen that we get a constant c times the same vector. If so, ψ is said to be an eigenvector of L and c is the eigenvalue. It is easy to prove that the eigenvectors of a hermitian operator are real and that its eigenvectors belonging to different eigenvalues are orthogonal. If the eigenvectors of a hermitian operator form a complete set or basis, then the operator is said to be *self-adjoint.* We shall restrict ourselves to hermitian operators which are also self-adjoint.

In quantum mechanics, a self-adjoint operator is an *observable.* There may be a limitation on the observability of certain self-adjoint operators because of the existence of so-called superselection rules postulated by Wick *et al.* (1952). (See, however, Aharonov and Susskind, 1967.) However, these considerations are outside the scope of our work.

Every linear operator can be written as a sum of a hermitian and an antihermitian operator. This can be seen as follows: Any linear operator L can be written

$$L = \tfrac{1}{2}(L + L^\dagger) + \tfrac{1}{2}(L - L^\dagger).$$

Since $(L^\dagger)^\dagger = L$, it follows that $\frac{1}{2}(L + L^\dagger)$ is hermitian and $\frac{1}{2}(L - L^\dagger)$ is antihermitian.

A linear operator is *unitary* if

$$U^\dagger = U^{-1}. \tag{3.28}$$

For U to be unitary, both the following conditions must hold

$$U^\dagger U = U U^\dagger = 1. \tag{3.29}$$

Transformations of vectors by unitary operators leave the scalar product invariant. Let $\psi' = U\psi$ and $\phi' = U\phi$. Then

$$(\psi', \phi') = (U\psi, U\phi) = (U^\dagger U\psi, \phi) = (\psi, \phi). \tag{3.30}$$

The converse is not necessarily true. If an operator U satisfies

$$(U\psi, U\phi) = (\psi, \phi),$$

U does not have to be unitary. It is true that $U^\dagger U = 1$, but not necessarily that $UU^\dagger = 1$. We shall not consider this possibility further.

Any unitary operator can be written as follows:

$$U = e^{iaL}, \tag{3.31}$$

where L is a hermitian operator and a is a real number. The operator L is said to be the *generator* of U. An *antiunitary* operator K is an antilinear operator satisfying

$$(K\psi, K\phi) = (\phi, \psi) = (\psi, \phi)^*. \tag{3.32}$$

In quantum mechanics, if ϕ and ψ describe a possible state of a physical system, the quantity $|(\psi, \phi)|^2$ is interpreted as a transition probability from the state ϕ to the state ψ. It is therefore a quantity closely related to the results of an experimental measurement.

For this reason, operators which transform ψ and ϕ but leave $|(\psi, \phi)|$ invariant have a special importance in quantum mechanics. Unitary and antiunitary operators are among the operators having this property.

Wigner has proved an important theorem about operators which leave $|(\psi, \phi)|$ invariant. The theorem states that in considering operators which leave $|(\psi, \phi)|$ invariant, we lose no generality if we restrict ourselves to unitary and antiunitary operators.

Unitary and antiunitary operators as we have seen, are special members of the class which leave $|(\psi, \phi)|$ invariant. In other words, if an operator A leaves $|(\psi, \phi)|$ invariant, i.e.,

$$|(A\psi, A\phi)| = |(\psi, \phi)|, \tag{3.33}$$

we can find a unitary or antiunitary operator B which, when operating on any vector χ in the space, gives a result $B\chi$ which differs from $A\chi$ only by a constant of magnitude unity. The constant may, of course, depend on the vector χ. The proof of the theorem is given by Wigner (1959). Since an antiunitary operator can be written as the product of a unitary operator times complex conjugation, we shall usually confine our attention to unitary operators. Antiunitary operators are important in considering such transformations as time reversal.

We can use an orthonormal basis ϕ_k to write an operator in the form of a matrix. If we operate on one of the vectors ϕ_j with an operator L, we must obtain a vector which can be written as a linear combination of the vectors ϕ_k

$$L\phi_j = \sum_k L_{kj}\phi_k, \tag{3.34}$$

where the coefficients L_{kj} are complex numbers. Taking the scalar product of ϕ_i with $L\phi_j$, we obtain

$$(\phi_i, L\phi_j) = \sum_k L_{kj}(\phi_i, \phi_k) = \sum_k L_{kj}\delta_{ki} = L_{ij}. \tag{3.35}$$

The L_{ij} can be written as a matrix array, or simply matrix, and the L_{ij} are called matrix elements. The number of basis vectors ϕ_i is the same as the dimensionality of the matrix. The matrix is square if the dimensionality is finite. If the dimensionality is infinite, the matrix is square in the sense that there is a one-to-one correspondence between its rows and columns.

The matrix corresponding to the product of two operators is that matrix obtained by ordinary matrix multiplication of the matrices corresponding to the two operators. This is seen as follows: Let $C = AB$, where A, and B, and C are operators. Now, we have

$$B\phi_j = \sum_k B_{kj}\phi_k.$$

Then

$$AB\phi_j = \sum_k AB_{kj}\phi_k = \sum_k B_{kj}A\phi_k = \sum_k B_{kj}\sum_l A_{lk}\phi_l. \tag{3.36}$$

But

$$C\phi_j = \sum_l C_{lj}\phi_l. \tag{3.37}$$

Comparing (3.36) and (3.37), we get

$$C_{lj} = \sum_k A_{lk}B_{kj}, \tag{3.38}$$

which is the definition of matrix multiplication.

The *hermitian conjugate* or *hermitian adjoint* $L^\dagger$ of a matrix L has matrix elements given by

$$(L^\dagger)_{ij} = L^*_{ji}. \tag{3.39}$$

The notion of the hermitian conjugate of a matrix applies even if the matrix is not square. If L has m columns and n rows, $L^\dagger$ has n columns and m rows.

In matrix notation, a vector is a column matrix of real or complex numbers, and the scalar product of two vectors ψ and ϕ is given by

$$(\psi, \phi) = \psi^\dagger\phi = \Sigma\psi_i{}^*\phi_i, \tag{3.40}$$

where

$$\phi = \begin{pmatrix}\phi_1\\ \vdots\end{pmatrix}, \qquad \psi^\dagger = (\psi_1{}^*, \psi_2{}^*, \ldots). \tag{3.41}$$

The numbers $\phi_1, \phi_2, \ldots, \psi_1{}^*, \psi_2{}^*, \ldots$ refer to the set of scalar products of ϕ and ψ with a set of given basis vectors $f_1, f_2, \ldots$, i.e.,

$$\phi_i = (\phi, f_i), \qquad \psi_i{}^* = (f_i, \psi), \qquad i = 1, 2, \ldots.$$

An important transformation on a set of operators and vectors is a *similarity transformation.* This is a transformation by means of a linear operator S which leaves unaltered the algebra in the transformed system. All operators L are transformed by

$$L' = SLS^{-1}, \tag{3.42}$$

and all vectors ϕ by

$$\phi' = S\phi. \tag{3.43}$$

Then, if $\phi + \psi = \chi$, it follows that $\phi' + \psi' = \chi'$ and if $L_1 L_2 \phi = L\phi$, it follows that $L_1' L_2' \phi' = L'\phi'$. Note that S must have an inverse. A unitary transformation is a special case of a similarity transformation.

3.3 Some Properties of Representations

We next come to the important concept of a representation of a group. Physicists deal more with group representations than with the properties of abstract groups.

A *representation* of a group G is a homomorphism of G onto a group of linear operators acting on a linear vector space. If the linear operators are matrices, the representation is called a *matrix representation.* Unless otherwise specified, by a representation we shall mean a matrix representation.

We now consider some properties of representations. If, to every member g_i of a group G, we associate a matrix $D(g_i)$, then, in order for the matrices to form a representation of G, we must have the following:

$$D(g_i)D(g_j) = D(g_i g_j). \tag{3.44}$$

If the group is continuous, there is the additional requirement that if $g_i g_j$ is a neighboring element of g_i, then all the numbers of the matrix $D(g_i g_j)$ must differ by only a small amount from the corresponding numbers of $D(g_i)$. We denote the set of all matrices of a representation of a group G by $D(G)$. If we consider more than one representation, we can distinguish between them by superscript $D^{(i)}(G)$.

As a simple example, we consider four representations of the two-element group e, a.

$$\begin{aligned}
&1.\quad D^{(1)}(e) = 1, && D^{(1)}(a) = -1,\\
&2.\quad D^{(2)}(e) = 1, && D^{(2)}(a) = 1,\\
&3.\quad D^{(3)}(e) = \begin{pmatrix} 1 & 0 \\ 0 & 1 \end{pmatrix}, && D^{(3)}(a) = \begin{pmatrix} 1 & 0 \\ 0 & -1 \end{pmatrix},\\
&4.\quad D^{(4)}(e) = \begin{pmatrix} 1 & 0 \\ 0 & 0 \end{pmatrix}, && D^{(4)}(a) = \begin{pmatrix} -1 & 0 \\ 0 & 0 \end{pmatrix}.
\end{aligned} \tag{3.45}$$

All of these representations are by square matrices (including one-dimensional matrices or numbers). We shall always restrict ourselves to this case, with the understanding that a square infinite matrix is one in which the rows and columns can be put in one-to-one correspondence. The matrices of the representations $D^{(4)}$ have vanishing determinants and therefore do not have inverses. Unless otherwise specified, we shall subsequently restrict ourselves to representations by matrices with nonvanishing determinants. The representations $D^{(1)}$, $D^{(3)}$, and $D^{(4)}$ are isomorphic to the group; $D^{(2)}$ is not. If a representation is isomorphic to the group it is said to be *faithful*. All groups of matrices are faithful representations of themselves.

A representation is often called a *vector* representation to distinguish it from a *ray* representation in which, if a group element g_i is represented by the matrix $D(g_i)$, then it is also represented by all the matrices $e^{i\alpha_i}D(g_i)$, where α_i is real. Then the multiplication law becomes

$$D(g_i)D(g_j) = e^{i\alpha_{ij}}D(g_i g_j), \tag{3.46}$$

where α_{ij} is an arbitrary real number which can depend on g_i and g_j. Sometimes α_{ij} is restricted to take on only a finite number of values, in which case the representation is called multiple-valued. Of special importance are the double-valued representations in which $\alpha_{ij} = 0$ or $\alpha_{ij} = \pi$, so that any group element g_i is represented by two matrices $\pm D(g_i)$. Then

$$D(g_i)D(g_j) = \pm D(g_i g_j). \tag{3.47}$$

Unless otherwise specified, we mean, by representation, a single-valued or vector representation.

Two representations are *equivalent* if one can be transformed into the other by a similarity transformation. By this we mean that all matrices of one of the representations can be transformed into their corresponding matrices by the same similarity transformation. A representation by unitary matrices is called a *unitary* representation.

An important theorem states that a representation of a finite or compact Lie group by matrices with nonvanishing determinants can be transformed into a unitary representation by a similarity transformation. For the proof of the theorem see Wigner (1959). Thus, in considering representations of finite or compact Lie groups, we lose no generality in restricting ourselves to unitary representations.

If a representation can be brought into the following form by a similarity transformation:

$$D(g) = \begin{pmatrix} D_1(g) & X(g) \\ 0 & D_2(g) \end{pmatrix}, \tag{3.48}$$

for all $g \in G$, then the representation is called *reducible*. Otherwise it is called *irreducible*. By the representation being brought into a certain form by a transformation, we mean that every matrix of the representation is brought to that form by the *same* transformation. If the representation can be brought into the following form by a similarity transformation:

$$D(A) = \begin{pmatrix} D_1(g) & 0 \\ 0 & D_2(g) \end{pmatrix}, \tag{3.49}$$

then the representation is called *fully reducible* or *decomposable.*

A theorem states that if a representation by unitary matrices is reducible, then it is fully reducible. From this theorem and the fact that representations of a finite or compact Lie group are equivalent to unitary representations, we get the following:

THEOREM. *If a representation of a finite or compact Lie group is reducible, then it is fully reducible.*

The representation is said to be decomposable into a direct sum

$$D = D^{(1)} + D^{(2)} + \cdots \tag{3.50}$$

of irreducible representations. This sum is unique except for the order in which the irreducible representations are written down. Usually, when referring to such representations, we shall just call them reducible.

An important theorem about group representations is known as

SCHUR'S LEMMA. *Let G be a finite or compact Lie group and let D and D' be irreducible representations of G of dimension n and n', respectively. Let M be an arbitrary $n' \times n$ matrix (n' rows and n columns) which is independent of the elements g of G, and let*

$$MD(g) = D'(g)M,$$

for all g in G. Then if $n' \neq n$ it follows that $M = 0$, while if $n' = n$, either $M = 0$ or M is nonsingular and D and D' are equivalent.

[Note: A corollary, *also called Schur's lemma,* is: *A matrix which commutes with all matrices of an irreducible representation is a multiple of the unit matrix.*]

The *trace* of a matrix is the sum of its diagonal elements. A matrix with trace equal to zero is often said to be traceless. The trace of a matrix $D(g)$ belonging to a representation $D(G)$ is called the *character* of g in the representation D, and is denoted by $\chi(g)$. The set of characters of all members of the representation D is called the character of the representation D and it is denoted by $\chi(D)$.

The importance of character lies in the following:

THEOREM. *All equivalent representations have the same character.*

This theorem is useful in enabling one to decide whether two representations are equivalent.

3.4 Unitary Representations, Multiplets, and Conservation Laws

We now discuss the reasons for the importance of group representations in quantum mechanics. Suppose a group of transformations $T(a)$ leave a physical system invariant. If this is the case, the quantities $|(\phi, \psi)|$, where ϕ and ψ are any vectors (or *functions* or *states*) of the system, are left invariant by the operators representing the transformation. Let us restrict ourselves to the case in which the transformations are represented by unitary operators $U(a)$. Furthermore, consider only those cases in which the Hamiltonian of the system is left invariant by the transformations.

Now consider an eigenfunction ϕ_n of the Hamiltonian H with energy eigenvalue E_n. The eigenvalue equation is

$$H\phi_n = E_n \phi_n. \tag{3.51}$$

If we operate on this equation with the unitary operator $U(a)$, we obtain (omitting the argument a)

$$UH\phi_n = UHU^{-1}U\phi_n = E_n U\phi_n. \tag{3.52}$$

Letting

$$H' = UHU^{-1} \qquad \text{and} \qquad \phi_n' = U\phi_n, \tag{3.53}$$

we get

$$H'\phi_n' = E_n \phi_n'. \tag{3.54}$$

By assumption, the transformation U leaves the Hamiltonian invariant; that is, we have $H' = H$. Therefore, from (3.54) we see that the transformed state ϕ_n' is an eigenstate of the Hamiltonian with the same energy as the state ϕ_n. If we operate with $U(b)$, where b is another value of the parameter, on the state ϕ_n we will obtain, in general, still another eigenstate of H with the same energy E_n. All the states which can be obtained by operating with all U on a given state can be written as linear combinations of a set of basis vectors which span the subspace of eigenstates of the Hamiltonian with the given energy. These vectors are the basis vectors of the unitary representation U of the group of transformations. Furthermore, in general the vectors are the basis vectors of an *irreducible* representation. This can be seen as follows:

Suppose the representation is reducible. Then, since the representation is unitary, it is fully reducible and can be decomposed by a similarity transformation into the form

$$U = \begin{pmatrix} U_1 & 0 \\ 0 & U_2 \end{pmatrix}, \tag{3.55}$$

where U_1 and U_2 are themselves unitary matrices. Likewise the vectors take the form

$$\phi_n = \begin{pmatrix} \phi_{1n} \\ \phi_{2n} \end{pmatrix}, \tag{3.56}$$

where ϕ_{1n} and ϕ_{2n} are themselves column matrices. Operating on ϕ_n with U we obtain

$$\phi_n' = \begin{pmatrix} U_1\phi_{1n} \\ U_2\phi_{2n} \end{pmatrix}. \tag{3.57}$$

Thus the transformed vectors of either subspace never involve a linear combination of vectors from the other subspace. Therefore, the vectors in the two subspaces may have different energies. However, if the vectors are the basis vectors of an *irreducible* representation, any one can be transformed into any of the others (or into linear combinations of them) by the unitary transformations of the group. Therefore, all must have the same energy eigenvalue.

The basis vectors of an irreducible unitary representation of a symmetry transformation denote a set of quantum mechanical states. These states are said to constitute a *multiplet*. Since all of the states of a multiplet are eigenstates of the Hamiltonian with the same energy eigenvalue, the states are said to be *degenerate* in the energy. The degeneracy is the number of states with a given energy. If there is only one state corresponding to a given energy, the state is said to be singly degenerate, or more often, nondegenerate.

As an example of a multiplet, we can consider the different states of a particle with spin J. There exist $2J + 1$ vectors representing different orientations of the spin of the particle with respect to an arbitrary z axis. The unitary operators representing rotations transform these states into linear combinations of one another. Because of the rotational invariance of the Hamiltonian, the $2J + 1$ states all have the same energy and constitute a multiplet of degeneracy $2J + 1$. If $J = 0$, the state is nondegenerate.

As another example, the $2I + 1$ different charge states of a particle with isospin I constitute a multiplet. However, since isospin is not an exact symmetry, the different states are not exactly degenerate in energy. It is often said under such circumstances that different members of the multiplet are different particles, rather than different states of the same particle.

Next we point out the relation between symmetry transformations and conservation laws. If $T(a)$ is a symmetry transformation which can be represented by the operator $U(a)$, then we have (again omitting the argument a):

$$H = UHU^{-1}. \tag{3.58}$$

Operating on this equation from the right with U, we obtain $HU = UH$ or

$$[U, H] = 0, \tag{3.59}$$

where

$$[U, H] = UH - HU, \tag{3.60}$$

is the *commutator* of U and H. But in quantum mechanics, the time dependence of an operator F is given by (see, e.g., Schiff, 1968)

$$\frac{dF}{dt} = \frac{\partial F}{\partial t} + \frac{1}{i\hbar}[F, H]. \tag{3.61}$$

Now since the operator U does not depend explicitly on the time, the fact that it commutes with H shows that it is a conserved quantity (a constant of the motion). But any unitary operator U can be written in terms of a hermitian operator A as

$$U = e^{iaA},$$

where a is real. Since U is conserved, so is the operator A. But since A is hermitian it is an observable. Therefore, if a symmetry transformation can be represented by a unitary operator, there exists a conserved observable quantity. The existence of a conserved observable is said to be a conservation law.

As an example, consider a translation operator $U(a)$ which is defined by

$$U(a)\psi(x) = \psi(x + a) \tag{3.62}$$

We expand $\psi(x + a)$ in a Taylor series about x:

$$\begin{aligned} \psi(x + a) &= \psi(x) + a\frac{d}{dx}\psi(x) + \frac{a^2}{2!}\frac{d^2}{dx^2}\psi(x) + \cdots \\ &= \sum_{n=0}^{\infty}\frac{a^n}{n!}\frac{d^n}{dx^n}\psi(x) \\ &= \exp[ia(-id/dx)]\psi(x). \end{aligned} \tag{3.63}$$

Thus $U(a)$ is given by

$$U(a) = e^{iap_x} \tag{3.64}$$

where p_x is just the ordinary linear momentum operator[3] $-id/dx$. Thus if the Hamiltonian of the system is invariant under translations, linear momentum is conserved.

[3] We let $\hbar = 1$.

CHAPTER 4

THE SYMMETRIC GROUP AND IDENTICAL PARTICLES

4.1 Two- and Three-Particle States

Although we are primarily interested in the unitary group $SU(n)$, especially when $n = 2$, 3, or 6, it is convenient to discuss a few properties of the symmetric or permutation group S_n. We mention two reasons for this. The first is that we shall consider a number of systems of several identical particles, and the symmetric group is the relevant group in this regard. The second reason is that we shall introduce the concept of Young diagrams or Young tableaux, which will prove very useful when we discuss the basis functions of the unitary irreducible representations of $SU(n)$. In this and later chapters, we shall place our primary emphasis on the basis functions or multiplets rather than on the representations of the underlying groups of transformations.

Of all the permutations, those that interchange two objects, called *transpositions* are of special importance. This is because any permutation can be written as a product of transpositions. We denote by P_{ij} the transposition which interchanges the object in the ith and jth box. Note that $P_{ji} = P_{ij}$ and $P_{ij}^2 = 1$. A permutation is even or odd according to whether it is a product of an even or odd number of transpositions.

We now wish to make a correspondence between objects in boxes and quantum mechanical states of identical particles. Suppose we have n nonidentical objects, each of which is in a separate identical box. The boxes are labeled by the numbers 1, 2, ..., n. A permutation

$$\begin{pmatrix} 1, & 2, & 3, & \dots, n \\ a_1, & a_2, & a_3, & \dots, a_n \end{pmatrix},$$

consists of putting the objects originally in box number 1 into box number a_1, etc. Now, suppose we have n identical particles labeled by 1, 2, ..., n, and each one is in a distinct quantum-mechanical state. Then the identical particles correspond to the identical boxes, and the nonidentical objects correspond to the nonidentical states or wave functions of the particles. This is sometimes a bit confusing, since the objects are in the boxes, while the particles are in states. Nevertheless, the boxes correspond to the particles.

We can use a number of different notations to denote a particular arrangement of objects in boxes or particles in states. Let us denote by a, b, and c three nonidentical objects or three nonidentical single-particle states or wave functions. Then the notation

$$a(1)b(2)c(3),$$

means that object a is in box 1, object b is in box 2, and object c is in box 3. Alternatively, this notation means that particle 1 is in state a, particle 2 is in state b, and particle 3 is in state c. From now on, we shall use the language of quantum-mechanical states. If we perform the permutation $\binom{123}{231}$ on the state $a(1)b(2)c(3)$, we obtain the state $a(2)b(3)c(1)$:

$$\begin{pmatrix} 123 \\ 231 \end{pmatrix} a(1)b(2)c(3) = a(2)b(3)c(1). \tag{4.1}$$

There is a simpler notation for these states of identical particles, based on the convention of always writing first the state of particle 1, second the state of particle 2, etc. Then (4.1) becomes

$$\begin{pmatrix} 123 \\ 231 \end{pmatrix} abc = cab. \tag{4.2}$$

We observe that the permutation $\binom{123}{231}$ can be written as the product of two transpositions

$$P_{13}P_{12} = \begin{pmatrix} 123 \\ 321 \end{pmatrix}\begin{pmatrix} 123 \\ 213 \end{pmatrix} = \begin{pmatrix} 213 \\ 231 \end{pmatrix}\begin{pmatrix} 123 \\ 213 \end{pmatrix} = \begin{pmatrix} 123 \\ 231 \end{pmatrix}. \tag{4.3}$$

Note also that we always operate first with the permutation furthermost on the right. Still another notation for the state $a(1)b(2)c(3)$ is $\psi(123)$. This notation has the advantage that the state $\psi(123)$ may be more general than a product of three one-particle states. In this notation $\psi(231)$ corresponds to $a(2)b(3)c(1)$ or cab.

We now wish to consider states of two and three identical particles as examples. We begin with S_2. Suppose we have a state of two identical particles and want to consider the symmetry of that state under the interchange of the two particles. Let the vector describing this state be $\psi(12)$ where 1 and 2 stand for all the coordinates of the first and second particles, respectively. If the original state $\psi(12)$ has no particular symmetry under the interchange of the two particles, we can construct a symmetric state ψ_s or and antisymmetric state ψ_a as follows:

$$\begin{aligned}\psi_s &= \psi(12) + \psi(21),\\ \psi_a &= \psi(12) - \psi(21),\end{aligned} \qquad (4.4)$$

where

$$\psi(21) = P_{12}\psi(12),$$

and P_{12} is a transposition. The state ψ_s is a basis function for a one-dimensional representation of the permutation group. This follows because ψ_s is an eigenstate of any permutation applied to it, as can be directly verified. Thus, ψ_s is a nondegenerate multiplet or singlet. Similarly, ψ_a is an eigenstate of all permutations of S_2 and is a singlet.

Another notation for the two-particle states ψ_s and ψ_a is by means of two boxes, one for each particle. Two boxes in a row denote a symmetric state; two boxes in a column denote an antisymmetric state.

$$\psi_s = \square\square \qquad \psi_a = \begin{array}{c}\square\\ \square\end{array} \qquad (4.5)$$

A collection of boxes to represent a state with a particular symmetry property is known as a *Young diagram* or *Young tableau.*[1] For a one-particle state, there is only one diagram, consisting of a single box.

For a three-particle state, we have the following three possible Young tableaux

$$\psi_s = \square\square\square \qquad \psi_a = \begin{array}{c}\square\\ \square\\ \square\end{array} \qquad \psi_n = \begin{array}{l}\square\square\\ \square\end{array} \qquad (4.6)$$

where we adopt the convention that no row is longer than the row above it. Again the row tableau

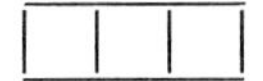

[1] Some authors distinguish between the terminology *Young diagram* and *Young tableau.* To those authors, a tableau is a diagram with a positive integer in each box.

represents a state which is totally symmetric under the interchange of any two of the particles, and the column tableau

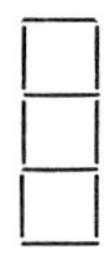

represents a state which is antisymmetric under the interchange. However, there is the possibility of having states of mixed symmetry denoted by the tableau

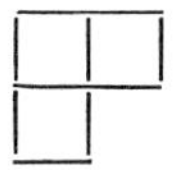

This tableau represents all states which are first made symmetric under the interchange of two of the particles and then made antisymmetric between one of these and the third particle. If we denote the identity operator by 1, then the operator S_{ij} which symmetrizes a state with respect to particles i and j is given by

$$S_{ij} = 1 + P_{ij}. \tag{4.7}$$

Similarly the antisymmetrizing operator A_{ij} is given by

$$A_{ij} = 1 - P_{ij}. \tag{4.8}$$

Then, starting from the state $\psi(123)$, we obtain four independent states ψ_i of mixed symmetry

$$\begin{aligned} \psi_1 &= A_{13} S_{12} \psi(123), \qquad & \psi_2 &= A_{23} S_{12} \psi(123), \\ \psi_3 &= A_{23} S_{13} \psi(123), \qquad & \psi_4 &= A_{12} S_{13} \psi(123). \end{aligned} \tag{4.9}$$

These four mixed states, plus the totally symmetric state ψ_s and the totally antisymmetric state ψ_a, are six linearly independent combinations of the six permutations of $\psi(123)$, including the identity. We could obtain two additional states by using the symmetrizer S_{23}, but these would not be independent of the states (4.9). Alternatively, the mixed tableau

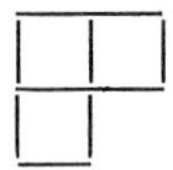

represents the states $\psi_n = S_{ij} A_{ik} \psi(123)$ formed by first antisymmetrizing and then symmetrizing, but again these states are not independent of the states in (4.9). Note, for example, that a state formed by first symmetrizing with respect to particles i and j and then antisymmetrizing with respect to j and k, loses its symmetry under the interchange of i and j. The last-performed symmetrizer or antisymmetrizer is controlling.

The six possible states $\psi(ijk)$ constructed by applying the permutation operators to $\psi(123)$ are the basis functions of a *reducible* representation of S_3. The basis functions for the irreducible representations are those states with definite symmetry properties. Since there is only one symmetric state and one antisymmetric state, the remaining four irreducible states must be of mixed symmetry. However, these states do not form the basis functions of a four-dimensional irreducible representation of S_3, but rather are linear combinations of two two-dimensional irreducible representations. In the next section we shall show how to obtain the dimensionality of the irreducible unitary representations of S_n.

4.2 Standard Arrangements of Young Tableaux

It is true, in general, that the totally symmetric and totally antisymmetric states of S_n are basis functions for one-dimensional irreducible unitary representations. However, all other (inequivalent) irreducible representations of S_n are multidimensional. The dimensionality of these representations can be obtained with the use of Young tableaux. To do this, we must consider so-called standard arrangements of Young tableaux.

Let the number of boxes in the ith row of a tableau be λ_i. We adopt the convention that $\lambda_i \geq \lambda_{i+1}$. Thus a typical tableau looks as follows:

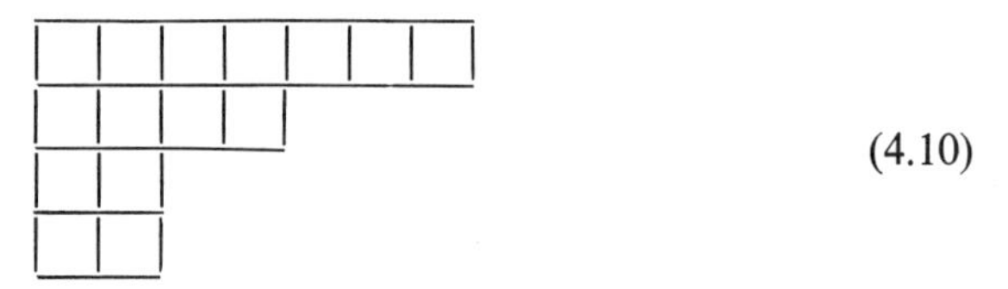

(4.10)

Here $\lambda_1 = 7$, $\lambda_2 = 4$, $\lambda_3 = \lambda_4 = 2$. The tableau corresponds to permutations in which we first symmetrize with respect to each row and then antisymmetrize with respect to each column (or conversely). The tableau (4.10) corresponds to all the possible states of 15 particles with the given symmetry.

It is often more convenient to label a tableau with integers p_i rather than with λ_i. The p_i are defined by

$$p_i = \lambda_i - \lambda_{i+1}. \tag{4.11}$$

For the tableau (4.10) the p_i are given by

$$p_1 = 3, \qquad p_2 = 2, \qquad p_3 = 0, \qquad p_4 = 2.$$

It is easily seen that p_1 is the number of columns with one box, p_2 the number of columns with two boxes, etc. We denote the tableau (4.10) by

$$\lambda = (\lambda_1 \lambda_2 \lambda_3 \lambda_4) = (7422), \qquad \text{or} \qquad p = (p_1 p_2 p_3 p_4) = (3202).$$

Hereafter we shall use only the integers p_i to label a tableau.

We now define a *standard arrangement* of a tableau to be one in which positive integers are placed in the boxes such that the numbers *do not decrease* in going from left to right in a row, and *increase* in going from top to bottom in a column. Each number stands for a possible state in which a single particle can be. If n states are available to a single particle, these states are ordered in some manner, each being assigned a number between 1 and n. Then the number j in each box satisfies $1 \leq j \leq n$.

As examples of the standard arrangements of tableaux we consider two-particle tableaux with three allowed single-particle states. The symmetric tableau is $\begin{array}{|c|c|}\hline \; & \; \\ \hline\end{array}$ and the standard arrangements are

$$\begin{array}{|c|c|}\hline 1 & 1 \\ \hline\end{array} \quad \begin{array}{|c|c|}\hline 1 & 2 \\ \hline\end{array} \quad \begin{array}{|c|c|}\hline 1 & 3 \\ \hline\end{array} \quad \begin{array}{|c|c|}\hline 2 & 2 \\ \hline\end{array} \quad \begin{array}{|c|c|}\hline 2 & 3 \\ \hline\end{array} \quad \begin{array}{|c|c|}\hline 3 & 3 \\ \hline\end{array} \qquad (4.12)$$

The antisymmetric tableau is

$$\begin{array}{|c|}\hline \; \\ \hline \; \\ \hline\end{array}$$

and the standard arrangements are

$$\begin{array}{|c|}\hline 1 \\ \hline 2 \\ \hline\end{array} \quad \begin{array}{|c|}\hline 1 \\ \hline 3 \\ \hline\end{array} \quad \begin{array}{|c|}\hline 2 \\ \hline 3 \\ \hline\end{array} \qquad (4.13)$$

Note that the same number can appear more than once in a given row of a standard arrangement but the numbers in a column must all be different. This is because if two particles are in the same state, this state is symmetric and cannot be antisymmetrized.

We can use the standard arrangements of tableaux to obtain the dimensionality of the irreducible representations of the symmetric group S_n. To do this, we must consider all tableaux with n boxes, each different tableau standing for a different permutation symmetry and hence for a different representation. To obtain the dimensionality of the representation associated with a given tableau of S_n, we count the number of standard arrangements with n states available, and with each particle in a *different* state.

As an example, consider three identical particles, each in a different state, with only three single-particle states available. The standard arrangements of the Young diagrams corresponding to this situation are

$$\begin{array}{|c|c|c|}\hline 1 & 2 & 3 \\ \hline\end{array} \quad \begin{array}{|c|}\hline 1 \\ \hline 2 \\ \hline 3 \\ \hline\end{array} \quad \begin{array}{|c|c|}\hline 1 & 2 \\ \hline 3 \\ \cline{1-1}\end{array} \quad \begin{array}{|c|c|}\hline 1 & 3 \\ \hline 2 \\ \cline{1-1}\end{array} \qquad (4.14)$$

We see there is only one way to make a standard arrangement of the tableau

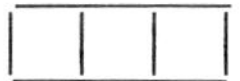

and similarly for the tableau

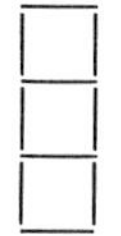

This corresponds to the fact that there is only one symmetric and one antisymmetric combination of three particles. (This is, of course, true for any number of particles.) The nonstandard arrangements correspond to the same states as the standard arrangements after properly symmetrizing, and so must not be counted. For example, the nonstandard arrangement

1	3	2

is the same as the standard arrangement

1	2	3

after symmetrizing with respect to all the particles.

Since there is only one standard arrangement of the symmetric and antisymmetric tableaux, each of them stands for a basis vector of a one-dimensional irreducible unitary representation of S_3. Each is a nondegenerate multiplet or singlet. On the other hand, there are two standard arrangements of the tableau

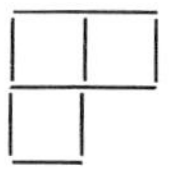

This means that the tableau of mixed symmetry can be used as a shorthand notation for the two functions which are the basis vectors of a two-dimensional irreducible representation of S_3. In other words, the tableau denotes a doublet. So far we have considered only four functions. But we have seen that there are 3! or six functions corresponding to $\psi(123)$ and its permutations. The remaining two functions form the basis vectors of *another* doublet of S_3 represented by the *same* tableau. In general, it can be shown that if an irreducible representation of S_n has N dimensions, there are N such representations of S_n.

As another example, we consider the irreducible representations of S_4. The Young diagrams are

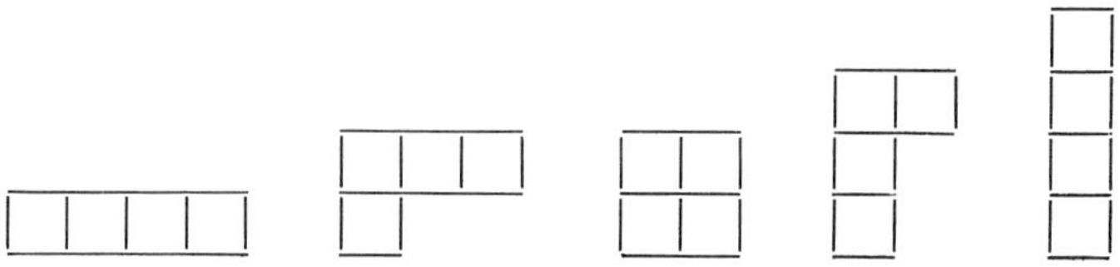

(4.15)

The standard arrangements of these tableaux are

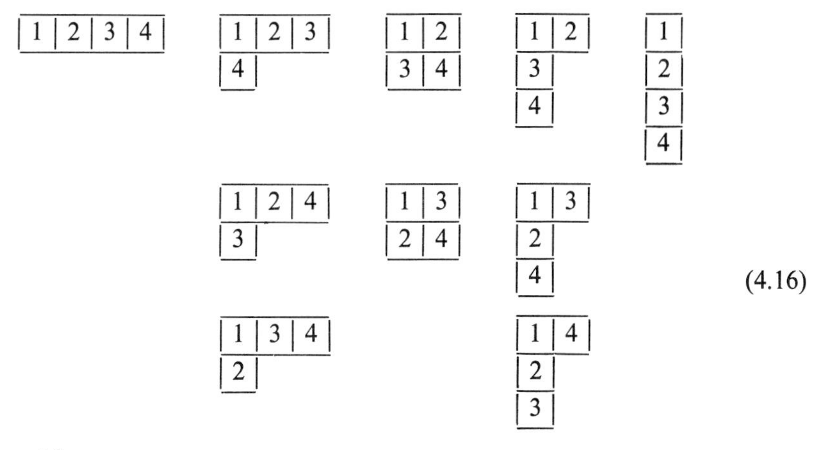

(4.16)

The tableaux

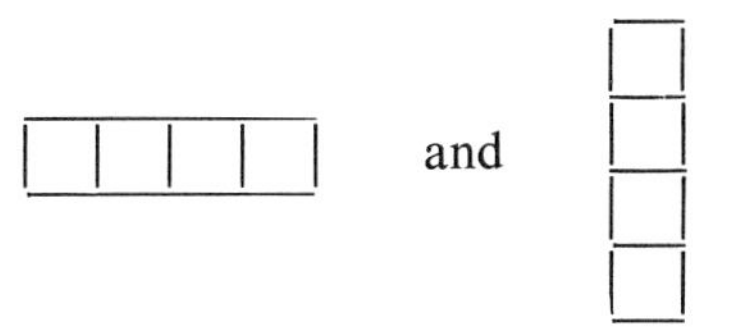

denote the basis vectors of one-dimensional representations. Likewise each of the tableaux

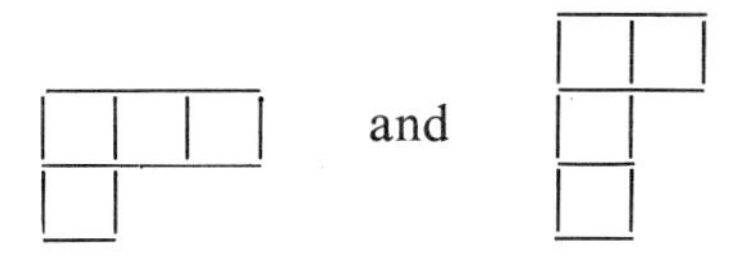

denotes the basis vectors of three three-dimensional representations. Finally, the tableau

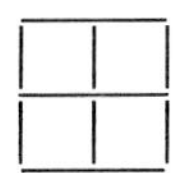

denotes the vectors of two two-dimensional representations. This gives a total of

$$1 + 3^2 + 2^2 + 3^2 + 1 = 24$$

basis vectors. But S_4 has $4! = 24$ elements, and therefore 24 functions can be formed from $\psi(1234)$ and its permutations. We have obtained the symmetry properties of the basis functions of the irreducible representations which can be obtained from a reducible 24-dimensional representation of S_4.

In general, a Young diagram with n boxes represents the basis functions for an irreducible representation of S_n. The basis functions are often called *basis tensors*, rather than basis vectors since they depend on n indices.

We now introduce the notion of a Young diagram which is *conjugate* to a given tableau. There are two ways to define conjugation: The first is for the symmetric group and the second is for $SU(n)$. Here we shall define conjugation only for the symmetric group and postpone defining the second kind of conjugation to Section 7.1.

Suppose we consider a given Young tableau with n boxes representing n identical particles, each in a different state. Then the *conjugate tableau* for the symmetric group is that diagram in which each row becomes a column and each column becomes a row. For example, if we consider four-particle states, we have

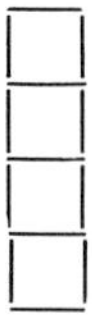 is conjugate to 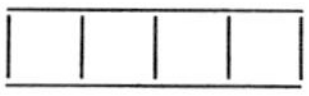

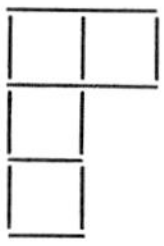 is conjugate to 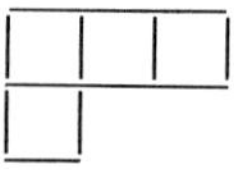

and

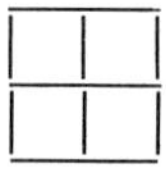 is self conjugate.

Conjugate diagrams belong to different inequivalent representations of the symmetric group with the same dimensionality.

We now reiterate the importance of finding the dimensionality of the irreducible representations. All states which are basis functions of the same irreducible representation of the group (in this case S_n) have the same energy, provided the Hamiltonian of the system is invariant under the group (in this case, permutations of the identical particles). Thus, the dimensionality of an irreducible representation gives the degeneracy of the state, barring any additional accidental degeneracy. The degeneracy associated with the dimensionality of an irreducible representation is often called *essential* degeneracy.

Now suppose the Hamiltonian of a system is invariant under two or more different groups of transformations. These transformations taken together form a larger group, which may or may not be a direct product group, depending on whether or not all the transformations of each group commute with all the transformations of all the others. In any case, the degenerate multiplets are the basis vectors of the irreducible representations of the larger group. If the group is a direct product group, the multiplicities are just the products of multiplicities of the groups comprising the direct product.

All particles observed in nature so far appear to be either bosons or fermions. State vectors describing n identical bosons belong to the symmetric representation of S_n, while the state vectors of n fermions belong to the antisymmetric representation. Thus all particles observed so far appear to belong to one-dimensional representations of the symmetric group, or are singlets.

Why, then, do we study the higher multiplets of S_n? One possible reason is that at some future time particles may be discovered which are neither bosons nor fermions. However, the principal reason for studying higher multiplets of S_n is as follows: Although states of several identical particles are singlets under any permutation of *all* the coordinates of identical particles (i.e., of the particles themselves), nevertheless, the states often exhibit higher multiplicities under permutations of only *some* of the coordinates of the particles. For example, consider three identical spin-1 particles. The overall state of such a system is symmetric. However, the state may have mixed symmetry under the permutations of the spin or space coordinates alone such that the state is symmetric overall.

4.3 Basis Functions of S_3

We construct here a set of orthogonal basis functions for the irreducible unitary representations of S_3 to illustrate how the multiplets are obtained. The state functions of mixed symmetry are not unique, but can be related to others by a similarity transformation. The totally symmetric function is

$$\psi_s = \boxed{1 \mid 2 \mid 3} = S_{123}\, abc$$

where the symmetrizer S_{123} is given by

$$S_{123} = 1 + P_{12} + P_{13} + P_{23} + P_{13}P_{12} + P_{12}P_{13}, \tag{4.17}$$

and the P_{ij} are transpositions. Then

$$\psi_s = abc + bac + cba + acb + cab + bca. \tag{4.18}$$

The totally antisymmetric function is

$$\psi_a = \begin{array}{|c|} \hline 1 \\ \hline 2 \\ \hline 3 \\ \hline \end{array} = A_{123}\, abc$$

where the antisymmetrizer A_{123} is

$$A_{123} = 1 - P_{12} - P_{13} - P_{23} + P_{13}P_{12} + P_{12}P_{13}. \tag{4.19}$$

Then

$$\psi_a = abc - bac - cba - acb + cab + bca. \tag{4.20}$$

Four functions ψ_i of mixed symmetry can be written using the symmetrizers S_{ij} and antisymmetrizers A_{ij} defined in (4.7) and (4.8). We obtain for ψ_1

$$\psi_1 = \begin{array}{|c|c|}\hline 1 & 2 \\ \hline 3 & \\ \cline{1-1}\end{array} = A_{13}\, S_{12}\, abc$$

or

$$\psi_1 = abc + bac - cba - cab. \tag{4.21}$$

For ψ_2 we obtain

$$\psi_2 = \begin{array}{|c|c|}\hline 1 & 2 \\ \hline 3 & \\ \cline{1-1}\end{array} = A_{23}\, S_{12}\, abc,$$

or

$$\psi_2 = abc + bac - acb - bca. \tag{4.22}$$

For ψ_3 we obtain

$$\psi_3 = \begin{array}{|c|c|}\hline 1 & 3 \\ \hline 2 & \\ \cline{1-1}\end{array} = A_{12}\, S_{13}\, abc,$$

or

$$\psi_3 = abc + cba - bac - bca. \tag{4.23}$$

For ψ_4 we obtain

$$\psi_4 = \begin{array}{|c|c|}\hline 1 & 3 \\ \hline 2 & \\ \cline{1-1}\end{array} = A_{23}\, S_{13}\, abc,$$

or

$$\psi_4 = abc + cba - acb - cab. \tag{4.24}$$

The functions ψ_1, ψ_2, ψ_3, ψ_4 are not orthogonal to one another. We construct a function ψ which is a linear combination of ψ_1 and ψ_2 and which is orthogonal to ψ_1. It is given by $\psi = \psi_1 - 2\psi_2$, or

$$\psi = 2acb + 2bca - abc - bac - cba - cab. \tag{4.25}$$

It can be seen (by directly operating on ψ_1 and ψ with all six permutations) that any permutation of either of them is a linear combination of the two functions. Thus, ψ_1 and ψ are an orthogonal basis for a two-dimensional representation of S_3. In other words ψ_1 and ψ form a doublet.

We then construct a function ϕ which is a linear combination of ψ_1, ψ, and ψ_3 and which is orthogonal to ψ_1 and ψ. It is given by

$$\phi = 2\psi_3 + \tfrac{1}{2}\psi_1 + \tfrac{1}{2}\psi,$$

or

$$\phi = 2abc - 2bac + cba - cab + acb - bca. \tag{4.26}$$

Finally we construct a linear combination χ of ψ_1, ψ, ϕ, and ψ_4 which is orthogonal to ψ_1, ψ, and ϕ. It is

$$\chi = \tfrac{1}{3}(4\psi_4 - \psi_1 + \psi - \phi),$$

or

$$\chi = -acb + bca - cab + cba. \tag{4.27}$$

It can be directly verified that the functions ϕ and χ also form a doublet of mixed symmetry. All the functions ψ_s, ψ_a, ψ_1, ψ, ϕ, and χ can be easily normalized.

This technique can be generalized to enable us to obtain the basis functions of S_n. To do so, we must make use of symmetrizers and antisymmetrizers of up to n particles. This method is not often used, since it is possible to obtain more useful basis functions which are simultaneously basis functions of $SU(n)$.

For further discussion of the symmetric group, see, for example, Wigner (1959).

CHAPTER 5

LIE GROUPS AND LIE ALGEBRAS

5.1 Some Definitions and Examples

A continuous group with a finite number of parameters is sometimes called a *finite continuous group*. The number of parameters is the *order* of the group. Let us consider a finite continuous group of order r. Denote a group element by $g(a)$, where now a stands for r parameters, namely $a_1, a_2, \ldots, a_r$. The identity element of the group is characterized by the parameter a^0. Without loss of generality we sometimes take $a^0 = 0$. The group elements have the following properties:

$$g(a^0)g(a) = g(a) \qquad \text{or} \qquad g(0)g(a) = g(a), \tag{5.1}$$

$$g(\bar{a})g(a) = g(a^0) \qquad \text{or} \qquad g(\bar{a})g(a) = g(0), \tag{5.2}$$

where we have defined

$$g(\bar{a}) = [g(a)]^{-1} = g^{-1}(a). \tag{5.3}$$

Also if

$$g(c) = g(a)g(b), \tag{5.4}$$

then c is a continuous function of a and b

$$c = \phi(a, b). \tag{5.5}$$

This notation stands for

$$c_i = \phi_i(a_1, \ldots, a_r, b_1, \ldots, b_r), \qquad i = 1, 2, \ldots, r.$$

If further, c is an analytic function of a and b, and $\bar{a}$ is an analytic function of a, then the group is said to be an *r-parameter Lie group*. Instead of calling a group element $g(a)$, we can simply call it a. The domain over which the parameters vary may comprise a number of disconnected regions. If so, the group is said to be a mixed continuous group.

We now consider groups of transformations on an n-dimensional space by means of an r-parameter Lie group. We assume that the r parameters are essential. If the parameters are not essential, we can find a formal procedure for obtaining the essential parameters (for example, see Hamermesh, 1962). Our treatment follows in part that of Racah (1965) and Behrends *et al.* (1962), but is somewhat less rigorous in the interest of simplicity. We also omit the proofs of a number of theorems. Some of these proofs can be found in Racah (1965).

Suppose we have an n-dimensional vector space with basis vectors $x_1, \ldots, x_n$. A transformation on that space can be written

$$x_i' = f_i(x_1, \ldots, x_n, a_1, \ldots, a_r), \qquad i = 1, 2, \ldots, n, \tag{5.6}$$

or in simpler notation:

$$x' = f(x, a). \tag{5.7}$$

The f_i are all analytic functions of their arguments. Suppose we have a set of transformations $x' = f(x, a)$. We assume that an identity transformation exists and that the transformations are associative. Then, in order for the transformations to form a Lie group, the following requirements must hold.

(1) If $x' = f(x, a)$ and $x'' = f(x', b)$, there exists a parameter c satisfying

$$c = \phi(a, b) \tag{5.8}$$

where ϕ is an analytic function of a and b such that

$$x'' = f(x, c) = f(x, \phi(a, b)). \tag{5.9}$$

(2) For every a there exists a unique $\bar{a}$ such that if $x' = f(x, a)$ then

$$x = f(x', \bar{a}). \tag{5.10}$$

As an example of a Lie group we consider the group of transformations

$$x' = a_1 x + a_2, \qquad a_1 \neq 0. \tag{5.11}$$

Identity: $a_1{}^0 = 1, \qquad a_2{}^0 = 0,$

Inverse: $\bar{a}_1 = 1/a_1, \qquad \bar{a}_2 = -a_2/a_1,$

Product: $c_1 = b_1 a_1, \qquad c_2 = b_1 a_2 + b_2.$

Alternatively, a group element can be written as an ordered pair (a_1, a_2). Then the identity is $(1, 0)$, the inverse is $(1/a_1, -a_2/a_1)$ and the product is given by

$$(b_1, b_2)(a_1, a_2) = (b_1 a_1, b_1 a_2 + b_2).$$

If a_1 and a_2 are real, this is a two-parameter group; otherwise it is a four-parameter group. If we restrict the domain of variation of the parameters, the transformations do not necessarily form a group at all. For example, if a_2 is restricted to be between zero and one, the product of two transformations may be outside the interval.

Two Lie groups are isomorphic if they have the same number of parameters, the same domain of variation of the parameters, and the same law of combination. As an example we note that the one-dimensional group

$$x' = ax, \qquad a \neq 0,$$

is isomorphic to the two-dimensional group

$$x_1{}' = ax_1, \qquad x_2{}' = ax_2, \qquad a \neq 0.$$

Both of these groups are Abelian Lie groups. It should be clear that we can have a one-dimensional group isomorphic to a group of transformations in any number of dimensions.

We shall enlarge the definition of a Lie group to include a mixed analytic group which has a finite number of discontinuous pieces. For example, the orthogonal group $O(3)$ divides into two pieces, one piece consisting of those elements with determinant 1 and the other piece consisting of those elements with determinant -1. Only the elements of the first piece can be reached continuously from the identity.

5.2 Generators of Lie Groups

The great contribution of Sophus Lie (Lie and Scheffers, 1893) to the theory of continuous analytic groups (Lie groups) was to consider those elements which differ infinitesimally from the identity, and to show that from them one can obtain most of the properties of the group in the large. Specifically, Lie showed that the properties of the elements of a Lie group which can be reached continuously from the identity are determined from the elements lying in neighborhood of the unit element. In view of the restrictive nature of analytic functions, this result is not surprising.

Consider a Lie group of transformations which take x into x':

$$x' = f(x, a). \tag{5.12}$$

The vectors x and x' are n-dimensional and a stands for r parameters. In other words, the transformation $x' = f(x, a)$ is shorthand notation for

$$x_i' = f_i(x_1, \ldots, x_n, a_1, \ldots, a_r), \qquad i = 1, \ldots, n.$$

Let us use the convention that the identity element is given by $a = 0$. Then

$$x = f(x, 0). \tag{5.13}$$

We can make a small change in x by varying the identity element by a small amount da, obtaining

$$x + dx = f(x, da). \tag{5.14}$$

Then to lowest order in da we obtain

$$dx = f(x, da) - f(x, 0) = \frac{\partial f(x, 0)}{\partial a}\, da. \tag{5.15}$$

Define $u(x)$ by

$$u(x) = \frac{\partial f(x, 0)}{\partial a}. \tag{5.16}$$

Then

$$dx = u(x)\ da. \tag{5.17}$$

This equation is shorthand for

$$dx_i = \sum_{\nu}^{r} u_{i\nu}(x)\, da_\nu, \qquad i = 1, 2, \ldots, n, \tag{5.18}$$

where

$$u_{i\nu} = \left.\frac{\partial f(x_i, a_\nu)}{\partial a_\nu}\right|_{a_\nu = 0} \equiv \frac{\partial f(x_i, 0)}{\partial a_\nu}.$$

We are using the convention that Greek indices go from 1 to the number of parameters r, and Latin indices go from 1 to the dimension of the space n.

Now consider the change in a function $F(x)$ under an infinitesimal transformation

$$dF = (\partial F/\partial x)\, dx. \tag{5.19}$$

But $dx = u(x)\, da$. Therefore (5.19) becomes

$$dF = (\partial F/\partial x)u(x)\, da = da\{u(x)(\partial/\partial x)\}F. \tag{5.20}$$

Now define the quantity X by

$$X = u(x)\frac{\partial}{\partial x} = \frac{\partial f(x, 0)}{\partial a}\frac{\partial}{\partial x}. \tag{5.21}$$

The operator X is an r-dimensional vector operator with components

$$X_\nu = \sum_{i=1}^{n} u_{i\nu}(\partial/\partial x_i), \qquad \nu = 1, 2, \ldots, r. \tag{5.22}$$

The component operators X_ν are called the *generators* of the group. There is one generator for each essential parameter.

With a suitable choice of parametrization, these generators can be made hermitian. From a set of hermitian generators we can obtain a representation of the group by means of unitary operators $U(a_\nu)$ which are given by

$$U(a_\nu) = e^{ia_\nu X_\nu} \equiv U_\nu \tag{5.23}$$

where the a_ν are real parameters which vary continuously from the identity. An arbitrary element of the group which can be reached continuously from the identity can be written as a product of operators U_ν. Therefore, to obtain that portion of a Lie group which is connected to the identity element, we need only to study the properties of the generators. All the Lie groups with a given set of generators X_ν are *locally isomorphic*, that is isomorphic in a neighborhood of the identity. Furthermore, there exist homomorphic mappings from each of these groups onto the piece which can be reached continuously from the identity, which constitutes an invariant subgroup.

An important theorem is that the commutators of the generators of a Lie group are linear combinations of the generators. See, for example, Racah (1965) for the proof of the theorem. The statement of the theorem can be written

$$[X_\mu, X_\nu] = \sum_{\lambda=1}^{r} c_{\mu\nu\lambda} X_\lambda \tag{5.24}$$

where the $c_{\mu\nu\lambda}$ are complex numbers called *structure constants*. The algebra of the generators is called a *Lie algebra*. Since commutators are linearly independent and satisfy the relation

$$[X_\nu, X_\mu] = -[X_\mu, X_\nu], \tag{5.25}$$

it follows that the structure constants satisfy

$$c_{\mu\nu\lambda} = -c_{\nu\mu\lambda}. \tag{5.26}$$

Also, since commutators satisfy the Jacobi identity

$$[[X_\nu, X_\mu], X_\lambda] + [[X_\mu, X_\lambda], X_\nu] + [[X_\lambda, X_\nu], X_\mu] = 0, \tag{5.27}$$

we obtain

$$\sum_{\beta=1}^{r} (c_{\nu\mu\beta} c_{\beta\lambda\delta} + c_{\mu\lambda\beta} c_{\beta\nu\delta} + c_{\lambda\nu\beta} c_{\beta\mu\delta}) = 0. \tag{5.28}$$

Since the foregoing is somewhat abstract, we shall give some examples of how to obtain the generators of Lie groups and how to obtain the Lie algebras. As a first example, consider the two-parameter group of transformations on a one-dimensional vector

$$x' = \alpha_1 x + \alpha_2, \qquad \alpha = (\alpha_1, \alpha_2).$$

We write this in another way so that the identity element is zero. This is

$$x' = (1 + a_1)x + a_2, \qquad a = (a_1, a_2)$$

where $a_1 = \alpha_1 - 1$ and $a_2 = \alpha_2$. The infinitesimal transformation now is

$$x + dx = (1 + da_1)x + da_2$$

or

$$dx = x\ da_1 + da_2.$$

Then, comparing this expression for dx with the expression in Eq. (5.18), we obtain

$$u_{11} = x, \qquad u_{12} = 1.$$

Then, using (5.22), we obtain the generators

$$X_1 = x(\partial/\partial x), \qquad X_2 = (\partial/dx).$$

We can work out the commutator by operating with it on a function $F(x)$:

$$\begin{aligned}[X_1, X_2]F &= x\frac{\partial^2 F}{\partial x^2} - \frac{\partial}{\partial x}\left(x\frac{\partial F}{\partial x}\right) \\ &= x\frac{\partial^2 F}{\partial x^2} - \frac{\partial F}{\partial x} - x\frac{\partial^2 F}{\partial x^2} = -\frac{\partial F}{\partial x} = -X_2 F.\end{aligned}$$

Thus the commutator is

$$[X_1, X_2] = -X_2,$$

and the structure constants are

$$c_{121} = 0, \qquad c_{122} = -1.$$

It should be clear that the way we parametrize a group is, to some extent, arbitrary. For example the transformation $x' = (1 + a_1)x + a_2$ which we have just considered may be written in terms of new parameters b_1 and b_2 as

$$x' = 1 + (b_1 + b_2)x + b_1 - b_2.$$

If we obtain the generators of the group with this parametrization, they will be different from the generators we previously obtained, and the structure constants will be different. In fact, any linear combination of the generators of a Lie group is also a generator. This means that there is a continuum of different sets of structure constants for a given Lie group. However, all the sets of structure constants correspond to only one Lie algebra.

As another example, consider the two-dimensional, two-parameter Abelian group given by the following transformations

$$x_1' = \alpha_1 x_1,$$
$$x_2' = \alpha_2 x_2.$$

We write this so the identity is zero.

$$x_1' = (1 + a_1)x_1,$$
$$x_2' = (1 + a_2)x_2.$$

Then for an infinitesimal transformation

$$x_1 + dx_1 = (1 + da_1)x_1$$
$$x_2 + dx_2 = (1 + da_2)x_2$$

or

$$dx_1 = x_1\, da_1, \qquad dx_2 = x_2\, da_2.$$

Thus

$$X_1 = x_1(\partial/\partial x_1), \qquad X_2 = x_2(\partial/\partial x_2).$$

In this case the commutator is

$$[X_1, X_2] = 0.$$

In fact, all the structure constants are zero for any Abelian Lie group. Therefore, the Lie algebra for an r-parameter Abelian group is trivial, consisting of r operators X_μ, all of which commute. The Lie algebra is then said to be Abelian.

Still another example is given by the two-dimensional, one-parameter group

$$x_1' = (1 + a)x_1$$
$$x_2' = (1 + a)^{-1}x_2.$$

We have written this transformation so that the identity element is zero. The infinitesimal transformation is

$$x_1 + dx_1 = (1 + da)x_1,$$
$$x_2 + dx_2 = (1 + da)^{-1}x_2 = (1 - da)x_2.$$

Then

$$dx_1 = x_1\, da$$
$$dx_2 = -x_2\, da.$$

Then the single generator is

$$X = x_1(\partial/\partial x_1) - x_2(\partial/\partial x_2).$$

Now let us consider the special orthogonal group in three dimensions $SO(3)$ or $R(3)$. We use matrix notation

$$x' = Ax \tag{5.29}$$

where x and x' are three-dimensional vectors and A is a real three-dimensional matrix satisfying $A\tilde{A} = 1$; where $\tilde{A}$ means the transpose of A, that is $\tilde{A}_{ij} = A_{ji}$. Writing the transformation in the form $x' = (1 + a)x$ so that the identity $a^0 = 0$, we obtain

$$x + dx = (1 + da)x, \qquad dx = x\, da.$$

Now

$$A\tilde{A} = 1 \Rightarrow (1 + da)(1 + d\tilde{a}) = 1,$$

or to lowest order in a

$$1 + da + d\tilde{a} = 1, \qquad \text{or} \qquad da = -d\tilde{a}.$$

Thus da must be of the form

$$da = \begin{pmatrix} 0 & da_{12} & -da_{13} \\ -da_{12} & 0 & da_{23} \\ da_{13} & -da_{23} & 0 \end{pmatrix}. \tag{5.30}$$

Thus the group has three essential parameters. Letting $da_{23} = a_1$, $da_{13} = a_2$, $da_{12} = a_3$ we get

$$\begin{aligned} dx_1 &= x_2 a_3 - x_3 a_2, \\ dx_2 &= -x_1 a_3 + x_3 a_1, \\ dx_3 &= x_1 a_2 - x_2 a_1. \end{aligned}$$

Then the operators are

$$\begin{aligned} X_1 &= x_3 \frac{\partial}{\partial x_2} - x_2 \frac{\partial}{\partial x_3}, \\ X_2 &= -x_3 \frac{\partial}{\partial x_1} + x_1 \frac{\partial}{\partial x_3}, \\ X_3 &= x_2 \frac{\partial}{\partial x_1} - x_1 \frac{\partial}{\partial x_2}. \end{aligned} \tag{5.31}$$

We can obtain the commutation relations in a straightforward way:

$$[X_1, X_2] = X_3, \qquad [X_2, X_3] = X_1, \qquad [X_3, X_1] = X_2. \tag{5.32}$$

In parametrizing a Lie group, we have taken all the essential parameters to be real. Therefore, in general the Lie algebra is an algebra with real scalars. However, we can take certain complex linear combinations of the generators and still preserve the algebra. We shall often have occasion to do this. For example, in the case of $R(3)$, we can let $J_\nu = iX_\nu$. Then the commutation relations of Eq. (5.32) become

$$[J_\mu, J_\nu] = i\varepsilon_{\mu\nu\lambda} J_\lambda, \tag{5.33}$$

where $\varepsilon_{\mu\nu\lambda}$ is antisymmetric under odd permutations and $\varepsilon_{123} = 1$. A comparison of (5.32) and (5.33) again shows that the structure constants multiplying the operators depend on the choice of operators.

We next consider the group $SU(2)$ and show that it has the same Lie algebra as $R(3)$. The importance of this is that if two Lie groups have the same Lie algebra they are locally isomorphic; that is, they are isomorphic in a neighborhood of the identity. Furthermore, if the two groups are not isomorphic in the large, there exists a homomorphic mapping of the larger one on to the smaller one.

The group $SU(2)$ consists of the unitary matrices of determinant unity (denoted by A) acting on a complex two-dimensional vector space. The transformation is given by

$$x' = Ax. \tag{5.34}$$

We write the infinitesimal transformations as

$$x' = (1 + da)x, \qquad dx = dax \tag{5.35}$$

Since $AA^\dagger = 1$, we have

$$(1 + da)(1 + da^\dagger) = 1 = 1 + da + da^\dagger$$

or

$$da + da^\dagger = 0. \tag{5.36}$$

Adding the requirement that

$$\det(1 + da) = 1, \tag{5.37}$$

we find we can write da in the form

$$da = \begin{pmatrix} ia_1 & a_2 + ia_3 \\ -a_2 + ia_3 & -ia_1 \end{pmatrix}. \tag{5.38}$$

where the a_i are infinitesimal real numbers.

$$dx_1 = ia_1x_1 + (a_2 + ia_3)x_2,$$
$$dx_2 = (-a_2 + ia_3)x_1 - ia_1x_2.$$

The generators are

$$X_1 = ix_1 \frac{\partial}{\partial x_1} - ix_2 \frac{\partial}{\partial x_2},$$
$$X_2 = x_2 \frac{\partial}{\partial x_1} - x_1 \frac{\partial}{\partial x_2}, \tag{5.39}$$
$$X_3 = ix_2 \frac{\partial}{\partial x_1} + ix_1 \frac{\partial}{\partial x_2}.$$

The commutation relations satisfied by these operators are

$$[X_1, X_2] = -2X_3, \tag{5.40}$$

and cyclic permutations. If we let $X_\mu = -2iJ_\mu$, we find that the J_μ satisfy the commutation relations (5.33). This shows that $SU(2)$ has the same Lie algebra as $R(3)$. The two groups are not isomorphic, but there is a homomorphic mapping of one onto the other. In this case there is a two-to-one mapping of $SU(2)$ onto $R(3)$.

Equations (5.34) through (5.37) hold for the special unitary group in any number of dimensions. For $SU(n)$, however, the infinitesimal element da, instead of being given by (5.38) is given in terms of $n^2 - 1$ parameters. For $SU(3)$, da is given in terms of eight real infinitesimal parameters c_ν.

$$da = \begin{pmatrix} ic_1 & c_2 + ic_3 & c_4 + ic_5 \\ -c_2 + ic_3 & c_6 & c_7 + ic_8 \\ -c_4 + ic_5 & -c_7 + ic_8 & -ic_1 - ic_6 \end{pmatrix}.$$

With this parametrization, the generators are

$$\begin{aligned} X_1 &= ix_1 \frac{\partial}{\partial x_1} - ix_3 \frac{\partial}{\partial x_3}, \\ X_2 &= x_2 \frac{\partial}{\partial x_1} - x_1 \frac{\partial}{\partial x_2}, \\ X_3 &= ix_2 \frac{\partial}{\partial x_1} + ix_1 \frac{\partial}{\partial x_2}, \\ X_4 &= x_3 \frac{\partial}{\partial x_1} - x_1 \frac{\partial}{\partial x_3}, \\ X_5 &= ix_3 \frac{\partial}{\partial x_1} + ix_1 \frac{\partial}{\partial x_3}, \\ X_6 &= ix_2 \frac{\partial}{\partial x_2} - ix_3 \frac{\partial}{\partial x_3}, \\ X_7 &= x_3 \frac{\partial}{\partial x_2} - x_2 \frac{\partial}{\partial x_3}, \\ X_8 &= ix_3 \frac{\partial}{\partial x_2} + ix_2 \frac{\partial}{\partial x_3}. \end{aligned} \tag{5.41}$$

If a group has r generators, then the number of commutation relations which must be worked out is $\frac{1}{2}(r-1)r$. This can be seen as follows: Every generator commutes with itself. Thus, there are $r - 1$ commutation relations between the first generator and the others. Likewise, there are $r - 2$ commutation relations between the second and all the others except the first, etc. The total number of commutation relations is then

$$(r - 1) + (r - 2) + \cdots + 1 = \tfrac{1}{2}(r - 1)r. \tag{5.42}$$

Since $SU(n)$ has $r = n^2 - 1$ parameters it has $n^2 - 1$ generators, and there are $\frac{1}{2}(n^2 - 2)(n^2 - 1)$ commutation relations to be worked out. In the next section we shall see how to simplify the problem of obtaining the commutation relations.

5.3 Simple and Semisimple Lie Algebras

Consider a subspace B of a Lie algebra A. If the commutator of any operator in A with any operator in B also lies in B, then B is said to be an *ideal.* A Lie algebra is *simple* if it is not Abelian and its only nonzero ideal is A itself. A Lie algebra is *semisimple* if it has no nonzero Abelian ideals. Since all the generators of an Abelian Lie group commute, the Lie algebra of an Abelian Lie group is not semisimple. Likewise, the Lie algebra of a one-parameter group is not semisimple. We can now give a definition of a simple Lie group which is more precise than that given in Section 2.4. If a Lie algebra is simple or semisimple, then so are all the Lie groups corresponding to the algebra.

The Lie algebra of a direct product of two groups is just the algebra of the generators of both. The generators of one of the groups in the direct product, of course, all commute with the generators of the other group. The Lie algebra of the direct product of two simple groups is semisimple.

It is often convenient to study Lie algebras rather than Lie groups. The reason is that by considering a single Lie algebra, one is, in effect, considering a whole class of Lie groups, all of which are locally isomorphic. However, the Lie algebra of a group does not tell about the global properties of the group. Thus, the question arises as to the importance of the global properties for physics. As we have remarked previously, we are more interested in the representations of Lie groups than the abstract groups themselves. Now, in general, if two groups differ globally, then one of them has more vector representations than the other. However, in quantum mechanics we should not exclude multivalued representations. When these are included, then all groups with the same Lie algebra have the same representations.

5.4 Standard Form of Lie Algebras

Let us consider further properties of the generators X_μ of a given Lie group. The number of generators which are mutually commuting operators depends on the particular choice of X_μ. For example, suppose we have three operators X_1, X_2, and X_3 such that

$$[X_1, X_2] = 0, \qquad [X_1, X_3] \neq 0, \qquad [X_2, X_3] \neq 0.$$

If we define

$$X_1' = X_1 + X_3, \qquad X_2' = X_2, \qquad X_3' = X_3,$$

then none of the operators X_1', X_2', X_3' commute. Yet this is just another way to describe the same Lie algebra.

An important question is what is the *maximum* number of commuting generators of the algebra. In other words, we want to know how many generators can be made to commute by choosing new generators to be arbitrary linear combinations of the old. If the group is a symmetry group of the system, all commuting generators will also commute with the Hamiltonian. If the generators are chosen to be hermitian, the commuting set of generators will be simultaneously observable constants of the motion.

We now discuss the question of how to find the maximum number of commuting generators of a Lie group, given a particular set of generators X_μ. Let us choose a linear combination A of the X_μ which can be diagonalized. We let

$$A = \sum_\mu a_\mu X_\mu, \tag{5.43}$$

where the sum goes from 1 to r, the number of generators of the group. Consider the "eigenvalue" problem

$$[A, X] = \rho X, \tag{5.44}$$

where the unknown X is again a linear combination of the X_μ:

$$X = \sum_\mu b_\mu X_\mu, \tag{5.45}$$

and ρ is an eigenvalue to be determined. But we know

$$[X_\mu, X_\nu] = \sum_\lambda c_{\mu\nu\lambda} X_\lambda. \tag{5.46}$$

Substituting these expressions for A and X into Eq. (5.44) and using Eq. (5.46), we obtain

$$\left[\sum_\mu a_\mu X_\mu, \sum_\nu b_\nu X_\nu\right] = \sum_{\mu\nu\lambda} a_\mu b_\nu c_{\mu\nu\lambda} X_\lambda = \rho \sum_\lambda b_\lambda X_\lambda,$$

or

$$\sum_\lambda \left[\sum_{\mu\nu} a_\mu b_\nu c_{\mu\nu\lambda} - \rho b_\lambda\right] X_\lambda = 0. \tag{5.47}$$

Since the X_λ are linearly independent, each coefficient must vanish:

$$\sum_{\mu\nu} a_\mu b_\nu c_{\mu\nu\lambda} - \rho b_\lambda = 0,$$

or

$$\sum_\nu \Big(\sum_\mu a_\mu c_{\mu\nu\lambda} - \rho\delta_{\nu\lambda}\Big) b_\nu = 0. \qquad (5.48)$$

This is a set of homogeneous linear equations for b. For a solution to exist, the determinant of the coefficients of b_ν must vanish:

$$\det\Big(\sum_\mu a_\mu c_{\mu\nu\lambda} - \rho\delta_{\nu\lambda}\Big) = 0. \qquad (5.49)$$

This equation is called the secular equation for the eigenvalues ρ. These eigenvalues are called *roots*.

If we have an r-parameter group, there may be fewer than r linearly independent eigenvectors if the secular equation has degenerate roots. By a degenerate root, we mean the following: The secular equation is an rth order polynomial in ρ, and thus has r roots. However, all these roots need not be different; if two or more are the same, the root is said to be *degenerate*. The *multiplicity* of a root is the number of times it occurs.

For the groups of interest in most physical problems, Eq. (5.44) does have r independent eigenvectors independently of whether any root is degenerate. In particular, this is true for a semisimple Lie group. An important theorem about semisimple Lie groups has been proven by Cartan.

THEOREM. *If A is chosen so that the secular equation has the maximum number of roots, then only the root $\rho = 0$ is degenerate. Furthermore, if l is the multiplicity of this root, there are l linearly independent eigenvectors H_i $(i = 1, 2, \ldots, l)$ which commute with one another*

$$[H_i, H_j] = 0. \qquad (5.50)$$

The number l is called the *rank* of the group, and is the maximal number of mutually commuting generators. Since A commutes with itself, the rank of a semisimple Lie group is $l \geq 1$. We shall use Latin letters for indices which can vary from 1 to l, Greek letters α, β, γ for indices which can vary from 1 to $r - l$, and other Greek letters for indices which can vary from 1 to r.

We denote the noncommuting eigenvectors (corresponding to nonzero nondegenerate values of ρ) by the symbols E_α. Thus, for a rank l semisimple Lie group, we have the following eigenvectors.

$$H_i, \quad i = 1, 2, \ldots, -l, \qquad E_\alpha, \quad \alpha = 1, 2, \ldots, r - l. \qquad (5.51)$$

The H_i and E_α are, of course, generators of the group.

Let us use $SU(2)$ as an example: We take for the original X_μ the J_μ which satisfy the commutation relations of Eq. (5.33). Let $A = J_3$. Then if we solve the equation

$$[A, X] = \rho X,$$

we obtain for the three values of X

$$\begin{aligned} X &= J_3 \equiv H_1 = A, \qquad \rho = 0, \\ X &= (J_1 + iJ_2)/\sqrt{2} \equiv E_1, \qquad \rho = 1, \\ X &= (J_1 - iJ_2)/\sqrt{2} \equiv E_2, \qquad \rho = -1, \end{aligned} \tag{5.52}$$

or

$$[H_1, H_1] = 0, \qquad [H_1, E_1] = E_1, \qquad [H_1, E_2] = -E_2. \tag{5.53}$$

The operators E_1 and E_2 are just the usual raising and lowering operators $J_\pm$ of angular momentum except for normalization

$$\sqrt{2}\,E_1 = J_+, \qquad \sqrt{2}\,E_2 = J_-. \tag{5.54}$$

Note that the nonvanishing values of ρ are $\rho = \pm 1$. It is true for any semisimple compact group that the nonvanishing roots come in pairs equal in magnitude and opposite in sign.

Let the roots, denoted by $\rho_i(\alpha)$, be solutions to the equations

$$[H_i, E_\alpha] = \rho_i(\alpha) E_\alpha, \qquad i = 1, 2, \ldots, l. \quad \alpha = 1, 2 \ldots, r - l. \tag{5.55}$$

Then the l values of ρ_i for a given α can be considered to be the components of a *root vector* (or just *root*) of dimensionality equal to the rank of the group. The l operators H_i can also be considered as the components of a vector $\mathbf{H}$, and we can write

$$[\mathbf{H}, E_\alpha] = \boldsymbol{\rho}(\alpha) E_\alpha. \tag{5.56}$$

We now quote without proof two theorems for semisimple compact groups.

THEOREM. *If* $\boldsymbol{\rho}(\alpha)$ *is a root, then* $-\boldsymbol{\rho}(\alpha)$ *is also a root.*

We shall now use the notation $\boldsymbol{\rho}(-\alpha) \equiv -\boldsymbol{\rho}(\alpha)$ and let α take on $\frac{1}{2}(r - l)$ values rather than $r - l$ values.

THEOREM. *The* E_α *satisfy the following commutation relations*

$$[E_\alpha, E_{-\alpha}] = \sum_i c_{\alpha, -\alpha, i} H_i, \qquad \alpha = 1, 2, \ldots, \pm\tfrac{1}{2}(r - l), \tag{5.57}$$

$$[E_\alpha, E_\beta] = \begin{cases} c_{\alpha\beta\gamma} E_\gamma, & \textit{if} \quad \boldsymbol{\rho}(\gamma) = \boldsymbol{\rho}(\alpha) + \boldsymbol{\rho}(\beta) \neq 0 \\ 0 & \textit{otherwise}. \end{cases} \tag{5.58}$$

We can normalize the H_i such that

$$\sum_{\alpha=1}^{(r-l)/2} \rho_i(\alpha)\rho_j(\alpha) = \delta_{ij}. \tag{5.59}$$

Then it can be shown that we can normalize the E_α such that

$$[E_\alpha, E_{-\alpha}] = \sum_i \rho_i(\alpha)H_i, \tag{5.60}$$

or in vector notation

$$[E_\alpha, E_{-\alpha}] = \boldsymbol{\rho}(\alpha) \cdot \mathbf{H}. \tag{5.61}$$

We introduce the notation $N_{\alpha\beta} = c_{\alpha\beta, \alpha+\beta}$ and collect Eqs. (5.50), (5.56), (5.58), and (5.61). Then we have the Lie algebra in standard form:

$$\begin{aligned} [H_i, H_j] &= 0, \\ [\mathbf{H}, E_\alpha] &= \boldsymbol{\rho}(\alpha)E_\alpha, \\ [E_\alpha, E_{-\alpha}] &= \boldsymbol{\rho}(\alpha) \cdot \mathbf{H}, \end{aligned} \tag{5.62}$$

$$[E_\alpha, E_\beta] = \begin{cases} N_{\alpha\beta} E_{\alpha+\beta} & \text{if } \boldsymbol{\rho}(\alpha) + \boldsymbol{\rho}(\beta) \text{ is a nonvanishing root} \\ 0 & \text{otherwise.} \end{cases}$$

The $N_{\alpha\beta}$ can be directly computed once **H** and **E** are known. These equations are called the *standard form* of the commutation relations. For $SU(2)$, the standard form of the commutation relations is given by

$$\begin{aligned} [H_1, E_1] &= E_1, \\ [H_1, E_{-1}] &= -E_{-1}, \\ [E_1, E_{-1}] &= H_1, \end{aligned} \tag{5.63}$$

where

$$H_1 = J_3, \qquad \sqrt{2}E_1 = J_+ = J_1 + iJ_2, \qquad \sqrt{2}E_{-1} = J_- = J_1 - iJ_2. \tag{5.64}$$

Since there is only one positive value of α, there are no $N_{\alpha\beta}$ to be computed in this case.

Next we shall quote two important theorems about Lie groups, and then discuss the roots in some more detail.

THEOREM. *From the members of a Lie algebra of a semisimple Lie group of rank l, one can construct l nonlinear invariant operators, called Casimir operators, which commute with every member of the algebra.*

For example, the group $SU(2)$ (rank $l = 1$) has one invariant operator $J^2 = J_1{}^2 + J_2{}^2 + J_3{}^2$.

THEOREM. *There are only a finite number of simple Lie algebras of any rank l.*

The importance of this theorem for applications in elementary particle physics is that there appear to be two relevant additive conserved internal quantum numbers specifying elementary particles: the hypercharge Y and the third component of the isospin I_3. If I_3 and Y are considered to be commuting operators of a simple Lie algebra of rank 2, then we need consider only a finite number of such algebras. By comparing the predictions of all these algebras with experiment, we can pick out the one which agrees best with the facts. This happens to be the algebra of $SU(3)$.

We now return to a discussion of the roots of a Lie group. We consider a *root diagram*, which is the graphical representation of the root vectors in an l-dimensional space. Since it is hard to draw root diagrams of three dimensions on a two-dimensional surface and even harder to visualize a space of more than three dimensions, we shall restrict much of our discussion to groups of rank $l \leq 2$.

THEOREM. *If* $\boldsymbol{\rho}(\alpha)$ *and* $\boldsymbol{\rho}(\beta)$ *are two roots then*

$$\frac{2\boldsymbol{\rho}(\alpha)\cdot\boldsymbol{\rho}(\beta)}{\rho^2(\alpha)} \quad \text{and} \quad \frac{2\boldsymbol{\rho}(\alpha)\cdot\boldsymbol{\rho}(\beta)}{\rho^2(\beta)}, \tag{5.65}$$

are integers, and

$$\boldsymbol{\rho}(\beta) - 2\boldsymbol{\rho}(\alpha)\frac{\boldsymbol{\rho}(\alpha)\cdot\boldsymbol{\rho}(\beta)}{\rho^2(\alpha)} = \boldsymbol{\rho}(\gamma) \tag{5.66}$$

is a root.

The geometrical interpretation of the theorem is that the root $\boldsymbol{\rho}(\gamma)$ is the reflection of an l dimensional root $\boldsymbol{\rho}(\beta)$ with respect to a hyperplane (of $l-1$ dimensions) perpendicular to $\boldsymbol{\rho}(\alpha)$.

For a rank 1 group, the root diagram is one-dimensional, and all the roots lie on the same line. The hyperplane degenerates to a point in this case, and we obtain no new information from the theorem. However, a previous theorem, that if $\boldsymbol{\rho}(\alpha)$ is a root, then $\boldsymbol{\rho}(-\alpha) = -\boldsymbol{\rho}(\alpha)$ is also a root, is useful. Using this theorem, we obtain the root diagram of Fig. 5.1. In fact this root diagram is the only root diagram for a rank one simple group. The Lie algebra of this group is the Lie algebra of $SU(2)$.

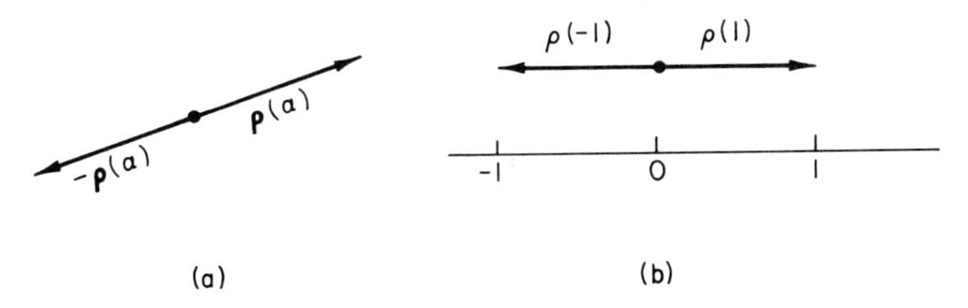

Fig. 5.1. (a) Root diagram obtained from the theorem that states if $\boldsymbol{\rho}(\alpha)$ is a root, then $-\boldsymbol{\rho}(\alpha)$ is a root. (b) Normalized root diagram for all rank 1 simple Lie groups. All these groups [e.g., $SU(2)$, $O(3)$] have the same Lie algebra.

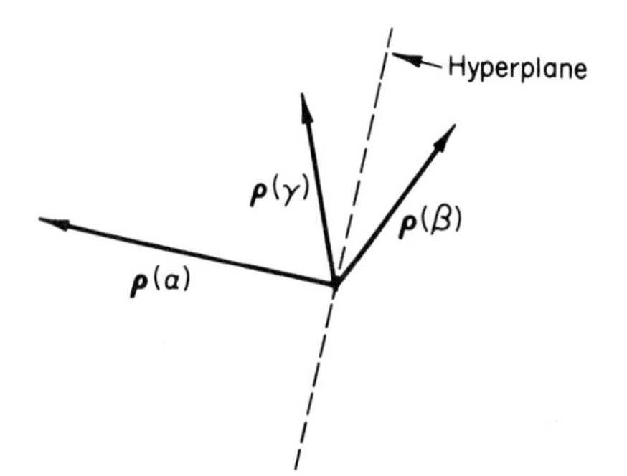

Fig. 5.2. Given two roots $\boldsymbol{\rho}(\alpha)$ and $\boldsymbol{\rho}(\beta)$, we construct a third root $\boldsymbol{\rho}(\gamma)$ by drawing a hyperplane (in two dimensions a line) perpendicular to $\boldsymbol{\rho}(\alpha)$. Then the reflection of $\boldsymbol{\rho}(\beta)$ with respect to this hyperplane is the third root.

We now consider simple rank 2 groups. The hyperplane in this case is a line. If we have two roots $\boldsymbol{\rho}(\alpha)$ and $\boldsymbol{\rho}(\beta)$ making an angle not equal to 90° or 180°, we obtain a third distinct root [not obtainable from the knowledge that $-\boldsymbol{\rho}(\alpha)$ and $-\boldsymbol{\rho}(\beta)$ are roots]. Figure 5.2 shows the construction. The theorem also states that

$$\frac{\boldsymbol{\rho}(\alpha)\cdot\boldsymbol{\rho}(\beta)}{\rho^2(\alpha)} = \frac{1}{2}m, \qquad \frac{\boldsymbol{\rho}(\alpha)\cdot\boldsymbol{\rho}(\beta)}{\rho^2(\beta)} = \frac{1}{2}n,$$

where m and n are integers. Letting the angle between $\boldsymbol{\rho}(\alpha)$ and $\boldsymbol{\rho}(\beta)$ be ϕ, and multiplying the two equations together, we obtain

$$\cos^2\phi = \tfrac{1}{4}mn.$$

The solutions to this equation are

$$\phi = 0°, \quad 30°, \quad 45°, \quad 60°, \quad 90°.$$

This makes plausible the theorem that there are only a finite number of simple compact groups of a given rank. There is another relevant theorem due to Cartan (1933):

THEOREM. *Every semisimple compact group can be written as a direct product of simple compact groups.*

This theorem implies that there are only a finite number of semisimple compact algebras of a given rank. Also we need consider only the simple algebras to obtain all the relevant information about the semisimple ones. All the Lie algebras of simple compact groups have been classified by Cartan (1933). We shall give here only the root diagrams for the ones of rank 2. The first is $SU(3)$ (or A_2 in Cartan's classification). The groups A_l of Cartan are of order $(l+1)^2 - 1$ and of rank l. They are just the groups $SU(l+1)$. The root diagram of $SU(3)$ is shown in Fig. 5.3. With the aid of the root diagram of Fig. (5.3) and Eqs. (5.62), we can write down the commutation

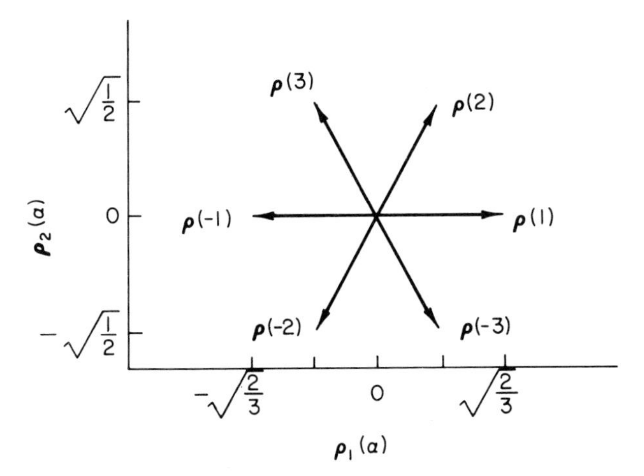

Fig. 5.3. Normalized root diagram for $SU(3)$, or A_2 according to Cartan's classification.

relations of $SU(3)$ in standard form. However, the normalization constants $N_{\alpha\beta}$ are not given in terms of the roots but must be computed separately. In Fig. 5.4 are given the only other root diagrams of simple rank 2 groups. See Cartan (1933) or Racah (1965) for further information about these groups.

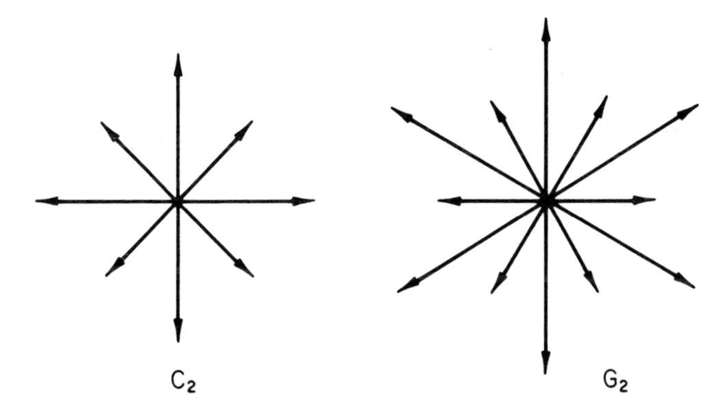

Fig. 5.4. Root diagrams of the rank 2 groups C_2 and G_2 according to the classification of Cartan (1933). See also Racah (1965). (The group B_2 of Cartan's classification has the same diagram as C_2 rotated by 45°.)

CHAPTER 6

MULTIPLETS

6.1 Diagonal Generators and Weights

We shall now discuss the states which are the basis vectors of irreducible unitary representations of Lie groups. The basis vectors of such a representation constitute a multiplet.

As an example, we first consider the multiplets of the rotation group in three dimensions $R(3)$. The basis vectors for the representations of this group are the ordinary spherical harmonics[1] $Y_j^m(\theta\phi)$. These functions are eigenstates of the generator J_3 and of the invariant operator

$$J^2 = J_1^{\,2} + J_2^{\,2} + J_3^{\,2}, \tag{6.1}$$

which commutes with all the generators. The quantum number j labels the representation and the quantum number m labels the different vectors belonging to the representation, i.e., the different members of the multiplet. Since m takes on all integral values from $-j$ to j, the dimensionality of the representation is $2j + 1$.

If we are interested merely in how the Y_j^m transform among themselves when operated on by the generators, we can represent the Y_j^m by column vectors and J_i $(i = 1, 2, 3)$ by square matrices in a space of $2j + 1$ dimensions.

[1] It is customary to use the notation Y_l^m for the spherical harmonics, but we reserve the letter l for the rank of a group.

If double-valued representations are included, j can take on both integral and half-integral values. The multiplicity N of a state belonging to the representation specified by j is

$$N = 2j + 1. \tag{6.2}$$

We shall adopt the notation that the quantum numbers labeling a representation are written as superscripts, and the quantum numbers distinguishing the different vectors belonging to the representation by subscripts. We sometimes omit the superscripts when this can be done without confusion. Thus we write $\psi_m^{(j)}$ for the basis vectors of the representations of $R(3)$. When the double-valued representations are included, the multiplets of $R(3)$ are the same as the multiplets of $SU(2)$.

Now it is well known that the J_i can be represented in terms of the Pauli spin matrices σ_i $(i = 1, 2, 3)$:

$$J_1 = \tfrac{1}{2}\sigma_1, \qquad J_2 = \tfrac{1}{2}\sigma_2, \qquad J_3 = \tfrac{1}{2}\sigma_3, \tag{6.3}$$

where

$$\sigma_1 = \begin{pmatrix} 0 & 1 \\ 1 & 0 \end{pmatrix}, \qquad \sigma_2 = \begin{pmatrix} 0 & -i \\ i & 0 \end{pmatrix}, \qquad \sigma_3 = \begin{pmatrix} 1 & 0 \\ 0 & -1 \end{pmatrix}. \tag{6.4}$$

It can be directly verified by matrix multiplication that the J_i of Eq. (6.3) satisfy the commutation relations of Eq. (5.33).

The Pauli matrices are sometimes said to be a *representation* of the Lie algebra for $SU(2)$ [or $R(3)$]. The term "representation" may be confusing in this connection, for in different representations of a Lie algebra, the structure constants may be different.

The representation of the Lie algebra of $SU(2)$ by the Pauli matrices is the one lowest possible dimensionality. This is true because one-dimensional matrices (numbers) always commute. This fact that the lowest-dimensional representation of $SU(2)$ is by 2×2 matrices is a special case of a general theorem which we quote without proof. The theorem states that the lowest-dimensional representation of the Lie algebra of $SU(n)$ is by n-dimensional matrices. The matrices are traceless, and there are $n^2 - 1$ independent ones. These $n \times n$ matrices are called the *matrix generators* or more often the *generators* of the group. The generators of $SU(n)$ may also be represented by matrices of larger dimension. The n-dimensional representation of $SU(n)$ is called a fundamental representation. In Section 6.3, we shall give a more general definition of a fundamental representation of a simple Lie group. It will turn out that a rank l simple Lie group has l fundamental representations. Therefore $SU(n)$ has $n - 1$ fundamental representations, of which we have discussed only one.

If we write the Lie algebra of $SU(2)$ in standard form, the diagonal operator J_3 is denoted by H_1. The vector $\psi_m^{(j)}$ is an eigenstate of H_1:

$$H_1 \psi_m^{(j)} = m \psi_m^{(j)}.$$

The eigenvalue m is called a *weight.*

Now consider the general case of a compact semisimple Lie group of rank l in which there are l diagonal operators H_i, $i = 1, 2, \ldots, l$. Then we can write

$$H_i \psi_m^{(j)} = m_i \psi_m^{(j)}, \tag{6.5}$$

where here j denotes all the quantities necessary to specify a particular representation. The m_i can be considered as the components of an l-component vector called the *weight vector* or just the *weight* of the eigenstate $\psi_m^{(j)}$. Thus we can write

$$\mathbf{H} \psi_m^{(j)} = \mathbf{m} \psi_m^{(j)}. \tag{6.6}$$

The l-dimensional vector space spanned by the weights $\mathbf{m}$ is called the weight space. A drawing of all the weights of a representation is called the *weight diagram* of the representation.

The number of different eigenvectors with the same value of a weight is called the *multiplicity* of the weight. The multiplicity of a weight should not be confused with the multiplicity of a representation, which is the number of different eigenvectors of the representation. The term *multiplet* will be reserved for all the eigenstates of an irreducible unitary representation.

If a weight belongs to only one eigenvector of a representation, it is called *simple.* This definition should not be confused with the term *simple* as applied to a Lie algebra.

6.2 Generators of $SU(2)$ and $SU(3)$

Let us return to $SU(2)$ and consider the weights of the fundamental or two-dimensional representation. In this representation, we can express the Lie algebra in standard form in terms of the Pauli matrices. We obtain

$$H_1 = \tfrac{1}{2}\sigma_3, \qquad E_1 = \tfrac{1}{2}(\sigma_1 + i\sigma_2)/\sqrt{2}, \qquad E_{-1} = \tfrac{1}{2}(\sigma_1 - i\sigma_2)/\sqrt{2}. \tag{6.7}$$

The basis vectors of this representation are

$$\psi_{1/2}^{(1/2)} = \begin{pmatrix} 1 \\ 0 \end{pmatrix} = u_1, \qquad \psi_{-1/2}^{(1/2)} = \begin{pmatrix} 0 \\ 1 \end{pmatrix} = u_2 \tag{6.8}$$

Operating on these basis vectors with H_1, we obtain

$$H_1 u_1 = \tfrac{1}{2} u_1, \qquad H_1 u_2 = -\tfrac{1}{2} u_2. \tag{6.9}$$

Thus, for $SU(2)$ there are two one-dimensional weights in the fundamental representation given by

$$m(1) = \tfrac{1}{2}, \qquad m(2) = -\tfrac{1}{2}. \tag{6.10}$$

where the argument of $m(i)$, $i = 1, 2$, refers to the eigenvector to which it belongs. (We reserve the use of subscripts to the different components of a single weight.)

The states u_1 and u_2 can have a variety of physical interpretations. For example, they can represent the two possible spin eigenstates of a spin-$\frac{1}{2}$ particle (a doublet) such as the electron or proton. The weights or eigenvalues $m = \pm\frac{1}{2}$ then are the values of the z component of the spin (in units of $\hbar$) with respect to an arbitrary z axis. Alternatively, u_1 and u_2 can represent the two possible isospin eigenstates of an isospin-$\frac{1}{2}$ particle: for example, the nucleon. In this case the proton and neutron are regarded as the two possible orientations of the nucleon in isospin space, the proton having third component (or z component) of the isospin $m = I_3 = \frac{1}{2}$ and the neutron having third component $m = I_3 = -\frac{1}{2}$. The fact that the proton and neutron are not exactly degenerate in energy shows that isospin symmetry is approximate.

The situation is analogous to the case in which spin-$\frac{1}{2}$ particles are in a magnetic field pointed in the z direction. Then the degeneracy of the states with $m = \frac{1}{2}$ and $m = -\frac{1}{2}$ is broken. The crucial difference between the case of spin and isospin is of course that in the former case the symmetry-breaking magnetic field can be turned off, while in the latter case the symmetry-breaking interaction is always present. In the case of isospin, the proton and neutron differ in their electric charge, and the symmetry-breaking interaction is assumed to be the electromagnetic interaction. In fact, there is a relation between the charge of a state (in units of the proton charge) and the third component of the isospin I_3. This is the relation of Gell-Mann (1953) and Nishijima and Nakano (1953) which states that for all members of a given isospin multiplet, Q and I_3 are related by

$$Q = I_3 + Y/2, \tag{6.11}$$

where Y is a constant for all members of the isospin multiplet. The constant Y is known as the *hypercharge*.

Since I_3 and Y are additive quantum numbers, they can be interpreted as the diagonal generators of a group of rank 2. The group $SU(2)$ [or $R(3)$] must be a subgroup of this rank 2 group because of the known properties of isospin. Therefore we incorporate the nondiagonal generators of $SU(2)$, I_1 and I_2, into the algebra.

The smallest group of rank two which has the generators I_1, I_2, I_3, and Y is the group which has no other generators. The Lie algebra of the operators

I_1, I_2, I_3, and Y is not semisimple, since Y commutes with all the operators of the set. The group $U(1) \times SU(2)$ has the algebra of the operators I_μ and Y. This group is not semisimple, but is the direct product of the Abelian group $U(1)$ and the simple group $SU(2)$. The group $U(2)$ also has the same algebra, and therefore $U(2)$ and $U(1) \times SU(2)$ are locally isomorphic.

There is a two-to-one homomorphism of $U(1) \times SU(2)$ onto $U(2)$, the elements

$$e^{i\phi}\begin{pmatrix} 1 & 0 \\ 0 & 1 \end{pmatrix} \quad \text{and} \quad e^{i(\phi+\pi)}\begin{pmatrix} -1 & 0 \\ 0 & -1 \end{pmatrix},$$

of $U(1) \times SU(2)$ both corresponding to the element

$$\begin{pmatrix} e^{i\phi} & 0 \\ 0 & e^{i\phi} \end{pmatrix},$$

of $U(2)$. This is a trivial correspondence, and in the future we shall not distinguish between the groups $U(2)$ and $U(1) \times SU(2)$. Since analogous statements hold in n dimensions, we shall not distinguish between $U(n)$ and $U(1) \times SU(n)$.

To obtain the matrix generators of $U(2)$ we can add the unit matrix σ_0 given by

$$\sigma_0 = \begin{pmatrix} 1 & 0 \\ 0 & 1 \end{pmatrix},$$

to the Pauli matrices σ_ν, $\nu = 1, 2, 3$.

The physical interpretation of $U(2)$ for the strongly interacting particles is that the multiplets of $SU(2)$ are the isospin multiplets, such as proton and neutron, and the generator of $U(1)$ is the hypercharge, which is a constant for a given isospin multiplet.

If the symmetry group of the strongly interacting particles, or hadrons, is taken to be $U(2)$ rather than a larger group of rank 2, then no relation is predicted to exist between hypercharge and isospin. In other words, any isospin multiplet can have any hypercharge. However, if Y and I_3 are assumed to be the commuting generators of a *simple* rank 2 group, then there will be a relation between Y and I_3 for each member of a multiplet. In other words, more can be predicted. The Lie algebras of all the rank 2 simple groups contain the Lie algebra of $U(1) \times SU(2)$ as a subalgebra. This means that if any of the simple rank 2 groups is a symmetry of the hadrons, then isospin and hypercharge will be conserved. Thus, all of the algebras of the simple rank 2 groups should be examined to see how well their predictions agree with the experimental facts. This has been done by Ne'eman (1964), who found that the predictions of $SU(3)$ are in closest agreement with experiment.

We now consider the rank 2 simple group $SU(3)$. The hermitian generators [analogous to the Pauli spin matrices of $SU(2)$] are

$$\lambda_1 = \begin{pmatrix} 0 & 1 & 0 \\ 1 & 0 & 0 \\ 0 & 0 & 0 \end{pmatrix}, \quad \lambda_2 = \begin{pmatrix} 0 & -i & 0 \\ i & 0 & 0 \\ 0 & 0 & 0 \end{pmatrix}, \quad \lambda_3 = \begin{pmatrix} 1 & 0 & 0 \\ 0 & -1 & 0 \\ 0 & 0 & 0 \end{pmatrix},$$

$$\lambda_4 = \begin{pmatrix} 0 & 0 & 1 \\ 0 & 0 & 0 \\ 1 & 0 & 0 \end{pmatrix}, \quad \lambda_5 = \begin{pmatrix} 0 & 0 & -i \\ 0 & 0 & 0 \\ i & 0 & 0 \end{pmatrix}, \quad \lambda_6 = \begin{pmatrix} 0 & 0 & 0 \\ 0 & 0 & 1 \\ 0 & 1 & 0 \end{pmatrix}, \tag{6.12}$$

$$\lambda_7 = \begin{pmatrix} 0 & 0 & 0 \\ 0 & 0 & -i \\ 0 & i & 0 \end{pmatrix}, \quad \lambda_8 = \frac{1}{\sqrt{3}} \begin{pmatrix} 1 & 0 & 0 \\ 0 & 1 & 0 \\ 0 & 0 & -2 \end{pmatrix}.$$

These matrices are given with the notation and normalization used by Gell-Mann (1961). It is seen that λ_1, λ_4, and λ_6 are like σ_1 except that each has an extra row and column of zeros. Similarly, except for extra zeros, λ_2, λ_5, and λ_7 are like σ_2, and λ_3 is like σ_3. The matrix λ_8 has no analogy in $SU(2)$, as it is a matrix which commutes with λ_3. If we wish to consider the group $U(3)$ we can add the generator λ_0 given by

$$\lambda_0 = \sqrt{\frac{2}{3}} \begin{pmatrix} 1 & 0 & 0 \\ 0 & 1 & 0 \\ 0 & 0 & 1 \end{pmatrix}. \tag{6.13}$$

The group $U(3)$ is a rank 3 group and is not semisimple. The generator λ_0, which commutes with all the λ_ν, can be interpreted physically as associated with the baryon number B. In this case, I_3 and Y can vary within a given multiplet of $U(3)$ but all members of a multiplet have the same value of B.

We return to $SU(3)$. The generators λ_ν satisfy the Lie algebra given by the commutation relations

$$[\lambda_\kappa, \lambda_\mu] = 2if_{\kappa\mu\nu}\lambda_\nu \tag{6.14}$$

where the structure constants are denoted by $f_{\kappa\mu\nu}$ rather than $c_{\kappa\mu\nu}$. The nonzero $f_{\kappa\mu\nu}$ are given in Table 6.1.

TABLE 6.1

NONZERO INDEPENDENT STRUCTURE CONSTANTS $f_{\kappa\mu\nu}$ [a]

$\kappa\mu\nu$	$f_{\kappa\mu\nu}$	$\kappa\mu\nu$	$f_{\kappa\mu\nu}$
123	1	345	$\frac{1}{2}$
147	$\frac{1}{2}$	367	$-\frac{1}{2}$
156	$-\frac{1}{2}$	458	$\sqrt{3}/2$
246	$\frac{1}{2}$	678	$\sqrt{3}/2$
257	$\frac{1}{2}$		

[a] The $f_{\kappa\mu\nu}$ are antisymmetric under the permutation of any two indices.

The following properties of the λ_ν are also useful:

$$\begin{aligned} \mathrm{Tr}(\lambda_\mu \lambda_\nu) &= 2\delta_{\mu\nu}, \\ \{\lambda_\kappa, \lambda_\mu\} &= 2d_{\kappa\mu\nu}\lambda_\nu + \tfrac{4}{3}\delta_{\kappa\mu}, \\ \mathrm{Tr}(\lambda_\kappa[\lambda_\mu, \lambda_\nu]) &= 4if_{\kappa\mu\nu}, \\ \mathrm{Tr}(\lambda_\kappa\{\lambda_\mu, \lambda_\nu\}) &= 4\, d_{\kappa\mu\nu}. \end{aligned} \tag{6.15}$$

The $d_{\kappa\mu\nu}$ are given in Table 6.2. Relations (6.15) are not properties of the algebra but just of the three-dimensional λ-matrices.

TABLE 6.2

NONZERO INDEPENDENT ELEMENTS OF THE TENSOR $d_{\kappa\mu\nu}$ [a]

$\kappa\mu\nu$	$d_{\kappa\mu\nu}$	$\kappa\mu\nu$	$d_{\kappa\mu\nu}$
118	$1/\sqrt{3}$	355	$\frac{1}{2}$
146	$\frac{1}{2}$	366	$-\frac{1}{2}$
157	$\frac{1}{2}$	377	$-\frac{1}{2}$
228	$1/\sqrt{3}$	448	$-1/(2\sqrt{3})$
247	$-\frac{1}{2}$	558	$-1/(2\sqrt{3})$
256	$\frac{1}{2}$	668	$-1/(2\sqrt{3})$
338	$1/\sqrt{3}$	778	$-1/(2\sqrt{3})$
344	$\frac{1}{2}$	888	$-1\sqrt{3}$

[a] The $d_{\kappa\mu\nu}$ are symmetric under permutations of any two indices.

The λ-matrices operate on the basis vectors of the first fundamental representation of $SU(3)$. But $SU(3)$, being a simple rank-2 group, has a second fundamental representation. The generators of this second set λ_μ' are given by

$$\lambda_\mu' = -\lambda_\mu^* = -\tilde{\lambda}_\mu, \tag{6.16}$$

where the tilde over a symbol for a matrix denotes the transpose of the matrix. If we apply a similar transformation to the Pauli matrices of $SU(2)$

$$\sigma_\mu' = -\sigma_\mu^* = -\tilde{\sigma}_\mu,$$

we obtain the generators of an equivalent two-dimensional representation. In $SU(3)$, the new representation generated by the λ_μ' of Eq. (6.19) is inequivalent to the representation generated by the λ_μ.

When the Pauli matrices refer to isospin rather than spin, it is conventional to label them by τ_μ rather than σ_μ. The isospin generators I_μ are given by $I_\mu = \frac{1}{2}\tau_\mu$. Similarly, we can define $SU(3)$ generators F_μ of F-spin given by

$$F_\mu = \tfrac{1}{2}\lambda_\mu, \qquad \mu = 1, 2, \ldots, 8. \tag{6.17}$$

From Eqs. (6.14) and (6.15), we immediately obtain that the F_μ satisfy

$$\begin{aligned} [F_\kappa, F_\mu] &= if_{\kappa\mu\nu} F_\nu, \\ \mathrm{Tr}(F_\mu F_\nu) &= \tfrac{1}{2}\delta_{\mu\nu}, \\ \{F_\kappa, F_\mu\} &= d_{\kappa\mu\nu} F_\nu + \tfrac{1}{3}\,\delta_{\kappa\mu}, \\ \mathrm{Tr}(F_\kappa[F_\mu, F_\nu]) &= \tfrac{1}{2} i f_{\kappa\mu\nu}, \\ \mathrm{Tr}(F_\kappa\{F_\mu, F_\nu\}) &= \tfrac{1}{2}\, d_{\kappa\mu\nu}. \end{aligned} \tag{6.18}$$

We can also define a new set of operators D_μ, which are the following bilinear combinations of the F_μ:

$$D_\mu = \tfrac{2}{3} \sum_{\alpha\beta} d_{\mu\alpha\beta} F_\alpha F_\beta. \tag{6.19}$$

We shall now show that the D_μ satisfy the commutation relations

$$[D_\kappa, F_\mu] = if_{\kappa\mu\nu} D_\nu. \tag{6.20}$$

The proof makes use of the identity

$$[F_\mu, \{F_\gamma, F_\alpha\}] = \{[F_\mu, F_\gamma], F_\alpha\} + \{[F_\mu, F_\alpha], F_\gamma\},$$

which holds for any three linear operators. Making use of the relations of Eq. (6.18), this identity becomes

$$\sum_\beta d_{\gamma\alpha\beta}[F_\mu, F_\beta] = \sum_\beta (if_{\mu\gamma\beta}\{F_\beta, F_\alpha\} + if_{\mu\alpha\beta}\{F_\beta, F_\gamma\}).$$

For convenience, we have included a sum over the index β, even though, because of the properties of the $d_{\kappa\mu\nu}$ and $f_{\kappa\mu\nu}$, only one term in the sum is different from zero. We now multiply this expression by F_κ and take the trace, obtaining

$$\sum_\beta d_{\gamma\alpha\beta} f_{\kappa\mu\beta} = \sum_\beta (f_{\mu\gamma\beta} d_{\kappa\beta\alpha} + f_{\mu\alpha\beta} d_{\kappa\beta\gamma}). \tag{6.21}$$

We now substitute Eq. (6.19) into the left-hand side of Eq. (6.20), obtaining

$$[D_\kappa, F_\mu] = \tfrac{2}{3} \sum_{\alpha\beta} d_{\kappa\alpha\beta}[F_\alpha F_\beta, F_\mu].$$

Expanding the commutator, we obtain

$$[D_\kappa, F_\mu] = \tfrac{2}{3} \sum_{\alpha\beta} d_{\kappa\alpha\beta}([F_\alpha[F_\beta, F_\mu] + [F_\alpha, F_\mu]F_\beta).$$

Using (6.18), we get

$$[D_\kappa, F_\mu] = \tfrac{2}{3} i \sum_{\alpha\beta\gamma} d_{\kappa\alpha\beta}(f_{\beta\mu\gamma} F_\alpha F_\gamma + f_{\alpha\mu\gamma} F_\gamma F_\beta).$$

Again, for convenience, we have included a sum over the index γ. Relabeling the indices in the second term of the sum, we obtain

$$[D_\kappa, F_\mu] = \tfrac{2}{3}i \sum_{\alpha\beta\gamma} (d_{\kappa\alpha\beta} f_{\beta\mu\gamma} + d_{\kappa\beta\gamma} f_{\beta\mu\alpha}) F_\alpha F_\gamma.$$

Then, using the fact that the $d_{\kappa\mu\nu}$ are symmetric under all permutations of the indices and the $f_{\kappa\mu\nu}$ are antisymmetric under odd permutations, we can write this expression as

$$[D_\kappa, F_\mu] = \tfrac{2}{3}i \sum_{\alpha\beta\gamma} (d_{\kappa\beta\alpha} f_{\mu\gamma\beta} + d_{\kappa\beta\gamma} f_{\mu\alpha\beta}) F_\alpha F_\gamma.$$

Now, using Eq. (6.21), we obtain

$$[D_\kappa, F_\mu] = \tfrac{2}{3}i f_{\kappa\mu\beta} \sum_{\alpha\gamma} d_{\beta\alpha\gamma} F_\alpha F_\gamma,$$

where we have omitted the sum over the index β, since only one term contributes. Finally, using the definition of D_β from Eq. (6.19), we obtain Eq. (6.20).

Note that we have proved Eq. (6.20) only for 3×3 matrices. However, we can easily verify that these commutation relations are a property of the algebra which holds in any number of dimensions. To do this, we let

$$D_\kappa = D_\kappa^{(1)} + D_\kappa^{(2)}, \qquad F_\mu = F_\mu^{(1)} + F_\mu^{(2)}.$$

Then we have

$$[D_\kappa, F_\mu] = [D_\kappa^{(1)}, F_\mu^{(1)}] + [D_\kappa^{(2)}, F_\mu^{(2)}],$$

since generators with different superscripts commute, e.g.,

$$[F_\kappa^{(1)}, F_\mu^{(2)}] = 0.$$

Then, if Eq. (6.20) holds for $[D_\kappa^{(1)}, F_\mu^{(1)}]$ and $[D_\kappa^{(2)}, F_\mu^{(2)}]$ separately, it holds for $[D_\kappa, F_\mu]$. Since any D_κ and F_μ can be written as a sum of the corresponding three-dimensional generators, Eq. (6.20) holds in any number of dimensions.

From the λ_μ, we can obtain the generators in standard form. These are given by

$$H_1 = \frac{1}{\sqrt{6}}\begin{pmatrix}1 & 0 & 0\\ 0 & -1 & 0\\ 0 & 0 & 0\end{pmatrix}, \quad H_2 = \frac{1}{3\sqrt{2}}\begin{pmatrix}1 & 0 & 0\\ 0 & 1 & 0\\ 0 & 0 & -2\end{pmatrix},$$

$$E_1 = \frac{1}{\sqrt{3}}\begin{pmatrix}0 & 1 & 0\\ 0 & 0 & 0\\ 0 & 0 & 0\end{pmatrix}, \quad E_2 = \frac{1}{\sqrt{3}}\begin{pmatrix}0 & 0 & 1\\ 0 & 0 & 0\\ 0 & 0 & 0\end{pmatrix}, \quad E_3 = \frac{1}{\sqrt{3}}\begin{pmatrix}0 & 0 & 0\\ 0 & 0 & 1\\ 0 & 0 & 0\end{pmatrix},$$

$$E_{-1} = \frac{1}{\sqrt{3}}\begin{pmatrix}0 & 0 & 0\\ 1 & 0 & 0\\ 0 & 0 & 0\end{pmatrix}, \quad E_{-2} = \frac{1}{\sqrt{3}}\begin{pmatrix}0 & 0 & 0\\ 0 & 0 & 0\\ 1 & 0 & 0\end{pmatrix}, \quad E_{-3} = \frac{1}{\sqrt{3}}\begin{pmatrix}0 & 0 & 0\\ 0 & 0 & 0\\ 0 & 1 & 0\end{pmatrix}. \tag{6.22}$$

We see that

$$E_{-\alpha} = E_\alpha{}^\dagger. \tag{6.23}$$

If we operate on the three-dimensional basis vectors

$$u_1 = \begin{pmatrix}1\\0\\0\end{pmatrix}, \qquad u_2 = \begin{pmatrix}0\\1\\0\end{pmatrix}, \qquad u_3 = \begin{pmatrix}0\\0\\1\end{pmatrix}, \tag{6.24}$$

with the diagonal vector operator $\mathbf{H} = (H_1, H_2)$ we obtain

$$\begin{aligned}
\mathbf{H}u_1 &= \left(\frac{1}{\sqrt{6}}, \frac{1}{3\sqrt{2}}\right)u_1,\\
\mathbf{H}u_2 &= \left(-\frac{1}{\sqrt{6}}, \frac{1}{3\sqrt{2}}\right)u_2,\\
\mathbf{H}u_3 &= \left(0, -\frac{\sqrt{2}}{3}\right)u_3.
\end{aligned} \tag{6.25}$$

Thus, the three weights of the first fundamental representation of $SU(3)$ are

$$\mathbf{m}(1) = \left(\frac{1}{\sqrt{6}}, \frac{1}{3\sqrt{2}}\right), \qquad \mathbf{m}(2) = \left(-\frac{1}{6}, \frac{1}{3\sqrt{2}}\right), \qquad \mathbf{m}(3) = \left(0, -\frac{\sqrt{2}}{3}\right). \tag{6.26}$$

In this case we have three two-dimensional weights. For the basis vectors of any irreducible unitary representation of $SU(3)$ we have

$$\mathbf{H}\psi_m = \mathbf{m}\psi_m,$$

where $\mathbf{m} = (m_1, m_2)$.

From Eq. (6.16) we see that $-\mathbf{H}$ is the diagonal vector operator for the second fundamental representation. It follows that the weights of the second fundamental representation are just the negatives of the weights of the first.

The interpretation of the weights of $SU(3)$, as applied to the hadrons is that, except for normalization, m_1 is the third component of isospin I_3 and m_2 is the hypercharge Y. The exact relations are

$$I_3 = \tfrac{1}{2}\sqrt{6}\, m_1, \qquad Y = \sqrt{2}\, m_2. \tag{6.27}$$

Later we shall normalize the weights differently for convenience.

We have given a physical example of a multiplet belonging to the fundamental representation of $SU(2)$—the nucleon doublet. However, at the present time there is no firm evidence for the existence of any fundamental triplet of $SU(3)$. In Chapter 11 we shall consider a model in which an interpretation is given to this fundamental triplet—the quark model.

6.3 Properties of the Weights

We shall now give some of the properties of weights associated with the multiplets of compact simple Lie groups. For references, see Racah (1965) and Behrends *et al.* (1962).

First we note that there is only one simple Lie algebra of rank one, namely the algebra of $SU(2)$. For any representation j of this group, there are $2j + 1$ eigenvectors. Furthermore, there are $2j + 1$ different weights, each weight belonging to a different eigenvector. The weights are

$$m = -j, -j + 1, \ldots, j, \tag{6.28}$$

as is known from any book on quantum mechanics. Thus, all weights of simple rank-1 groups are simple. For groups of rank $l \geq 2$, the weights are not always simple.

Suppose we know the value of a weight $\mathbf{m}$ belonging to a particular representation of a simple rank l group. We can use this knowledge to obtain other weights of the representation if we also know the root diagram of the group. We obtain other weights of the representation by making use of the following:

THEOREM. *For any weight* $\mathbf{m}$ *and root* $\boldsymbol{\rho}(\alpha)$, *the quantity k defined by*

$$k = 2\mathbf{m} \cdot \boldsymbol{\rho}/\rho^2, \tag{6.29}$$

is an integer and

$$\mathbf{m}' = \mathbf{m} - \boldsymbol{\rho}(2\mathbf{m} \cdot \boldsymbol{\rho})/\rho^2, \tag{6.30}$$

is a weight with the same multiplicity as $\mathbf{m}$.

According to Eq. (6.30), if we have a weight $\mathbf{m}$, then by reflecting in the hyperplane perpendicular to $\boldsymbol{\rho}$, we obtain the new weight $\mathbf{m}'$. Weights related by reflections in hyperplanes are said to be *equivalent*. Equivalent weights should not be confused with equivalent representations, which are representations connected by a similarity transformation.

It is useful to be able to order the weights. The usual convention is that a weight $\mathbf{m}$ is higher than a weight $\mathbf{m}'$ if $\mathbf{m} - \mathbf{m}'$ has a positive number for its first nonvanishing component. However, with this ordering, the state vectors u_1, u_2, u_3 of Eq. (6.24) have weights which are not conveniently ordered. The vector u_1 has the highest weight, u_3 the next, and u_2 the lowest weight.

A more convenient convention for $SU(n)$ is that $\mathbf{m}$ is *higher* than $\mathbf{m}'$ if the *last* component of $\mathbf{m} - \mathbf{m}'$ is positive; if it is zero, we apply the criterion to the next to last component, etc. This is the convention we shall adopt for $SU(n)$. Then

$$\mathbf{H}u_i = \mathbf{m}(i)u_i.$$

The weights of Eq. (6.26) conform to this ordering. The highest weight of a set of equivalent weights is said to be *dominant.*

As we have remarked, the weights of the conjugate representation (the second fundamental representation) are negatives of the weights of the first. With our convention they are ordered as follows:

$$\mathbf{m}(1) = \left(0, \frac{\sqrt{2}}{3}\right), \qquad \mathbf{m}(2) = \left(\frac{1}{\sqrt{6}}, -\frac{1}{3\sqrt{2}}\right), \qquad \mathbf{m}(3) = \left(-\frac{1}{\sqrt{6}}, -\frac{1}{3\sqrt{2}}\right) \tag{6.31}$$

An important property of compact semisimple Lie groups is that there exists a highest weight for any irreducible representation. Furthermore this weight is simple. Another property is that two equivalent representations have the same highest weight, and two representations which have the same highest weight are equivalent.

Cartan (1933) has proved the following:

THEOREM. *For every simple group of rank l there are, l dominant weights called fundamental dominant weights denoted by* $\mathbf{M}^{(i)}$, $i = 1\ 2, \ldots, l$, *such that any other dominant weight* $\mathbf{M}$ *is a linear combination of the fundamental dominant weights*

$$\mathbf{M} = \sum_{i=1}^{l} p_i \mathbf{M}^{(i)} \tag{6.32}$$

where p_i are nonnegative integers. Furthermore, there exist l so-called fundamental irreducible representations which have the l fundamental dominant weights as their highest weights.

In general, to obtain the dimensionality of the fundamental representations of simple Lie groups, we need to discuss the characters of the representations. We shall not do this because the results happen not to be needed for the groups $SU(n)$. To find out how one uses the character to obtain the dimensionalities of the fundamental representations, see Racah (1965) and Behrends *et al.* (1962).

We now give some examples of how to obtain the weight diagrams of fundamental representations of simple Lie groups. First consider $SU(2)$. The root diagram consists of the roots $\rho(1) = 1$ and $\rho(-1) = -1$. Using Eq. (6.29) with $k = 1$, we find $m = \frac{1}{2}$ is a nonvanishing weight. Using Eq. (6.30), we obtain an equivalent weight $m' = -\frac{1}{2}$. These are the two weights of the fundamental representation of $SU(2)$, the weight $m = \frac{1}{2} = M^{(1)}$ being the fundamental dominant weight. The weight diagram of this representation is given in Fig. 6.1.

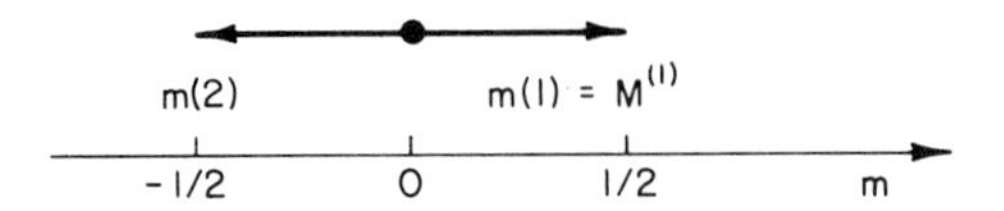

Fig. 6.1. Weight diagram of the fundamental representation of $SU(2)$.

Using Eq. (6.32), we see that the dominant weight M of any representation of $SU(2)$ is given by

$$M = pM^{(1)} = \tfrac{1}{2}p.$$

Thus, the dominant weight of a representation characterized by the integer p is either integral or half-integral according to whether p is even or odd.

Next we consider $SU(3)$. The root diagram is given in Fig. 5.3. From this figure we see that

$$\boldsymbol{\rho}(1) = (\sqrt{\tfrac{2}{3}}, 0).$$

Using this value of $\boldsymbol{\rho}(1)$ in Eq. (6.29) and letting $\mathbf{m} = (m_1 m_2)$, we obtain

$$k = \frac{2[m_1\rho_1(1) + m_2\rho_2(1)]}{\rho^2(1)} = \frac{2m_1\sqrt{\frac{2}{3}}}{\frac{2}{3}},$$

or

$$m_1 = k/\sqrt{6}.$$

Letting $k = 1$, we obtain the smallest nontrivial value of m_1:

$$m_1 = 1/\sqrt{6}.$$

Now consider another root from Fig. 5.3;

$$\boldsymbol{\rho}(2) = (1/\sqrt{6}, 1/\sqrt{2}),$$

and again take the scalar product with a weight $\mathbf{m}$ having the value $m_1 = 1\sqrt{6}$. We get

$$k = \frac{2[m_1\rho_1(2) + m_2\rho_2(2)]}{\rho^2(2)} = \frac{1}{2} + \frac{3}{\sqrt{2}}m_2.$$

Letting $k = 1$, we obtain $m_2 = 1/(3\sqrt{2})$, and letting $k = 0$, we obtain $m_2 = -1/(3\sqrt{2})$. Thus, we have obtained two weights

$$\mathbf{m} = (1/\sqrt{6}, 1/3\sqrt{2}) = \mathbf{M}^{(1)}, \tag{6.33}$$

and

$$\mathbf{m} = (1/\sqrt{6}, -1/3\sqrt{2}). \tag{6.34}$$

The weight of Eq. (6.33) is the fundamental dominant weight of the first fundamental representation of $SU(3)$, as can be seen by comparing with Eq. (6.26). Using this fundamental dominant weight and reflecting with respect to the lines (hyperplanes) perpendicular to the roots, we obtain the weight diagram of the first fundamental representation. This is shown in Fig. 6.2. The weights of Fig. 6.2 are just the ones we have obtained previously in Eq. (6.26).

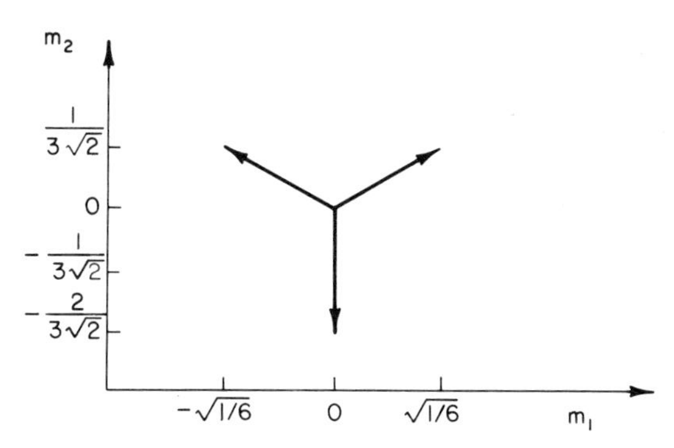

Fig. 6.2. Weight diagram for the first fundamental representation of $SU(3)$. This representation is denoted by (10) or **3**.

The weight of Eq. (6.34) belongs to the second fundamental representation, as can be seen from Eq. (6.31). Reflecting with respect to the lines perpendicular to the roots, we obtain the weight diagram of Fig. 6.3. The

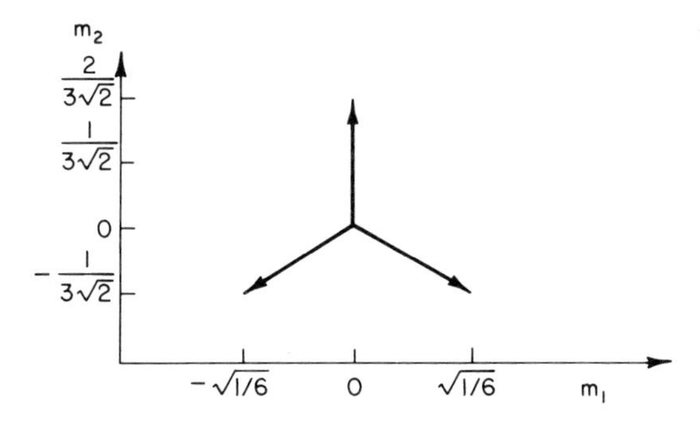

Fig. 6.3. Weight diagram of the second fundamental representation of $SU(3)$, denoted by (01) or $\bar{3}$.

fundamental dominant weight of this conjugate representation is (with our ordering convention)

$$\mathbf{M}^{(2)} = (0, \sqrt{2}/3). \tag{6.35}$$

A dominant weight M of any irreducible representation can be written

$$\mathbf{M} = p_1\mathbf{M}^{(1)} + p_2\mathbf{M}^{(2)}, \tag{6.36}$$

where p_1 and p_2 are nonnegative integers. The representation is fully characterized by $(p_1 p_2)$. Alternatively, the representation may be denoted by its dimensionality N, but this single number does not fully characterize the representation. The first fundamental representation is characterized by (10) or **3**, and the second by (01) or $\bar{\mathbf{3}}$. The bar on the number 3 is used to denote the conjugate of the representation **3**. Another notation for a representation of dimension N characterized by integers $(p_1 p_2)$ is $D^{(N)}(p_1 p_2)$.

We have obtained the weight diagrams of the two fundamental representations of $SU(3)$ from Eqs. (6.29) and (6.30) without a knowledge of the characters of the representations. This is because all weights of a fundamental representation of $SU(3)$ are simple and equivalent. However, the other rank two simple groups C_2 and G_2 do not have this property. Using Eqs. (6.29) and (6.30), we can obtain sets of equivalent weights, but cannot obtain the complete weight diagrams. Since we do not wish to discuss these groups in any detail, we shall just show the weight diagrams of the fundamental representations without illustrating here the procedure for obtaining them. The fundamental weight diagrams for C_2 are given in Fig. 6.4 and those for G_2 in Fig. 6.5. A method by which these weight diagrams (or the weight diagram of any simple compact group) can be obtained is described in Section 8.2. For a fuller treatment of C_2 and G_2, see, for example, Behrends *et al.* (1962).

The weight diagrams of C_2 and G_2 correspond to multiplets which do not have the properties of the known strongly interacting particles, as has been discussed by Ne'eman (1964). Therefore, if the hypercharge Y and third component of the isospin I_3 are to correspond to the diagonal operators of a compact simple Lie algebra, the only candidate is the Lie algebra of $SU(3)$. If we compare the predictions of $SU(3)$ with experiment, we find that the

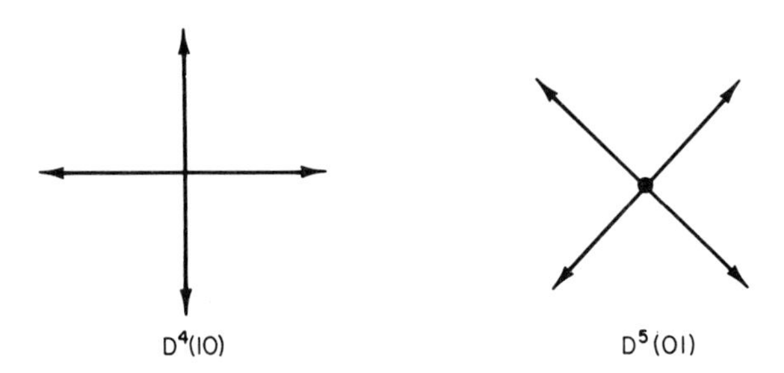

Fig. 6.4. Weight diagrams of the two fundamental representations of C_2. A solid small circle indicates a weight of zero length.

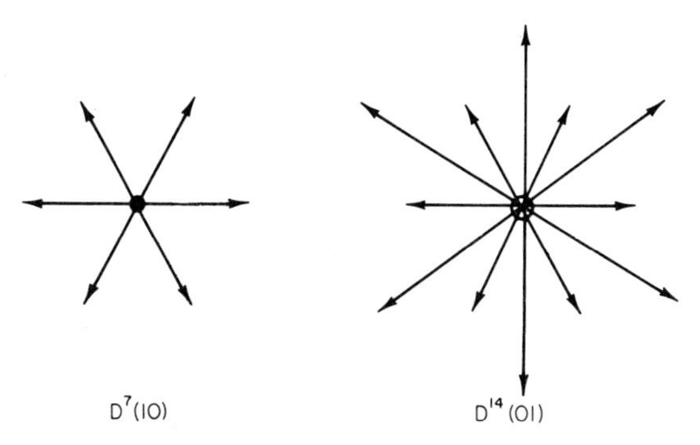

Fig. 6.5. Weight diagrams of the two fundamental representations of G_2. A solid circle indicates a simple weight of zero length and an open circle a double weight.

symmetry must hold only approximately. However, the evidence points very strongly toward $SU(3)$ being an underlying broken symmetry of the strong interactions of hadrons. We shall discuss this question further in Chapter 9.

6.4 Weight Diagrams of $SU(3)$

In Figs. 6.1, 6.2, and 6.3 are given the weight diagrams of the fundamental representations of $SU(2)$ and $SU(3)$. We wish to give a procedure for obtaining the weight diagram of any irreducible representation of $SU(3)$.

First, we make some preliminary remarks about $SU(2)$. The dimensionality N of a representation of $SU(2)$ associated with an integer p is

$$N = p + 1. \tag{6.37}$$

Since the dimensionality is also given in terms of the quantum number j as

$$N = 2j + 1, \tag{6.38}$$

we obtain a relation between p and j:

$$p = 2j. \tag{6.39}$$

In Eq. (6.28) are given the values of all the weights associated with the representation j. These values can be plotted on a one-dimensional weight diagram. A knowledge of the weights of $SU(2)$ is useful in obtaining the weight diagrams of $SU(3)$ because $SU(2)$ is a subgroup of $SU(3)$.

In $SU(3)$, there is also a relation between the dimensionality of a representation and the integers $(p_1 p_2)$ which denote it. It is given by

$$N = \tfrac{1}{2}(p_1 + 1)(p_1 + p_2 + 2)(p_2 + 1). \tag{6.40}$$

In Chapter 7 we shall use Young tableaux to prove Eq. (6.37) and to see how Eq. (6.40) arises.

It is convenient to change our notation slightly from the previous section. First, instead of drawing a vector from the origin to indicate a weight, we shall place a dot at the place where the end of the vector should go. If the weight has multiplicity n greater than one, we shall place $n - 1$ small concentric circles around the dot. These changes are to avoid cluttering up the weight diagram. We shall also draw dashed lines around the perimeter of a diagram to emphasize its shape. Another change is that we shall multiply all the weights of $SU(3)$ by a common factor $\sqrt{\frac{3}{2}}$ to conform to common useage. With this change in normalization, the first component of a weight of $SU(3)$ can be identified with the weight of an $SU(2)$ subgroup. If we relate the first component of a weight with I_3 and the second component with Y, we have the following relations

$$I_3 = m_1, \qquad Y = 2m_2/\sqrt{3}. \tag{6.41}$$

This normalization supersedes our previous one. With this normalization, m_1 is the eigenvalue of the generator F_3, and m_2 is the eigenvalue of F_8.

With these changes, the weight diagrams of the two fundamental representations of $SU(3)$ are given in Fig. 6.6. The dots in a horizontal row

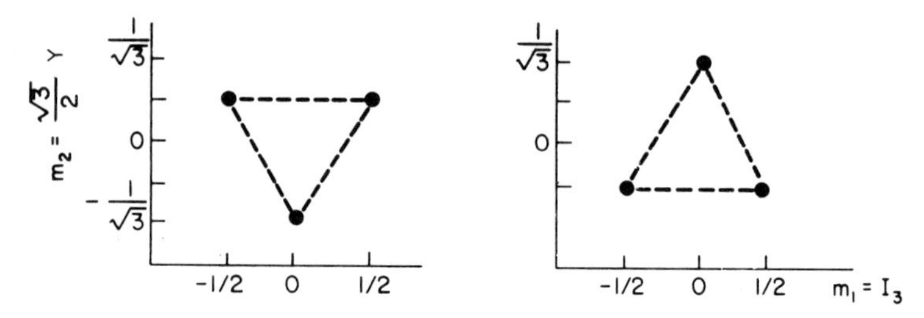

Fig. 6.6. Weight diagrams of the fundamental representations of $SU(3)$ normalized so that the values of m_1 are also weights of $SU(2)$.

of an $SU(3)$ weight diagram are the weights of $SU(2)$ multiplets. If the weights of any horizontal row are all simple, they constitute a single multiplet of $SU(2)$. Thus, the $SU(3)$ weight diagram of the representation (10) contains an $SU(2)$ doublet and an $SU(2)$ singlet, and likewise for the weight diagram of (01).

Note that the two fundamental weight diagrams of $SU(3)$ exhibit symmetry under rotations of 120° about the origin. This occurs because of the symmetry of the root diagram, and is therefore true for all representations. By reflecting a single nonvanishing weight with respect to lines perpendicular to the six roots, we obtain a set of three or six equivalent weights. If the original weight is vertical or makes an angle of 30° with the horizontal, then reflections produce a set of only three equivalent weights with a perimeter in the form of an equilateral triangle. This is the case for the two fundamental representations as well as many others. If the original weight is at another angle, reflections produce a set of six equivalent weights with a perimeter in the shape of a hexagon. This hexagon has the shape of an equilateral triangle with its corners removed by a horizontal line and lines making 60° with the horizontal. Note that all diagrams have an additional symmetry of reflections with respect to the vertical axis going through the origin.

Some $SU(3)$ diagrams have still greater symmetry. This can be seen as follows. If we rotate any weight diagram by 180°, we obtain the conjugate weight diagram. But a self-conjugate diagram must be unchanged by a rotation of 180°. Therefore such a diagram has both 120° and 180° symmetry, or 60° symmetry. Its perimeter is a regular hexagon.

Since all weight diagrams of $SU(3)$ have (at least) 120° symmetry, we must have $SU(2)$ multiplets, not only in horizontal rows, but in rows making angles of 120° with the horizontal. For example, *any* two weights of a three-dimensional representation of $SU(3)$ can be considered as a doublet of $SU(2)$ and the third weight can be considered an $SU(2)$ singlet. Thus, there are three different $SU(2)$ subgroups of $SU(2)$. We shall discuss these subgroups in more detail in Chapter 9.

With these preliminaries, we are ready to construct the weight diagram of any representation of $SU(3)$ specified by the integers $(p_1 p_2)$.

From Eq. (6.36), we can construct the highest weight

$$\mathbf{M} = p_1 \mathbf{M}^{(1)} + p_2 \mathbf{M}^{(2)},$$

where

$$\mathbf{M}^{(1)} = (1/2, 1/2\sqrt{3}), \qquad \mathbf{M}^{(2)} = (0, 1/\sqrt{3}), \tag{6.42}$$

with our new normalization. Making use of 120° symmetry and reflections through the vertical axis, we obtain a set of three or six equivalent weights. We may alternatively get a set of equivalent weights by reflecting with respect to the lines perpendicular to the roots. If the representation is $(p_1 0)$ or $(0 p_2)$, this procedure gives a triangular perimeter with three equivalent weights; otherwise it gives a set of six equivalent weights.

Next we may be able to obtain a number of other weights by making use of three different $SU(2)$ subgroups of $SU(3)$. We know the weights of $SU(2)$ decrease in integral steps along the horizontal and 120° lines. These are the only other weights of the diagram except for the problem of multiple weights.

All weights around the periphery of a diagram are simple by a theorem we have previously quoted; the others may be multiple. The maximum multiplicity v of any weight of a diagram is given by the formula

$$v = \tfrac{1}{2}(p_1 + p_2) - \tfrac{1}{2}|p_1 - p_2| + 1. \tag{6.43}$$

In the next chapter, we will indicate (but not prove) how the use of Young tableaux leads to this formula. Furthermore the multiplicities must not increase in going away from the center of a diagram, as otherwise we could not have complete multiplets of $SU(2)$. The actual rule of variation of multiplicities is somewhat stricter, as the multiplicity of two weights must either be the same or differ by unity. This can also be justified with Young tableaux.

These rules are sufficient to enable us to obtain any weight diagram of $SU(3)$. We shall illustrate the procedure with a few examples. First we shall obtain the weight diagram of the representation (30). The highest weight **M** is

$$\mathbf{M} = 3\mathbf{M}^{(1)} = (3/2, \sqrt{3}/2).$$

Using 120° symmetry we obtain the weights $(-3/2, \sqrt{3}/2)$ and $(0, -\sqrt{3})$. These weights are shown in Fig. 6.7a. Since the weights of the $SU(2)$ subgroups must differ by unity, we can add seven more weights as shown in

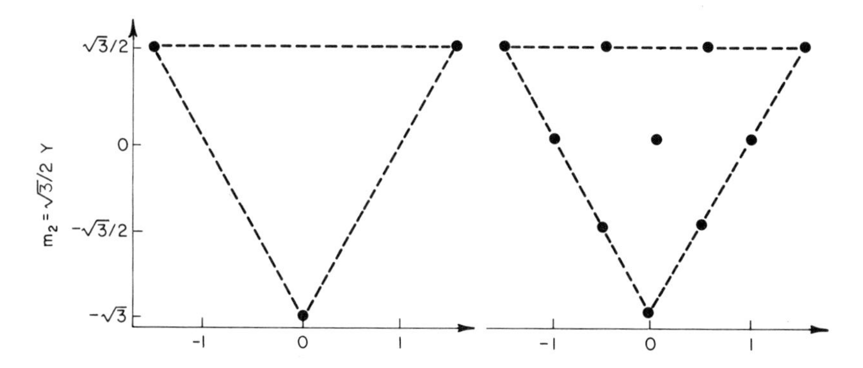

Fig. 6.7 (a) (left) A set of weights equivalent to the dominant weight of the representation (30) or **10**. (b) (right) The complete weight diagram of the representation (30).

Fig. 6.7b. From Eq. (6.43) we see that all the weights of any diagram $(p_1 0)$ are simple, so the diagram is complete. Furthermore, as a check we see from Eq. (6.40) that the dimensionality is $N = 10$.

As a second example consider the representation $p = (1, 1)$. The highest weight of this representation is

$$\mathbf{M} = (1/2, 1/2\sqrt{3}) + (0, 1/\sqrt{3}) = (1/2, \sqrt{3}/2).$$

The 60° symmetry of this self-conjugate representation gives a set of six equivalent weights which are shown in Fig. 6.8a. Since the weights of the $SU(2)$ subgroups must differ by unity, there must be a weight in the center. The maximum multiplicity of any weight of this diagram is 2 by Eq. (6.43).

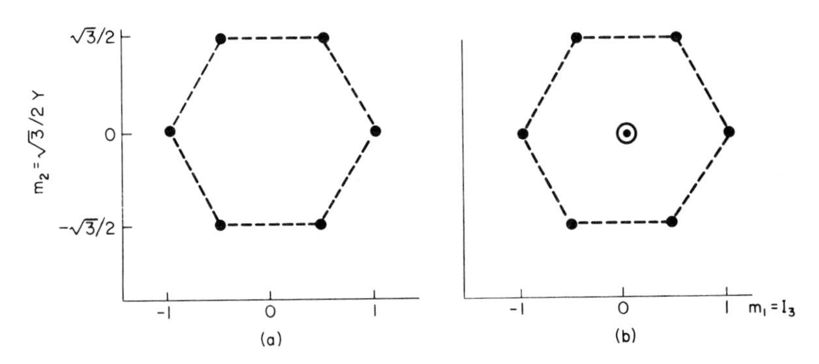

Fig. 6.8 (a) A set of weights equivalent to the dominant weight of (11) or **8**. (b) The complete weight diagram of (11).

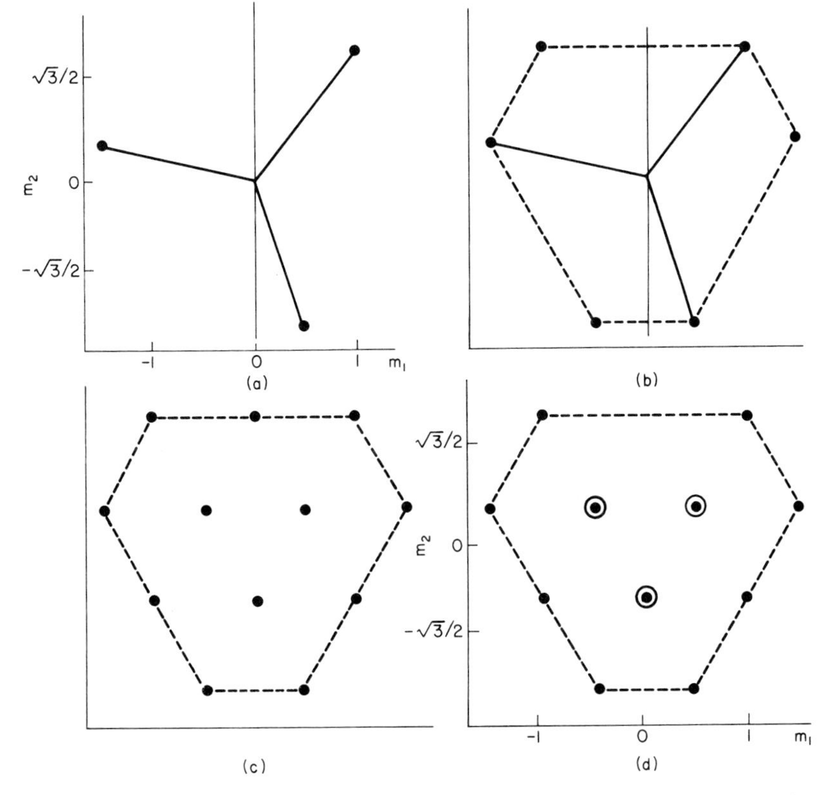

Fig. 6.9 (a) An incomplete set of weights equivalent to the dominant weight of (21) or **15**. (b) A complete set of equivalent weights formed by including the reflections of the weights of (a) with respect to the vertical axis. (c) Weights including those required by $SU(2)$ symmetry. (d) Complete weight diagram.

Furthermore, since the weights on the periphery are simple, the double weight must be in the center. The complete weight diagram is shown in Fig. 6.8b. As a check, we see from Eq. (6.40) that the dimensionality is $N = 8$.

As a third example, consider the representation $p = (2, 1)$. The highest weight of this representation is

$$\mathbf{M} = 2(1/2, 1/2\sqrt{3}) + (0, 1/\sqrt{3}) = (1, 2/\sqrt{3}).$$

Making use of 120° symmetry with respect to the origin, we obtain an incomplete set of three equivalent weights as shown in Fig. 6.9a. Reflecting these weights through the vertical axis, we obtain three additional equivalent weights, as shown in Fig. 6.9b. Making use of the fact that the weights of the $SU(2)$ subgroups differ in unit steps, we obtain Fig. 6.9c. Counting, we see the diagram has 12 weights thus far. From Eq. (6.40) we see that $N = 15$ for this weight diagram. By symmetry the three weights closest to the center must be double weights. This checks with the maximum multiplicity given in Eq. (6.43). The complete diagram is shown in Fig. 6.9d.

As a last example, consider the representation (2, 2) or **27**. The highest weight is $\mathbf{M} = (1, \sqrt{3})$. Making use of 60° symmetry, we obtain six equivalent weights in a regular hexagon. Making use of the $SU(2)$ subgroups, we obtain 13 additional weights. From Eq. (6.43) the maximum multiplicity of a weight for this diagram is three. Also since multiplicities cannot increase on out from the center, the central weight must be a triple weight. This gives two

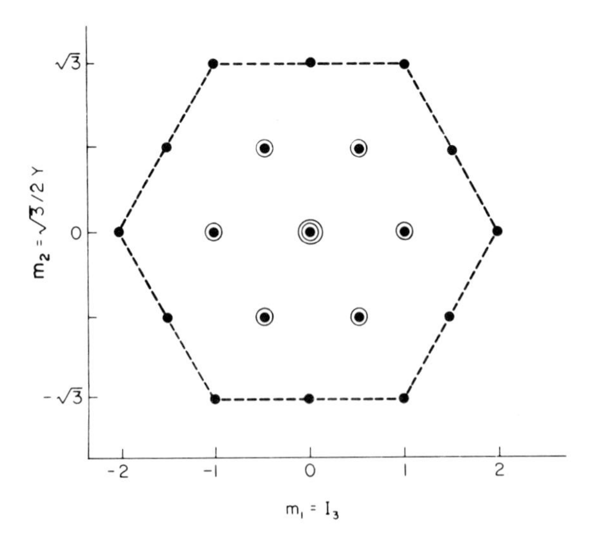

Fig. 6.10. Weight diagram of (22) or **27**.

more weights or 21 so far. But by Eq. (6.40) the dimensionality of the diagram is 27. By symmetry the six weights surrounding the center must be double weights. The complete weight diagram is shown in Fig. 6.10.

As an additional check on any weight diagram, we must see that we have complete multiplets of $SU(2)$. For example, in the diagram of Fig. 6.10, we have nine weights on the horizontal line with $m_2 = 0$. These weights have just the pattern to form a quintet, a triplet, and a singlet of $SU(2)$.

6.5 Casimir Operators and the Labeling of States

We shall now briefly discuss the invariant operators or so-called *Casimir operators* of simple compact groups, especially $SU(n)$. Let us begin with the example of $SU(2)$. This group has one nonlinear invariant operator which can be constructed from the members of the Lie algebra of the group, namely the square of the total angular momentum (or isospin)

$$J^2 = {J_1}^2 + {J_2}^2 + {J_3}^2.$$

This invariant operator commutes with all the generators and is a constant for all multiplets of a given representation. The quantum number j associated with the invariant operator J can be used to label a representation of $SU(2)$ instead of the integer p.

An important theorem states that a simple compact group of rank l has l Casimir or invariant operators which are nonlinear functions of the generators. From this theorem we see that there are two Casimir operators of $SU(3)$. We can label all the states of a given multiplet with the values of these Casimir operators, or alternatively we can label all members of the multiplet by the two numbers p_1 and p_2. Since $SU(n)$ has rank $n - 1$, we can label all members of a multiplet with the $n - 1$ p_i or Casimir operators C_i. A theorem states that the Casimir operators of $SU(n)$ can be written in the form

$$C_i = \sum_{\mu} a_{\mu i} X_{\mu}^{i+1}, \qquad i = 1, 2, \ldots, n-1, \tag{6.44}$$

where the X_μ are the generators of the group. By demanding that the C_i commute with all the generators, we can, in principle, determine the coefficients $a_{\mu i}$. However, it may be tedious to find the Casimir operators for $SU(n)$ if n is large. Furthermore, in an internal symmetry group like $SU(3)$ there does not seem to be any obvious physical significance to the Casimir operators.

Of course, if one wants to write down scalar or invariant interactions which depend on the internal symmetry coordinates, knowledge of the Casimir operators becomes quite useful. However, if one just wants to label the states of a multiplet, it is much more convenient to use the p_i.

We are now ready to consider the problem of how many numbers are necessary to label a state of $SU(n)$. First, we need to specify $n-1$ Casimir operators or p_i to distinguish which representation the state belongs to. Second, we need to specify the weight $\mathbf{m}$, which has $n-1$ components. For a multiplet in which all weights are simple, this is sufficient. However, for $n > 2$, we need more labels to distinguish the different members of a multiplet with the same weight.

For $SU(3)$, in which the weight signifies the value of the third component of isospin I_3 and the hypercharge Y, we solve this problem by also giving the value of the isospin I. Thus, for example, there is a double weight at the center of the octet weight diagram with $I_3 = Y = 0$. One of the states with this weight is an isospin singlet ($I = 0$) and the other is a member of an isospin triplet ($I = 1$). Thus, we complete the specification of the state by giving the value of the Casimir operator of one of the three $SU(2)$ subgroups. Alternatively, we could specify the value of p for the appropriate $SU(2)$ subgroup.

This solution to the labeling problem for $SU(3)$ can be generalized to $SU(n)$. In this case we can specify in addition to the p_i (or Casimir operators) and weights of $SU(n)$, the values of p_i (or Casimir operators) of $SU(n-1)$, $SU(n-2), \ldots, SU(2)$. This may not be a convenient way to label the states, but it is complete. Thus, a state of $SU(n)$ can be labeled by

$$2(n-1) + (n-2) + (n-3) + \cdots + 1 = \tfrac{1}{2}n(n+1) - 1$$

numbers. Two numbers are necessary to label a state of $SU(2)$ and five are necessary for a state of $SU(3)$.

The first of the two Casimir operators of $SU(3)$, denoted by F^2, can be given in terms of the F-spin generators defined in Eq. (6.17). It is

$$F^2 = \sum_{\mu=1}^{8} F_\mu{}^2. \tag{6.45}$$

The second Casimir operator G^3 is more conveniently given in terms of nine traceless operators A_{ij} defined by

$$(A_{ij})_{kl} = \delta_{ik}\,\delta_{jl} - \tfrac{1}{3}\delta_{ij}\,\delta_{kl} \qquad (i, j, k, l = 1, 2, 3). \tag{6.46}$$

Only eight of these operators are independent, since

$$A_{11} + A_{22} + A_{33} = 0.$$

Then G^3 is given by

$$G^3 = \sum_{ijk} A_{ij}\,A_{jk}\,A_{ki}\,. \tag{6.47}$$

See de Swart (1963) for further properties of the A_{ij} operators.

We can verify directly that F^2 and G^3 commute with every generator. It follows that matrix representations of F^2 and G^3 commute with all matrices of an irreducible representation, since all the matrices of an irreducible representation are functions of the generators. Therefore, by Schur's lemma (Section 3.3), F^2 and G^3 are constant multiples of the unit matrix. This proves that they are invariant operators. Thus, a given irreducible representation of $SU(3)$ can be labeled either by the two numbers p_1 and p_2 or by the quantum numbers f and g associated with the Casimir operators F^2 and G^3. Either pair of numbers completely specifies a representation. We shall use the numbers p_1 and p_2. The operator I^2 defined by

$$I^2 = \sum_{i=1}^{3} F_i^2, \tag{6.48}$$

is not an invariant operator of $SU(3)$, but it is an invariant operator of the isospin $SU(2)$ subgroup of $SU(3)$. A multiplet of $SU(3)$ can be chosen in a basis such that all members are eigenstates of I^2. The eigenvalues are $I(I+1)$, where I is integral or half integral.

6.6 Tensor Operators

Let us consider the matrix element of a self-adjoint operator T (an observable) between two states ϕ, ψ which are basis functions of a representation of a symmetry group which we take to be a compact Lie group. This matrix element M is given by

$$M = (\phi, T\psi). \tag{6.49}$$

We now ask how the matrix element M compares to the matrix element of the same observable between transformed states ϕ', ψ' given by

$$\phi' = U\phi, \qquad \psi' = U\psi \tag{6.50}$$

where U is a unitary operator of the symmetry group. The matrix element M' of T with respect to the transformed states is given by

$$M' = (\phi', T\psi') = (U\phi, T\psi), \tag{6.51}$$

or

$$M' = (\phi, U^{-1}TU\psi), \tag{6.52}$$

since U is unitary.

From Eq. (6.52), we see that the matrix element of T between the transformed states ϕ', ψ' is the same as the matrix element of a transformed observable T' given by

$$T' = U^{-1}TU, \tag{6.53}$$

between the original states. Thus, we are led to consider the transformation properties of operators. It is clear from Eq. (6.53) that if T commutes with U, then the transformed operator T' is equal to the original operator T. Such an operator is called a *scalar* under the transformation. However, in general, the operator will have more complicated transformation properties. We shall restrict ourselves to consideration of so-called tensor operators which have definite properties under the transformation of Eq. (6.53). Let us first consider the special case in which the operator T is X_μ, where X_μ is one of the generators of the symmetry group. We transform T by means of a unitary operator

$$U_\nu = \exp(ia_\nu X_\nu).$$

Then

$$T' = X_\mu' = \exp(-ia_\nu X_\nu) X_\mu \exp(ia_\nu X_\nu). \tag{6.54}$$

If the transformation is an infinitesimal one, then to lowest order

$$U_\nu = 1 + ia_\nu X_\nu. \tag{6.55}$$

Then Eq. (6.54) becomes

$$X_\mu' = (1 - ia_\nu X_\nu) X_\mu (1 + ia_\nu X_\nu) = X_\mu + ia_\nu [X_\mu, X_\nu]. \tag{6.56}$$

But the commutators of the generators are given as linear combinations of the generators

$$[X_\mu, X_\nu] = \sum_\lambda c_{\mu\nu\lambda} X_\lambda.$$

Then Eq. (6.56) becomes

$$X_\mu' - X_\mu = dX_\mu = ia_\nu \sum_\lambda c_{\mu\nu\lambda} X_\lambda. \tag{6.57}$$

If an operator T_μ has the same transformation properties as the generators, i.e.,

$$dT_\mu = ia_\nu \sum_\lambda c_{\mu\nu\lambda} T_\lambda, \tag{6.58}$$

the operator is said to be a vector operator. If we have an operator with n indices $T_{\mu_1, \mu_2, \ldots, \mu_n}$ which transforms like $X_{\mu_1} X_{\mu_2} \cdots X_{\mu_n}$, then $T_{\mu_1, \mu_2, \ldots, \mu_n}$ is called a tensor operator of rank n.

Of more interest to us are the so-called irreducible tensor operators. Let us consider the basis vectors $\psi_m^{(j)}$ of an irreducible representation of a Lie group. The index j identifies the irreducible representation and m distinguishes different vectors of the representation by some ordering process. If we operate on $\psi_m^{(j)}$ with a unitary operator U_ν of the symmetry group, we obtain

$$\psi_m'^{(j)} = U_\nu \psi_m^{(j)}$$

or, letting the matrix elements of U_ν be $D^{(j)}_{m'm}(a_\nu)$, we have

$$\psi'^{(j)}_m = \sum_{m'} D^{(j)}_{m'm}(a_\nu)\psi^{(j)}_{m'}. \tag{6.59}$$

We now define an irreducible tensor operator $T^{(j)}_m$ as an operator which transforms the same way as $\psi^{(j)}_m$ in Eq. (6.59). Thus we have

$$T'^{(j)}_m = \sum_{m'} D^{(j)}_{m'm}(a_\nu)T^{(j)}_{m'} = U_\nu^{-1}T^{(j)}_m U_\nu. \tag{6.60}$$

Other authors define a tensor operator by using the inverse or transpose of $D^{(k)}_{m'm}(a_\nu)$.

The *rank* of an irreducible tensor operator may also be defined. For $SU(2)$ if the number of operators $T^{(j)}_m$ is $2j + 1$, then j is the rank. For a more general simple compact Lie group, we may order the irreducible unitary representations in some manner and define the rank according to this ordering. Alternatively, we may regard the rank as a quantity specified by more than one number, for example by $\mathbf{p}$ for $SU(n)$.

We see from Eqs. (6.18) and (6.20) that for $SU(3)$ there are two sets of irreducible vector operators, the F_μ and D_μ. This is different from the case of $SU(2)$, for which there is only one set of irreducible vector operators associated with the generators.

CHAPTER 7

YOUNG TABLEAUX AND UNITARY SYMMETRY

7.1 Dimensionality of Multiplets of $SU(n)$

In the previous chapter, we represented the fundamental doublet of $SU(2)$ by the states

$$u_1 = \begin{pmatrix} 1 \\ 0 \end{pmatrix} \quad \text{and} \quad u_2 = \begin{pmatrix} 0 \\ 1 \end{pmatrix}. \tag{7.1}$$

Let us assume that these states denote states of a single particle of spin $\frac{1}{2}$. Another notation for these one-particle states is by means of a single-box Young tableau $\boxed{}$. We make the identification

$$u_1 = \boxed{1}, \qquad u_2 = \boxed{2}. \tag{7.2}$$

The single-box tableau without a number stands for both members of the doublet. Now suppose we have a two-particle state. We know that in order to be a multiplet of the symmetric group S_2, the state must be either symmetric, corresponding to the Young tableau $\begin{array}{|c|c|}\hline & \\ \hline\end{array}$, or antisymmetric, corresponding to the tableau

$$\begin{array}{|c|}\hline \\ \hline \\ \hline\end{array}$$

These same tableaux also represent multiplets of $SU(2)$. Consider first the symmetric state. If both particles are in the state u_1, the corresponding tableau is $\boxed{1}\boxed{1}$ whereas if both particles are in the state u_2, the tableau is $\boxed{2}\boxed{2}$. The only other symmetric possibility is for one of the particles to be in the state u_1 and the other in the state u_2. This is represented by the tableau $\boxed{1}\boxed{2}$. Thus, the symmetric two-particle state of $SU(2)$ is a triplet. The multiplicity is given by all the standard arrangements of the Young tableau $\boxed{}\boxed{}$ with integers restricted to be 1 or 2. The nonstandard arrangement $\boxed{2}\boxed{1}$ is obviously the same as the standard arrangement $\boxed{1}\boxed{2}$, since the state is symmetric. Therefore the nonstandard arrangement must not be counted. The only antisymmetric two-particle state is given by the tableau

$$\begin{array}{|c|}\hline 1 \\ \hline 2 \\ \hline\end{array}$$

This is a singlet state.

Next, consider multiplets composed of three spin-$\frac{1}{2}$ particles. The two possible tableaux are

$$\begin{array}{|c|c|c|}\hline & & \\ \hline\end{array} \quad \text{and} \quad \begin{array}{|c|c|}\hline & \\ \hline \\ \cline{1-1}\end{array}$$

This is because one cannot construct a totally antisymmetric state of three particles when only two states are available. The standard tableaux of the symmetric multiplet are

$$\begin{array}{|c|c|c|}\hline 1 & 1 & 1 \\ \hline\end{array} \qquad \begin{array}{|c|c|c|}\hline 1 & 1 & 2 \\ \hline\end{array} \qquad \begin{array}{|c|c|c|}\hline 1 & 2 & 2 \\ \hline\end{array} \qquad \begin{array}{|c|c|c|}\hline 2 & 2 & 2 \\ \hline\end{array}$$

Thus, this state is a quartet. The standard arrangements of the tableau of mixed symmetry are

$$\begin{array}{|c|c|}\hline 1 & 1 \\ \hline 2 \\ \cline{1-1}\end{array} \quad \text{and} \quad \begin{array}{|c|c|}\hline 1 & 2 \\ \hline 2 \\ \cline{1-1}\end{array}$$

Thus, this tableau

$$\begin{array}{|c|c|}\hline & \\ \hline \\ \cline{1-1}\end{array}$$

is a doublet, just like the single-box diagram $\boxed{}$. The reason that the tableaux

$$\begin{array}{|c|c|}\hline & \\ \hline \\ \cline{1-1}\end{array} \quad \text{and} \quad \begin{array}{|c|}\hline \\ \hline\end{array}$$

have the same multiplicity is that there is only one way to make an antisymmetric state of two particles, if only two different single-particle states are available. Thus, if we are interested in the multiplicity, but not in the number of particles, we can omit all columns with two boxes.

These examples illustrate the fact that a Young tableau can be used to denote any multiplet of $SU(2)$. The individual members of the multiplet are denoted by the standard arrangements and the multiplicity by the total number of standard arrangements.

An analogous result holds for $SU(n)$ with the numbers in each box restricted to be $1, 2, \ldots, n$. Consider a state of ν particles denoted by a Young tableau with ν boxes. Such a state is an irreducible tensor, (i.e., a basis tensor for an irreducible representation) of the symmetric group S_ν. We now state without proof an important theorem.

THEOREM. *If a ν-particle state is an irreducible tensor of S_ν and is constructed from one-particle states which are basis vectors of an irreducible n-dimensional representation of $SU(n)$, then the state is an irreducible tensor of $SU(n)$.*

A consequence of this theorem is that any Young tableau with n rows or less denotes the basis tensors of an irreducible representation of $SU(n)$. If there are more than n rows in a tableau, then at least one column must have more than n boxes. But a column with more than n boxes denotes a tensor which is identically zero. This follows because if the tensor is to be antisymmetric, no more than one particle can be in each state. But there are more than n particles and only n available states. When we try to antisymmetrize a function containing two particles in the same state we get zero.

To obtain the dimensionality of an irreducible representation of $SU(n)$, we simply count the number of standard arrangements of the corresponding Young tableau. When we put a number in a box, the number stands for one of the n possible states of a single particle. Therefore, the prescription for obtaining a standard arrangement of a tableau of $SU(n)$ is to put an integer from 1 to n in each box of the tableau such that the numbers increase from top to bottom in a column and do not decrease from left to right in a row. The number of possible standard arrangements is the number of components of the irreducible tensor which the tableau denotes. This, of course, equals the dimensionality of the corresponding irreducible representation. Furthermore, the dimensionality holds not only for the group $SU(n)$, but for the groups $GL(n)$, $SL(n)$, and $U(n)$. However, for the noncompact groups $GL(n)$ and $SL(n)$, the irreducible representations thus obtained are not unitary. In general the faithful unitary representations of noncompact groups are infinite dimensional. We shall not consider $GL(n)$ and $SL(n)$ further.

Each standard arrangement of a Young tableau stands for a particular irreducible tensor, and therefore has a weight. This weight can be obtained from the weights of the single-particle states by ordinary addition. Therefore, we can construct the weight diagram of a multiplet from its Young tableau.

As an example, we consider the one-particle vectors of $SU(n)$, which are denoted by a single box $\boxed{}$. This box represents the basis vectors for the first so-called *fundamental* representation of $SU(n)$. We can obtain the dimensionality of this fundamental representation by counting the number of standard arrangements. We have already pointed out that for $SU(2)$ we have two arrangements

$$\boxed{1} \quad \text{and} \quad \boxed{2}$$

with weights $m = \pm\frac{1}{2}$. Clearly, for $SU(n)$ there are n standard arrangements of the tableau $\boxed{}$, since any number from 1 to n can be put in the box. There are actually $n - 1$ different fundamental representations of $SU(n)$, at most two of which have dimension n. We shall define the others later in this section.

Next consider a two-particle state. There are two possible Young tableaux

$$\boxed{}\boxed{} \quad \text{and} \quad \begin{array}{c}\boxed{}\\\boxed{}\end{array}$$

The dimensionality of these irreducible tensors depends on n. We have already considered the number of standard arrangements for $SU(2)$. For $SU(3)$ the standard arrangements are

$$\boxed{1}\boxed{1} \qquad \boxed{1}\boxed{2} \qquad \boxed{1}\boxed{3} \qquad \boxed{2}\boxed{2} \qquad \boxed{2}\boxed{3} \qquad \boxed{3}\boxed{3}$$

$$\begin{array}{c}\boxed{1}\\\boxed{2}\end{array} \qquad \begin{array}{c}\boxed{1}\\\boxed{3}\end{array} \qquad \begin{array}{c}\boxed{2}\\\boxed{3}\end{array}$$

Thus the tableau $\boxed{}\boxed{}$, which we previously found was a 3-component irreducible tensor of $SU(2)$, is a 6-component irreducible tensor of $SU(3)$. Likewise the tableau

$$\begin{array}{c}\boxed{}\\\boxed{}\end{array}$$

which we found to be a scalar of $SU(2)$, is a 3-component tensor of $SU(3)$.

Since the weights are additive, it is easy to obtain the weights of the standard arrangements of the tableaux

$$\boxed{}\boxed{} \quad \text{and} \quad \begin{array}{c}\boxed{}\\\boxed{}\end{array}$$

starting with the weights of the first fundamental representation. From Fig. 6.6, these weights are

$$\begin{aligned} \boxed{1}&: \quad \mathbf{m}(1) = \mathbf{M}^{(1)} = (1/2,\ 1/2\sqrt{3}), \\ \boxed{2}&: \quad \mathbf{m}(2) = (-1/2,\ 1/2\sqrt{3}), \\ \boxed{3}&: \quad \mathbf{m}(3) = (0,\ -1/\sqrt{3}). \end{aligned} \tag{7.3}$$

Then the weights of the two-particle states are

$$\begin{aligned} \boxed{1\,|\,1}&: \quad \mathbf{m}(1) = \mathbf{M} = (1,\ 1/\sqrt{3}), \\ \boxed{1\,|\,2}&: \quad \mathbf{m}(2) = (0,\ 1/\sqrt{3}), \\ \boxed{2\,|\,2}&: \quad \mathbf{m}(3) = (-1,\ 1/\sqrt{3}), \\ \boxed{1\,|\,3}&: \quad \mathbf{m}(4) = (1/2,\ -1/2\sqrt{3}), \\ \boxed{2\,|\,3}&: \quad \mathbf{m}(5) = (-1/2,\ -1/2\sqrt{3}), \\ \boxed{3\,|\,3}&: \quad \mathbf{m}(6) = (0,\ -2/\sqrt{3}), \end{aligned} \tag{7.4}$$

where we have ordered the weights according to our convention that the last component is controlling. From these values, we obtain the weight diagram of Fig. 7.1. Because the weights are additive, the weight of a state does not depend on the configuration of its Young tableau, but only on the numbers in its boxes, i.e., a weight depends only on a set of integers. Thus, the weights of the antisymmetric states

$$\begin{array}{|c|}\hline 1\\\hline 2\\\hline\end{array} \qquad \begin{array}{|c|}\hline 1\\\hline 3\\\hline\end{array} \qquad \text{and} \qquad \begin{array}{|c|}\hline 2\\\hline 3\\\hline\end{array}$$

are given by $\mathbf{m}(2)$, $\mathbf{m}(4)$, and $\mathbf{m}(5)$, respectively, of (7.4). These weights are shown in Fig. 7.1 connected by a solid triangle which is the weight diagram of the $\overline{\mathbf{3}}$ representation. The use of Young tableaux also enables us to obtain the multiplicity of a given weight. Since the value of a weight depends only on a set of integers, all standard arrangements of a tableau with the same integers have the same weight. Thus, we need merely count the number of these arrangements to obtain the multiplicity of the weight.

The procedure for obtaining the dimensionality of an irreducible representation by counting the standard arrangements of a Young tableau can be simplified by noting that for $SU(n)$ [and also $U(n)$] any column with n boxes has only one standard arrangement. Therefore, so far as counting

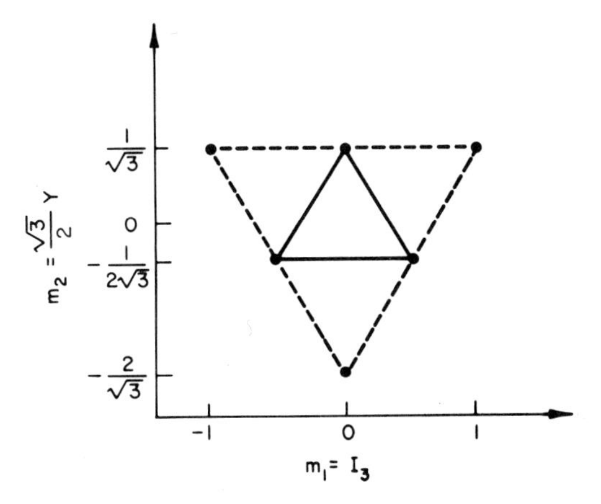

Fig. 7.1. Weight diagram of the representation (20) or **6**. The small solid triangle shows the part which is also the weight diagram of the representation (01) or $\bar{\mathbf{3}}$.

states is concerned, we can simplify a tableau by removing all such columns. For example, consider the tableau

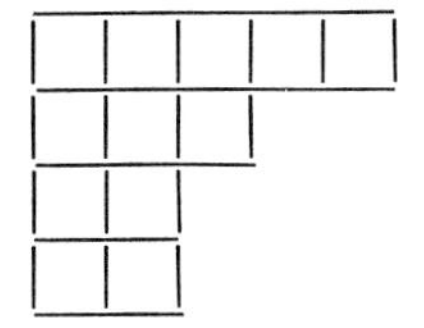

If $n = 4$, this tableau corresponds to the same irreducible representation as the tableau

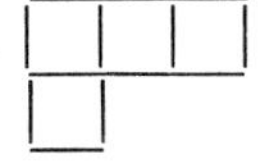

Thus, we can make the identification

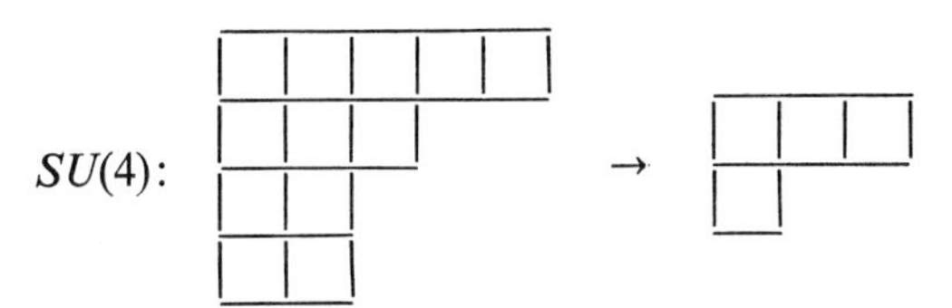

Of course, the first tableau denotes a 12-particle state while the second denotes a four-particle state; nevertheless, the dimensionalities of the representations are equal. Thus, if we are interested only in the dimensionality of a representation of $SU(n)$ associated with a particular Young tableau, but not in the number of particles, we can remove all columns with n boxes.

We need consider only tableaux specified by the $n - 1$ numbers

$$\mathbf{p} = (p_1, p_2, \ldots, p_{n-1}). \tag{7.5}$$

For $SU(n)$, the p_i are the same integers that appear in Eq. (6.32), which gives the dominant weight of the multiplet specified by $\mathbf{p}$.

We next consider the effect on the weights of omitting all columns with n boxes. It can be seen from Eqs. (6.4) and (6.12) that the Pauli matrices of $SU(2)$ and the matrices of $SU(3)$ are all traceless. It is generally true that the generators of $SU(n)$ can be represented by $n^2 - 1$ traceless matrices in n dimensions. Since all the matrices are traceless, this is true in particular of the diagonal matrices $\mathbf{H}$ whose elements are the weights. Therefore the sum of all the weights of the first fundamental representation is zero.

$$\sum_{i=1}^{n} \mathbf{m}(i) = 0. \tag{7.6}$$

This means that a column of n boxes has weight $\mathbf{m} = 0$. Thus, omitting columns of n boxes has no effect on the weights. This last statement is not true for $U(n)$ since $U(n)$ has an additional generator which has a nonzero trace. (In fact it is a multiple of the unit matrix.) The weight associated with this extra generator is proportional to the number of particles, i.e., to the number of boxes.

Thus, the group properties of a representation of $SU(n)$ [but not of $U(n)$] are completely specified by the numbers p_i $(i = 1, 2, \ldots, n - 1)$. All representations with the same values of p_i are equivalent. The p_i are just the numbers discussed in Section 6.3 in connection with constructing a dominant weight from the $n - 1$ fundamental dominant weights of $SU(n)$. [See Eq. (6.32)].

A Young tableau of $SU(n)$ containing only a single column of n boxes is a singlet of $SU(n)$ and is denoted by

$$\mathbf{p} = (00 \cdots 0), \qquad n - 1 \quad \text{zeros}. \tag{7.7}$$

In drawing such a tableau, if we remove the column, we have nothing left and might forget that we have a singlet. Thus, we shall either keep such a tableau or replace it by the symbol ①. The weight of any singlet of $SU(n)$ is

$$\mathbf{m} = (00 \cdots 0), \qquad n - 1 \quad \text{zeros}. \tag{7.8}$$

If the number of boxes in the longest column of a tableau is less than $n - 1$, we still use $n - 1$ integers to specify it. For example, suppose we have the four-particle tableau

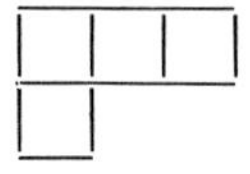

of $SU(6)$. This tableau is specified by the five integers

$$p_1 = 2, \quad p_2 = 1, \quad p_3 = 0, \quad p_4 = 0, \quad p_5 = 0,$$

or (21000). We can simplify this notation further by using a superscript for repeated integers $(21000) = (210^3)$.

Now suppose we have a tableau specified by the integers $(p_1, p_2, \ldots, p_{n-1})$. Then we define the *conjugate tableau* with respect to $SU(n)$ to be the one specified by the integers $(p_{n-1}, \ldots, p_2, p_1)$. As an example, the tableau conjugate to (210^3) is $(0^3 12)$ or

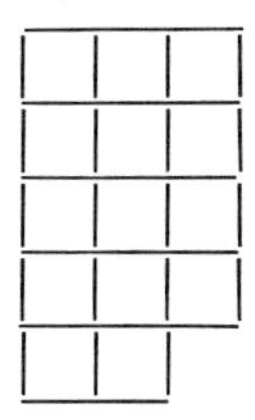

The number of boxes of a tableau is not equal, in general, to the number of boxes in its conjugate tableau. This is in contrast to conjugation with respect to S_n defined in Chapter 4. When we refer to a conjugate tableau, it should be clear from the context whether we mean with respect to $SU(n)$ or S_n.

If the same tableau is denoted by the integers $(p_1, \ldots, p_{n-1})$ and $(p_{n-1}, \ldots, p_n)$ it is said to be *self-conjugate*. If a tableau and its conjugate are not the same, they represent irreducible tensors of two *inequivalent* representations with the same dimensionality. In $SU(2)$, all tableaux are specified by a single integer p_1, and are therefore all self-conjugate. Thus, a representation of $SU(2)$ is fully specified by giving its dimensionality, but this is not true of $SU(n)$ with $n > 2$.

As we remarked previously, there are $n - 1$ fundamental representations of $SU(n)$. These correspond to all the Young tableaux with a single column with up to $n - 1$ boxes. The ith fundamental representation of $SU(n)$ is specified by the $n - 1$ integers

$$\begin{aligned} p_i &= 1, \\ p_j &= 0, \qquad j \neq i. \end{aligned} \tag{7.9}$$

Thus, for example, $SU(3)$ has the fundamental representations (10) and (01). Since they are conjugates of each other, they have the same dimensionality.

As another example, the three fundamental representations of $SU(4)$ have the tableaux

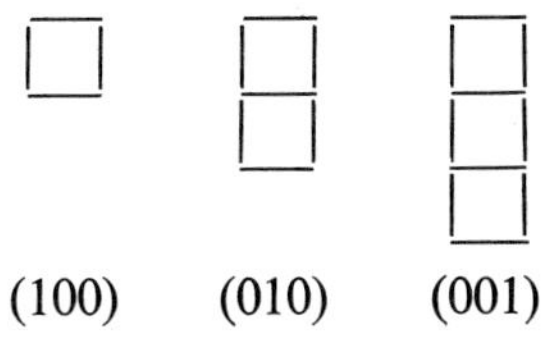

The representations (100) and (001) are conjugates of each other, and each has four dimensions. However the self-conjugate representation (010) has six dimensions. We have made particular use of the first fundamental representation of $SU(n)$, denoted by (10^{n-2}), as all others can be constructed from it.

7.2 Dimensionality Formulas

It can become very tedious to find the dimensionality of a Young tableau of $SU(n)$ by counting all its standard arrangements. However, we can simplify the problem by obtaining a general formula for the number of standard arrangements. This is a combinatorial problem with a straightforward solution. We shall derive the formula for $SU(2)$, and just state the result for $SU(n)$.

THEOREM. *For $SU(2)$, the dimensionality $N(p)$ of an irreducible representation corresponding to a tableau with p boxes is*

$$N(p) = p + 1. \tag{7.10}$$

Proof. Consider the tableau consisting of a single box $\square$. By direct counting, we find there are two standard arrangements $\boxed{1}$ and $\boxed{2}$. Therefore $N(1) = 2$, or $N(p) = p + 1$ if $p = 1$. We next show that if $N(p) = p + 1$, then $N(p + 1) = p + 2$. It then will follow by induction that $N(p) = p + 1$ holds for all p. Consider the standard arrangements with p boxes:

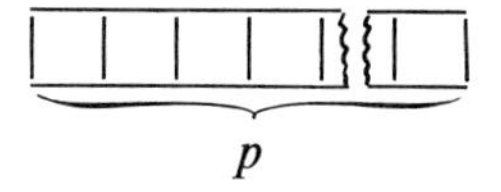

The last box may contain either a 1 or a 2. There is only one standard arrangement with the last box containing a 1: namely, the arrangement in which all boxes contain a 1. Now if we add another box, this last box can contain either a 1 or a 2:

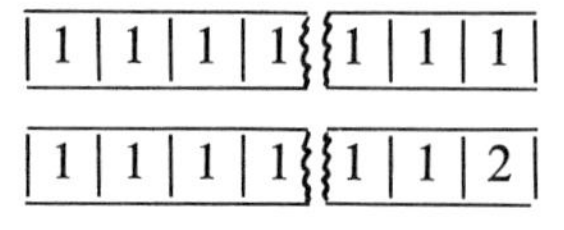

Therefore, we obtain two standard arrangements where we had one before. Now consider all other standard arrangements of p boxes in which the integer in the last box is 2. There must be p such arrangements, since by assumption, there are a total of $p + 1$ standard arrangements of p boxes, and only one of them ends with a 1. If we add one more box to these tableaux, we must put the integer 2 into this box. So the number of standard arrangements arising from the source remains at p. Finally, all together we have $p + 2$ standard arrangements of $p + 1$ boxes, which completes the proof.

The dimensionalities for $SU(3)$ and $SU(4)$ are given by

$$N_3(p_1 p_2) = \tfrac{1}{2}(p_1 + 1)(p_1 + p_2 + 2)(p_2 + 1), \tag{7.11}$$

or

$$N_3(p_1 p_2) = \tfrac{1}{2}(p_1 + p_2 + 2)(p_2 + 1)N_2(p_1). \tag{7.12}$$

$$N_4(p_1 p_2 p_3) = \frac{1}{2!3!}(p_1 + 1)(p_1 + p_2 + 2)(p_1 + p_2 + p_3 + 3)$$

$$\times (p_2 + 1)(p_2 + p_3 + 2)(p_3 + 1), \tag{7.13}$$

or

$$N_4(p_1 p_2 p_3) = (p_1 + p_2 + p_3 + 3)(p_2 + p_3 + 2)(p_3 + 1)N_3(p_1 p_2)/3! \tag{7.14}$$

The dimensionality for $SU(n + 1)$ is given by

$$N_{n+1}(p_1 \cdots p_n)$$

$$= \frac{1}{2!3! \cdots n!}(p_1 + 1)(p_1 + p_2 + 2) \cdots (p_1 + p_2 + \cdots + p_n + n)$$

$$\times (p_2 + 1)(p_2 + p_3 + 2) \cdots (p_2 + \cdots + p_{n-1} + n - 1) \cdots (p_n + 1) \tag{7.15}$$

or

$$N_{n+1}(p_1 \cdots p_n)$$

$$= (p_n + 1)(p_n + p_{n-1} + 2) \cdots (p_n + p_{n-1} + \cdots + p_1 + n)N_n(p_1 \cdots p_{n-1})/n!. \tag{7.16}$$

It is also a combinatorial problem to find the multiplicity of any weight associated with a given tableau of $SU(n)$. This is just the number of standard arrangements of a tableau with a given set of integers. For $SU(3)$, the solution is that the maximum multiplicity ν of any tableau is given by Eq. (6.43)

$$\nu = \tfrac{1}{2}(p_1 + p_2) - \tfrac{1}{2}|p_1 - p_2| + 1.$$

We shall not prove this, but shall prove the special case that all weights of a rectangular Young tableau of $SU(3)$ are simple.

First we prove that all weights of $SU(2)$ are simple. The proof follows at once from the fact that any tableau consists of a single row. If we have a given set of 1's and 2's, the only standard arrangement consists of all the 1's followed by all the 2's. The generalization to a single-row tableau of $SU(n)$ follows immediately, since all the 1's, 2's, 3's, ..., n's must go in consecutive order. For $SU(3)$, all rectangular Young tableaux have either one or two rows. The tableaux with two rows are conjugates of the one-row tableaux, and therefore have the same multiplicity. This completes the proof.

For tableaux of $SU(n)$, the combinational problem is harder. We shall not give the general result for the maximum multiplicity of a Young tableau of $SU(n)$. We merely state that even rectangular tableaux may have multiple weights for $n \geq 4$.

There is still another method for obtaining the dimensionality of a given Young tableau of $SU(n)$. For brevity, we write $\mathbf{p}$ for $p_1, p_2, \ldots, p_{n-1}$. Also we write the expression for the dimensionality $N_n(\mathbf{p})$ as a quotient:

$$N_n(\mathbf{p}) = a_n(\mathbf{p})/b(\mathbf{p}) \tag{7.17}$$

where $a_n(\mathbf{p})$ depends on the particular configuration of the Young tableau and on n, but $b(\mathbf{p})$ depends only on the configuration of the tableau.

We now give a prescription for obtaining $a_n(\mathbf{p})$ and $b(\mathbf{p})$ which, for large n, is more convenient than using the general formula (7.15). Suppose we have a tableau of $SU(n)$. We find $a_n(\mathbf{p})$ as follows: We write an n in the upper left corner of the tableau and then increase the numbers by unity in going from left to right in every row, and decrease them by unity in going down in every column. Then the product of the integers in the tableau is $a_n(\mathbf{p})$. To obtain $b(\mathbf{p})$, we proceed as follows. We put a point in the center of a given box of the tableau and draw two lines from the point, the first to the right and the second down. The number of boxes intersected by the lines is written in the given box, and the procedure is repeated for all other boxes. The product of these numbers is then $b(\mathbf{p})$.

We illustrate the method by an example. Consider the tableau

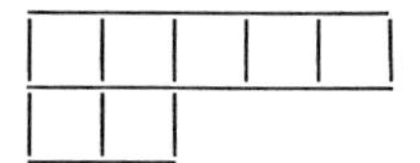

for $SU(3)$. The procedure to obtain the numerator $a_3(32)$ is to fill in the boxes of the tableau with numbers according to the prescription described for $a_n(\mathbf{p})$ with $n = 3$. We obtain the arrangement

3	4	5	6	7
2	3			

Then $a_3(320)$ is given by

$$a_3(32) = 3 \cdot 4 \cdot 5 \cdot 6 \cdot 7 \cdot 2 \cdot 3.$$

We illustrate the procedure to find $b(\mathbf{p})$ by drawing the two lines for one of the boxes. The lines intersect five boxes, so we put a 5 in the box

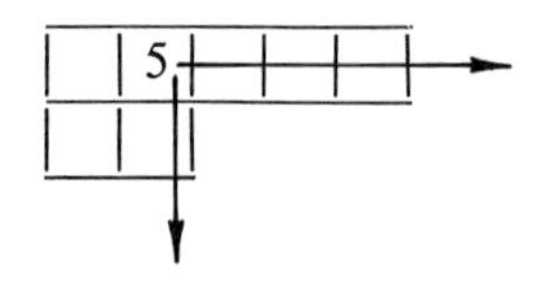

Repeating this procedure for the other boxes, we obtain

$$\begin{array}{|c|c|c|c|c|}\hline 6 & 5 & 3 & 2 & 1 \\ \hline 2 & 1 \\ \cline{1-2}\end{array}$$

Then $b(\mathbf{p})$ is given by

$$b(32) = 6 \cdot 5 \cdot 3 \cdot 2 \cdot 1 \cdot 2 \cdot 1.$$

Finally the dimensionality is given by

$$N_3(32) = \frac{a_3(32)}{b(32)} = \frac{3 \cdot 4 \cdot 5 \cdot 6 \cdot 7 \cdot 2 \cdot 3}{6 \cdot 5 \cdot 3 \cdot 2 \cdot 1 \cdot 2 \cdot 1} = 42.$$

To find the dimensionality of this tableau for $SU(4)$, we need only compute $a_4(320)$. This is

$$\begin{array}{|c|c|c|c|c|}\hline 4 & 5 & 6 & 7 & 8 \\ \hline 3 & 4 \\ \cline{1-2}\end{array}$$

Then $N_4(320)$ is given by

$$N_4(320) = \frac{4 \cdot 5 \cdot 6 \cdot 7 \cdot 8 \cdot 3 \cdot 4}{6 \cdot 5 \cdot 3 \cdot 2 \cdot 1 \cdot 2 \cdot 1} = 224,$$

since $b(\mathbf{p})$ for a given tableau is the same for all $SU(n)$.

7.3 Multiplets of the $SU(n-1)$ Subgroup of $SU(n)$

If we look at a weight diagram of a given multiplet of $SU(3)$, we see that it contains a number of multiplets of $SU(2)$. For example, the $SU(3)$ triplet contains a doublet and singlet of $SU(2)$, as can be seen from the weight diagram of Fig. 6.6. A more complicated case is that of the $SU(3)$ octet, which contains two doublets, a triplet, and a singlet of $SU(2)$, as can be seen from Fig. 6.8. These results can also be obtained directly from Young tableaux in a way that can be generalized to any multiplet of $SU(n)$.

We illustrate with the $SU(3)$ octet. The basis functions for the eight-dimensional representation of $SU(3)$ are summarized in the Young tableau

$$SU(3): \quad \begin{array}{|c|c|}\hline & \\ \hline \\ \cline{1-1}\end{array}$$

The eight standard arrangements of this tableau are

$$\begin{array}{|c|c|}\hline 1 & 1 \\ \hline 2 \\ \cline{1-1}\end{array} \quad \begin{array}{|c|c|}\hline 1 & 1 \\ \hline 3 \\ \cline{1-1}\end{array} \quad \begin{array}{|c|c|}\hline 1 & 2 \\ \hline 2 \\ \cline{1-1}\end{array} \quad \begin{array}{|c|c|}\hline 1 & 2 \\ \hline 3 \\ \cline{1-1}\end{array} \quad \begin{array}{|c|c|}\hline 1 & 3 \\ \hline 2 \\ \cline{1-1}\end{array} \quad \begin{array}{|c|c|}\hline 1 & 3 \\ \hline 3 \\ \cline{1-1}\end{array} \quad \begin{array}{|c|c|}\hline 2 & 2 \\ \hline 3 \\ \cline{1-1}\end{array} \quad \begin{array}{|c|c|}\hline 2 & 3 \\ \hline 3 \\ \cline{1-1}\end{array}$$

Now if a Young tableau refers to $SU(2)$, it can contain only the integers 1 and 2. So let us first look at those standard arrangements of the octet diagram in which 3 does not appear. There are two such arrangements

$$\begin{array}{|c|c|}\hline 1 & 1 \\ \hline 2 \\ \cline{1-1}\end{array} \qquad \begin{array}{|c|c|}\hline 1 & 2 \\ \hline 2 \\ \cline{1-1}\end{array}$$

The column

$$\begin{array}{|c|}\hline 1 \\ \hline 2 \\ \hline\end{array}$$

corresponds to no free indices, because in $SU(2)$ we can have only one antisymmetric combination of 1 and 2. Thus these arrangements correspond to an $SU(2)$ doublet:

$$\left.\begin{array}{c} \begin{array}{|c|c|}\hline 1 & 1 \\ \hline 2 \\ \cline{1-1}\end{array} \\ \\ \begin{array}{|c|c|}\hline 1 & 2 \\ \hline 2 \\ \cline{1-1}\end{array} \end{array}\right\} \to \left.\begin{array}{c} \begin{array}{|c|}\hline 1 \\ \hline\end{array} \\ \\ \begin{array}{|c|}\hline 2 \\ \hline\end{array} \end{array}\right\} \to \begin{array}{|c|}\hline \\ \hline\end{array}$$

Next let us consider the arrangements in which a 3 appears only in the right-hand box; there is only one such arrangement

$$\begin{array}{|c|c|}\hline 1 & 3 \\ \hline 2 \\ \cline{1-1}\end{array}$$

The only part of this tableau on which $SU(2)$ can act is the part where 3 does not occur, namely

$$\begin{array}{|c|}\hline 1 \\ \hline 2 \\ \hline\end{array},$$

which is a singlet. Next, consider the tableaux in which 3 appears only at the bottom; there are three such arrangements

$$\begin{array}{|c|c|}\hline 1 & 1 \\ \hline 3 \\ \cline{1-1}\end{array} \qquad \begin{array}{|c|c|}\hline 1 & 2 \\ \hline 3 \\ \cline{1-1}\end{array} \qquad \begin{array}{|c|c|}\hline 2 & 2 \\ \hline 3 \\ \cline{1-1}\end{array}$$

Again the box containing the 3 is irrelevant for $SU(2)$, and we can remove it. What is left are the arrangements

$$\begin{array}{|c|c|}\hline 1 & 1 \\ \hline\end{array} \qquad \begin{array}{|c|c|}\hline 1 & 2 \\ \hline\end{array} \qquad \begin{array}{|c|c|}\hline 2 & 2 \\ \hline\end{array}$$

which form a triplet. Finally, we consider the two tableaux containing two boxes with the number 3

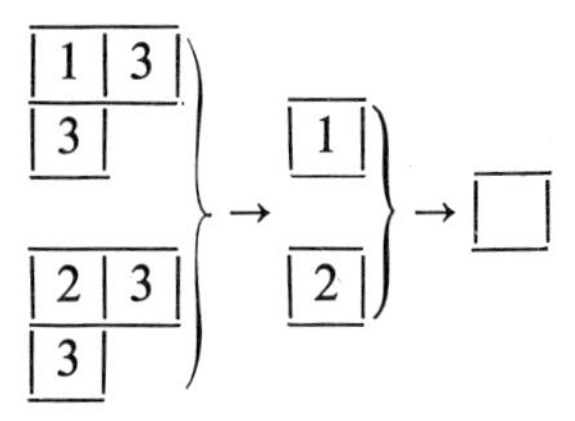

Since $SU(2)$ does not act on the index 3, this is another doublet. Thus, we find that the octet representation of $SU(3)$ contains two $SU(2)$ doublets, a triplet, and a singlet. It can be seen that we do not need to fill in the 1's and 2's to get this result, only the 3's. The formula for the multiplicity of a tableau of $SU(2)$ will then enable us to obtain the multiplicities.

By a generalization of this procedure, we can find the $SU(n-1)$ multiplets contained in any tableau of $SU(n)$. The prescription is to consider all possible ways of putting n's in the boxes consistent with standard arrangements from no n's to the maximum allowed number. We then remove all boxes with n's. The remaining tableaux give the $SU(n-1)$ multiplets.

For example suppose we have the tableau

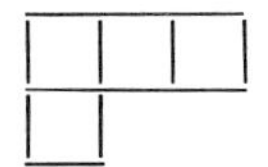

Following the prescription for putting n's in boxes, we obtain

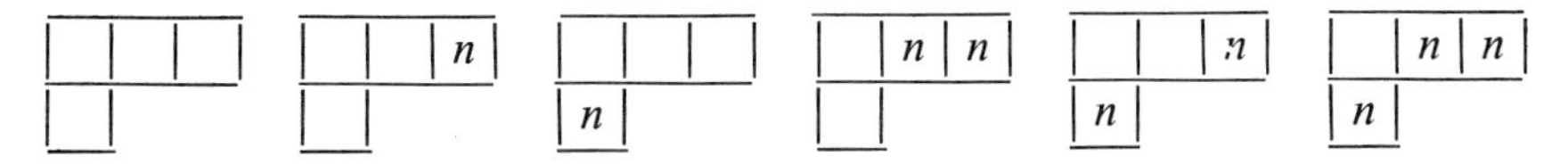

Each of these tableaux specifies a representation of $SU(n-1)$. Since $SU(-1)$ does not care about the index n, we drop the boxes which contain an n. We thus obtain

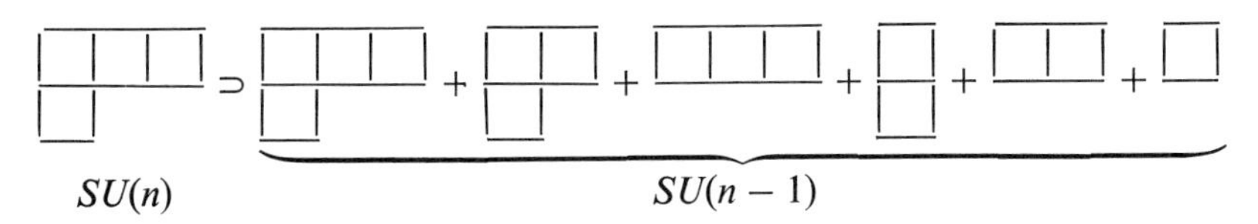

We can proceed further, obtaining the $SU(n-2)$ multiplets contained in these tableaux of $SU(n-1)$, and then continue the process until we have obtained the multiplets of all the $SU(n-n')$ subgroups of $SU(n)$, where $1 \leq n' \leq n-2$.

We can obtain the value of the weight m_2 which is characteristic of an entire $SU(2)$ multiplet contained in a multiplet of $SU(3)$. The single-particle states of $SU(3)$ have the following values of m_2

$$m_2(1) = m_2(2) = 1/(2\sqrt{3}), \qquad m_2(3) = -1/\sqrt{3}. \tag{7.18}$$

Then, since the weights are additive, a standard arrangement of a Young diagram with v boxes and v' 3's has a value of m_2 given by

$$m_2 = \tfrac{1}{2}(v - 3v')/\sqrt{3}, \qquad v \geq v' = 0, 1, 2, \ldots .$$

If m_2 is associated with the hypercharge as in the usual interpretation of $SU(3)$, then

$$Y = 2m_2/\sqrt{3} = \tfrac{1}{3}v - v'.$$

But the number of boxes of an $SU(3)$ Young diagram is given by $v = p_1 + 2p_2$. Therefore the hypercharge is given by

$$Y = 2m_2/\sqrt{3} = \tfrac{1}{3}(p_1 + 2p_2) - v'. \tag{7.19}$$

As an example of this procedure, let us find the hypercharge of each $SU(2)$ isospin multiplet contained in the $SU(3)$ decuplet or (30) representation. We have

$$\boxed{}\boxed{}\boxed{} \supset \boxed{}\boxed{}\boxed{} + \boxed{}\boxed{}\boxed{3} + \boxed{}\boxed{3}\boxed{3} + \boxed{3}\boxed{3}\boxed{3}$$

or

$$\boxed{}\boxed{}\boxed{} \supset \boxed{}\boxed{}\boxed{} + \boxed{}\boxed{} + \boxed{} + \textcircled{1}$$

Thus, the decomposition is

$$\mathbf{10} \supset \mathbf{4} \oplus \mathbf{3} \oplus \mathbf{2} \oplus \mathbf{1}.$$

Using Eq. (7.19), we have the following connection between $SU(2)$ multiplicity N_2 and hypercharge:

$$\begin{array}{llrrr} N_2: & 4 & 3 & 2 & 1, \\ Y: & 1 & 0 & -1 & -2. \end{array}$$

We can obtain the weight diagrams of $SU(3)$ in an easy way by making use of the $SU(2)$ subgroups of $SU(3)$. First we note that the isospin multiplet with the highest value of m_2 (or equivalently of the hypercharge Y) has Y given by Eq. (7.19) with $v' = 0$. Also, the isospin of this multiplet is given by the Young tableau obtained after removing the columns with two boxes. Thus this multiplet has isospin and hypercharge given by

$$I_1 = \tfrac{1}{2}p_1, \qquad Y = \tfrac{1}{3}(p_1 + 2p_2). \tag{7.20}$$

If the weight diagram is rotated by 60° in either direction, the multiplet at the top of the diagram in its new position has $SU(2)$ multiplicity N_2 given by

$$N_2 = p_2 + 1. \tag{7.21}$$

If we fill in that portion of the weight diagram which we obtain from Eqs. (7.20) and (7.21), we obtain either one side of a triangle (for a rectangular tableau) or three sides of a hexagon (for any other tableau). Then, using 120° rotational symmetry and the knowledge that $SU(2)$ subgroups have weights that differ by integral values, we can complete the diagram except for the problem of multiple weights. This problem can be solved as follows: If the weight diagram is a triangle, i.e., $N_2 = 1$, the diagram is finished. If the diagram is not a triangle, then to every weight of the interior, add an identical weight. If these double weights form a triangle, the diagram is finished. If not, then to every weight in the interior of the double weights, add an identical weight. Proceed until there is a triangle of weights or a central weight with the highest multiplicity. These rules can all be justified by counting standard arrangements of Young tableaux.

As an example, we consider the weight diagram of the representation (41) or **35**. From Eqs. (7.20) and (7.21), we obtain that portion of the weight

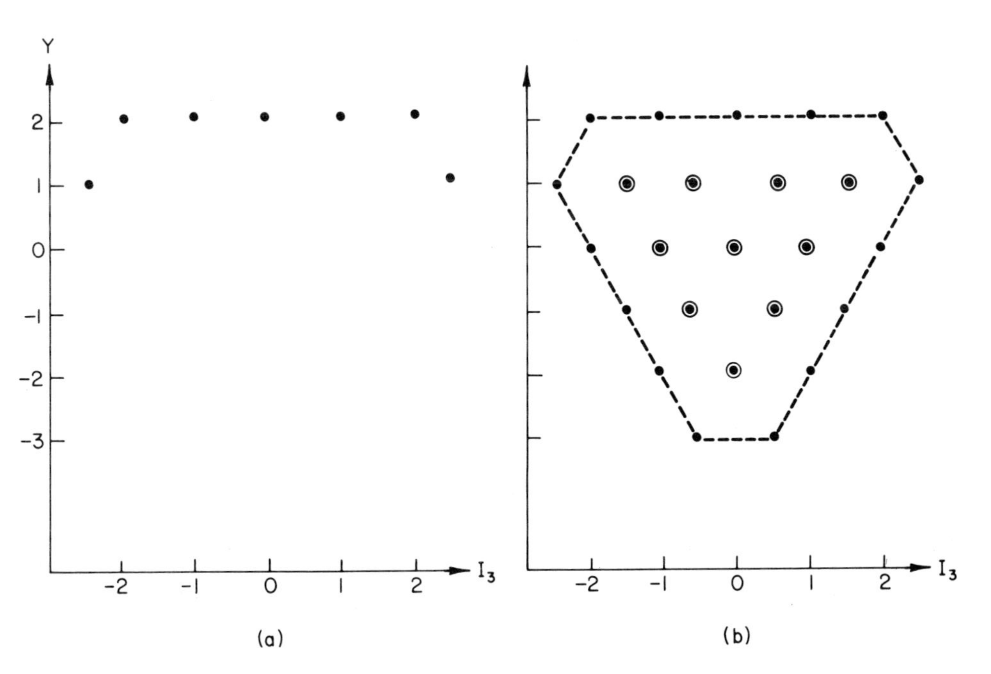

Fig. 7.2. Construction of the weight diagram of the (31) or **35** representation by a method given in Section 7.3.

diagram shown in Fig. 7.2a. Making use of 120° symmetry, plus the knowledge that $SU(2)$ weights differ by integers, and the rule for the multiplicities of weights, we complete the diagram as shown in Fig. 7.2b. The hypercharge Y, rather than m_2, is plotted as the ordinate in Fig. 7.2, but the vertical scale is shrunk so that the visual 120° symmetry remains.

7.4 Decomposition of Products of Irreducible Representations

Suppose we have two irreducible unitary representations of $SU(n)$ (or for that matter, of any compact simple group). Denote these representations by $D^{(\alpha)}$ and $D^{(\beta)}$, where $D^{(\alpha)}$ and $D^{(\beta)}$ have dimension n_α and n_β, respectively. Let the basis vectors for $D^{(\alpha)}$ and $D^{(\beta)}$ be x_i and y_i. Then the transformed basis vectors x_i' and y_i' are given by

$$x_i' = \sum_{j=1}^{n_\alpha} D_{ji}^{(\alpha)} x_j,$$

$$y_i' = \sum_{l=1}^{n_\beta} D_{li}^{(\beta)} y_l.$$

We now ask how the two-index quantity $x_i y_k$ transforms. From the transformations on the vectors we obtain

$$x_i' y_k' = \sum_{jl} D_{ji}^{(\alpha)} D_{lk}^{(\beta)} x_j y_l. \tag{7.22}$$

We can introduce the notation

$$D_{ji}^{(\alpha)} D_{lk}^{(\beta)} = D_{jl,\,ik}^{(\alpha\times\beta)}, \tag{7.23}$$

and obtain

$$x_i' y_k' = \sum_{jl} D_{jl,\,ik}^{(\alpha\times\beta)} x_j y_l. \tag{7.24}$$

The numbers $D_{jl,\,ik}^{(\alpha\times\beta)}$ are matrix elements of the representation $D^{(\alpha\times\beta)}$, which is called the *Kronecker product* of the representations $D^{(\alpha)}$ and $D^{(\beta)}$. This product representation is written

$$D^{(\alpha\times\beta)} = D^{(\alpha)} \times D^{(\beta)}. \tag{7.25}$$

We should note that the products $x_i y_k$ can be considered either as two-index tensors, or, by relabeling

$$z_j = x_i y_k, \qquad j = 1, 2, \ldots, n_\alpha n_\beta,$$

as basis vectors of a representation of $n_\alpha n_\beta$ dimensions. It is usually more convenient to consider the $x_i y_k$ as tensors. The representation $D^{(\alpha\times\beta)}$ is in general reducible. If the group is finite or simple and compact, $D^{(\alpha \times \beta)}$ decomposes into a direct sum of irreducible representations:

$$D^{(\alpha \times \beta)} = \sum_i \Gamma_i D^{(\gamma_i)}, \tag{7.26}$$

where the number Γ_i denotes how many times the irreducible representation $D^{(\gamma_i)}$ appears in the sum. This decomposition is called the *Clebsch–Gordan series* and is sometimes written

$$\alpha \times \beta = \sum_i \Gamma_i \gamma_i . \tag{7.27}$$

If none of the Γ_i is greater than unity, the product representation is said to be *simply reducible*. For a simple Lie group of rank one, a theorem says that the Kronecker product of any two irreducible representations is simply reducible. This is the principal requirement for calling a group simply reducible. Actually four requirements are necessary for a group to be simply reducible. They are:

(1) The direct sum contained in the Kronecker product of two irreducible representations contains each irreducible representation no more than once.

(2) Any representation D is either real (i.e., can be brought to real form by a similarity transformation), or is equivalent to D^*, the complex conjugate representation (i.e., the representation by matrices which are the complex conjugates of the original matrices).

(3) If a group element belongs to a certain class, then its inverse belongs to the same class.

(4) The characters (traces) of the matrices representing the group elements are real.

It is an important problem to construct from the $x_i y_k$ those *irreducible* tensors which form the basis functions of the irreducible unitary representations $D^{(\gamma_i)}$, as these tensors are the multiplets. We shall consider this problem for the group $SU(n)$, using the method of Young tableaux.

We consider as an example the tensors of a representation which is the Kronecker product of the fundamental representation of $SU(2)$ with itself. Suppose we have a system of two particles of spin $\frac{1}{2}$ (two doublets). Four linearly independent two-particle Kronecker product states can be constructed from the single-particle states u_1 and u_2. They are

$$u_1(1)u_1(2), \qquad u_1(1)u_2(2), \qquad u_2(1)u_1(2), \qquad u_2(1)u_2(2), \tag{7.28}$$

where the numbers in parentheses refer to particles number 1 and 2. If we adopt the convention of writing the state vector of the first particle to the left, we can omit the parentheses. Then the two-particle states of Eq. (7.28) are

$$u_1 u_1, \qquad u_1 u_2, \qquad u_2 u_1, \qquad u_2 u_2 . \tag{7.29}$$

The states $u_i u_j$ are the basis tensors of the Kronecker product representation. The irreducible tensors are the symmetric triplet states

$$u_1 u_1, \qquad (u_1 u_2 + u_2 u_1)/\sqrt{2}, \qquad \text{and} \qquad u_2 u_2, \tag{7.30}$$

and the antisymmetric singlet state

$$(u_1 u_2 - u_2 u_1)/\sqrt{2}. \tag{7.31}$$

Thus, the product representation of two doublets of $SU(2)$ decomposes into a triplet and a singlet. We can write this symbolically with Young tableaux as

$$\square \times \square = \square\square + \begin{array}{c}\square\\ \square\end{array} \tag{7.32}$$

This is the Clebsch–Gordan series for decomposition of the Kronecker product of two doublets. There are several other notations for this series. One notation is in terms of the dimensionality of the multiplets as follows:

$$\mathbf{2} \otimes \mathbf{2} = \mathbf{3} \oplus \mathbf{1}. \tag{7.33}$$

We put circles around the multiplication and addition signs to distinguish them from ordinary arithmetic operations.

We can also denote each representation by the integer p which stands for the configuration of its Young tableau. Then the Clebsch–Gordan series is

$$(1) \times (1) = (2) + (0). \tag{7.34}$$

Next consider an example from $SU(3)$. It is easy to find the Clebsch–Gordan series for the product of two triplets of $SU(3)$ with the aid of Young tableaux. In fact, it is just given by Eq. (7.32), which holds for any $SU(n)$. The dimensionalities are different, however, but can be found from any of the methods discussed in Sections 7.1 and 7.2. In particular, applying Eq. (7.11) to the tableaux of (7.32), we find that for $SU(3)$ we have

$$\mathbf{3} \otimes \mathbf{3} = \mathbf{6} \oplus \bar{\mathbf{3}}. \tag{7.35}$$

Using the p-notation, this is

$$(10) \times (10) = (20) + (01). \tag{7.36}$$

We shall now give the general rules for reducing the Kronecker product of two representations by means of Young tableaux in order to obtain the representations of the Clebsch–Gordan series. We draw the two Young tableaux of the representations, marking each box of the second diagram with the number of the row to which it belongs. We then attach the boxes of the second tableau in all possible ways to the first tableau, subject to the following rules of the combined tableaux:

(1) Each tableau should be a proper tableau; that is, no row is longer than any row above it.

(2) No tableau should have a column with more than n boxes if the group is $SU(n)$.

(3) We can make a path by counting each row from the right, starting with the top row. At each point of the path the number of boxes encountered with the number i must be less or equal to the number of boxes with $i - 1$.

(4) The numbers must not decrease in going from left to right across a row.

(5) The numbers must increase in going from top to bottom in a column.

As an example, let us obtain the reduction of the Kronecker product of the first and second fundamental representations of $SU(n)$. We obtain

$$\begin{array}{|c|}\hline \ \\\hline\end{array} \times \begin{array}{|c|}\hline 1\\\hline 2\\\hline\end{array} = \begin{array}{|c|c|}\hline \ & 1\\\hline 2 \\\cline{1-1}\end{array} + \begin{array}{|c|}\hline \ \\\hline 1\\\hline 2\\\hline\end{array}$$

or

$$(10^{n-2}) \times (010^{n-3}) = (110^{n-3}) + (0010^{n-4}).$$

For $SU(3)$ this is the reduction of the product of the fundamental representation and its conjugate. The $SU(3)$ numerology is

$$\mathbf{3} \otimes \bar{\mathbf{3}} = \mathbf{8} \oplus \mathbf{1}. \tag{7.37}$$

Note that the reduction of $\mathbf{3} \otimes \bar{\mathbf{3}}$ [Eq. (7.37)] is different from the reduction of $\mathbf{3} \otimes \mathbf{3}$ [Eq. (7.35)], since $\mathbf{3}$ and $\bar{\mathbf{3}}$ are inequivalent representations. In general, when we take the product of the first fundamental representation of any $SU(n)$ and its conjugate we get

$$\mathbf{n} \otimes \bar{\mathbf{n}} = (\mathbf{n}^2 - 1) \oplus \mathbf{1}. \tag{7.38}$$

The representation of dimension $n^2 - 1$ has the Young tableau specified by the integers $(10^{n-3}1)$ or $p_1 = p_{n-1} = 1$, all other $p_i = 0$. It is a self-conjugate representation and is often called the *adjoint representation*.

As another example, we find the irreducible representation contained in the Kronecker product of the representations (11) and (11) of $SU(3)$. Following the rules, we obtain

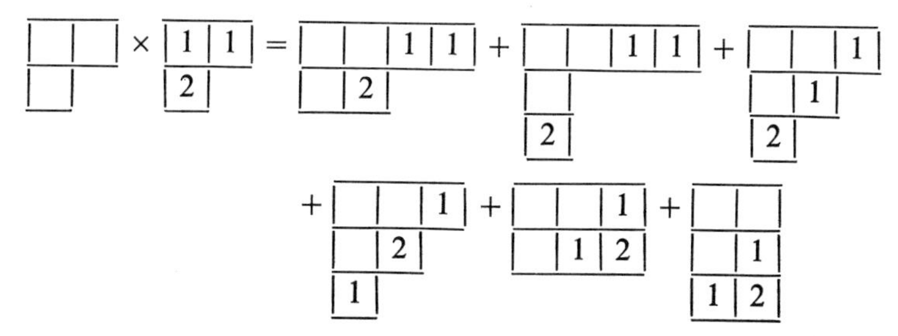

Omitting the superfluous columns with three boxes we get

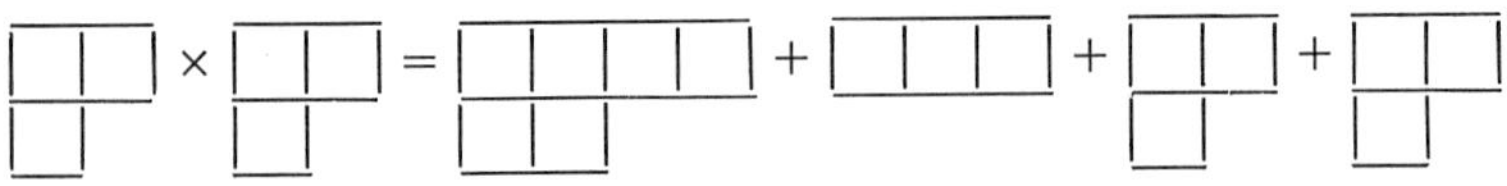

$$(11) \times (11) = (22) + (30) + (11) + (11)$$

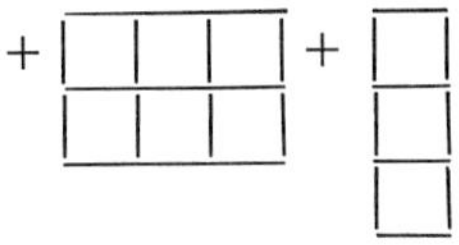

$$+ (03) + (00)$$

or

$$\mathbf{8} \otimes \mathbf{8} = \mathbf{27} \oplus \mathbf{10} \oplus \mathbf{8} \oplus \mathbf{8} \oplus \overline{\mathbf{10}} \oplus \mathbf{1}. \tag{7.39}$$

We see that the Clebsch–Gordan series for the Kronecker product $\mathbf{8} \otimes \mathbf{8}$ contains two equivalent representations, namely the eight-dimensional self-conjugate representations. Thus, this product is not simply reducible. This is in contrast to the case for $SU(2)$, in which the Kronecker product of any two representations is simply reducible. In fact the result for $SU(2)$ is a special case of a particular class of Kronecker products of $SU(n)$. The relevant theorem states that the Kronecker product of any representation and a representation specified by a rectangular tableau (i.e., a representation with only one p_i different from zero) is simply reducible. Since all tableaux of $SU(2)$ are rectangular (consisting of a single row), the theorem states that any Kronecker product representation of $SU(2)$ is simply reducible.

If two or more equivalent representations appear in the reduction of a Kronecker product of $SU(n)$, the use of group theory alone is not sufficient to enable us to label the states. We must know something of the physics of the problem in order to obtain a useful labeling. This problem has been encountered in $SU(2)$ in the Kronecker product of *three* representations. For example, consider the Kronecker product of three doublets of $SU(2)$

$$\mathbf{2} \otimes \mathbf{2} \otimes \mathbf{2} = (\mathbf{3} \oplus \mathbf{1}) \otimes \mathbf{2} = \mathbf{4} \oplus \mathbf{2} \oplus \mathbf{2}. \tag{7.40}$$

We see that in addition to a quartet two different doublets appear in the reduction. Let us assume that our example represents the coupling of three spin-$\frac{1}{2}$ particles. The first two particles can combine to form spin 1 or 0, and then the third particle can combine with either of these states to give a total spin of $\frac{1}{2}$. Group theory says nothing about whether the first two particles, in fact, combine to form spin 1 or 0 or a linear combination of the two. Only if we have a knowledge of the forces involved can we decide this question. At the level of $SU(n)$, with $n > 2$, the problem arises in the reduction of the Kronecker product of two representations.

7.5 Classes of Representations

A representation of $SU(2)$ can be put into one of two classes according to whether the number of boxes of its Young diagram is even or odd. The number of boxes p (all of which are in a single row) is related to the quantum number j by $p = 2j$. Thus, if p is even, j is an integer, while if p is odd, j is half of an odd integer. The representations of $SU(2)$ are called *integral* or *half integral*, according to whether j is integral or half integral.

We shall not prove the statement that a simply reducible group has only integral and half-integral unitary irreducible representations. The integral representations are those that can be made real by a similarity transformation, while the half-integral representations cannot be made real but are equivalent to their complex-conjugate representations. The group $R(3)$ has only integral vector representations, and they are just the same as the integral representations of $SU(2)$.

Suppose we have a product representation of $SU(2)$ whose states are denoted by the Young diagrams p and p'. We can use the method of Young diagrams to find the values of p'' of the irreducible representations contained in the product. Assuming $p > p'$, we have

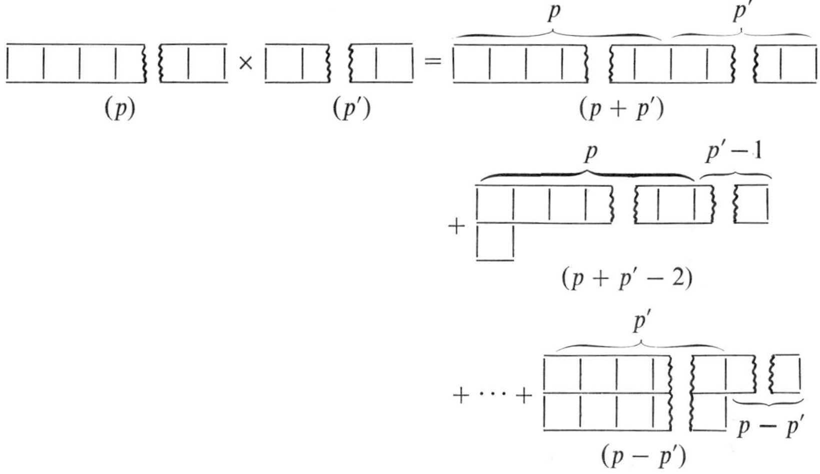

Removing all columns with two boxes, we obtain

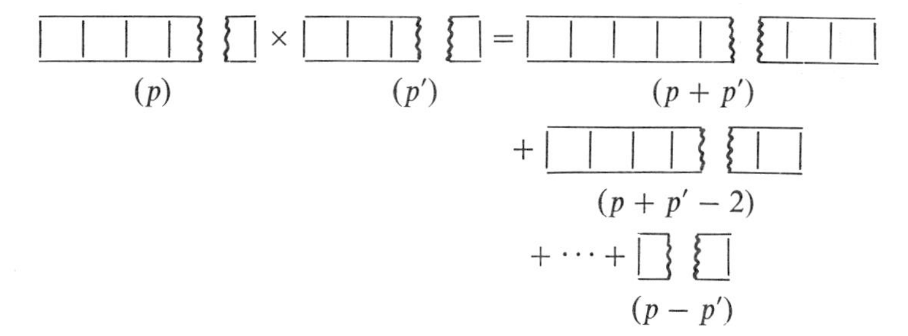

Denoting the representations by their values of j, we have

$$j'' = j + j', j + j' - 1, \ldots, j - j'. \tag{7.41}$$

According to Eq. (7.41), a product of two integral or two half-integral representations contains only integral irreducible representations in its direct sum, while a product of an integral and half-integral representation contains only half-integral irreducible representations.

Although the representations of $SU(n)$, $n > 2$, are not in general equivalent to their complex conjugates, we can nevertheless generalize the idea of the class of a representation. We subdivide the representations of $SU(n)$ into n classes according to the number of boxes contained in their Young tableaux. Let the number of boxes ν of any tableau be given by

$$\nu = ni + k, \tag{7.42}$$

where i and k are integers satisfying

$$i \geq 0, \qquad 0 \leq k \leq n - 1. \tag{7.43}$$

Each of the different possible values of k denotes a representation of a different class. Note that removing a column of n boxes from a tableau does not change the class. If we decompose the product of two representations of classes k_1 and k_2, then the irreducible representations contained in the product are of class k where

$$k = k_1 + k_2, \qquad \text{modulo } n. \tag{7.44}$$

From this formula it is clear that the Kronecker product of two representations of class $k = 0$ contains only irreducible representations of the same class.

In Chapter 2, we defined Z_2, a two-element invariant subgroup of $SU(2)$ [see Eq. (2.19)]. The group Z_2 consists of the two square roots of unity multiplied by the 2×2 unit matrix. Similarly, we can define Z_n as the group of the n nth roots of unity times the $n \times n$ unit matrix. Then Z_n is an invariant subgroup of $SU(n)$. The factor group $SU(n)/Z_n$ has vector representations only of class $k = 0$. However, if we also include multiple-valued representations of $SU(n)/Z_n$, then we include all the representations of $SU(n)$.

Baird and Biedenharn (1964) have introduced the term triality τ to distinguish the three classes of representations of $SU(3)$. They define

$$\begin{aligned} \tau &= 0 \qquad \text{if} \quad k = 0, \\ \tau &= 1 \qquad \text{if} \quad k = 1, \\ \tau &= -1 \qquad \text{if} \quad k = 2. \end{aligned} \tag{7.45}$$

The representations of zero triality are the only vector representations of the group $SU(3)/Z_3$. However, if we include the triple-valued representations of this group, we include all the representations of $SU(3)$.

7.6 Multiplets of $U(n)$

The Lie algebra of $U(n)$ is the same as the Lie algebra of $U(1) \times SU(n)$. Since $U(1)$ is Abelian, all its unitary irreducible representations are one-dimensional. Therefore, all the considerations of this chapter regarding the dimensionality of representations can be directly taken over to $U(n)$. However, in the case of $U(n)$, each Young tableau is associated with an additional quantum number. For example, in the case of $U(2)$, the $SU(2)$ subgroup can be interpreted as the isospin group and the $U(1)$ subgroup as the group of the hypercharge. In the case of $U(3)$, $SU(3)$ can be interpreted as a subgroup including both isospin and hypercharge, and $U(1)$ as the subgroup of the baryon number B.

The generator associated with $U(1)$ is an additive quantum number. Because of this, Young tableaux with different numbers of boxes correspond to different values of the quantum number, and are inequivalent. Thus, in considering the Young tableaux of $U(n)$, we can lose information if we omit columns of n boxes. However, it is convenient to remove these n-box columns in any case, keeping track of the $U(1)$ quantum number separately.

Another complication with $U(n)$ is that its different fundamental representations can have values of the extra quantum number which are not related in the usual way. Again we consider $U(3)$ as an example. Let the diagram $\square$ denote a one-particle state with baryon number $B = 1$ belonging to the first fundamental representation of $U(3)$. Then the two-particle states

$$\square\square \qquad \text{and} \qquad \begin{array}{c}\square\\\square\end{array}$$

both have $B = 2$. Now, however, consider the antiparticle to the state $\square$. The antiparticle is defined to be a state with the same energy as the particle but with its weight reversed. Thus, antiparticle of the state $\square$ has baryon number -1 and $SU(3)$ weight $-\mathbf{m}$, but the three states with weights $-\mathbf{m}$ belong to the conjugate $\bar{\mathbf{3}}$ representation of $SU(3)$ with Young tableau

$$\begin{array}{c}\square\\\square\end{array}$$

Therefore, starting from the state $\square$, we are led to two inequivalent representations of $SU(3)$, one with $B = 2$ and the other with $B = -1$, but both representations have the same Young tableau

$$\begin{array}{c}\square\\\square\end{array}$$

This creates no real difficulty provided we keep track of the quantum number B in a systematic way.

CHAPTER 8

CLEBSCH–GORDAN COEFFICIENTS

8.1 Some Properties of the Coefficients

Suppose we have the basis vectors of two unitary irreducible representations of a compact simple Lie group. We denote these vectors by $\psi_\mu^{(\alpha)}$ and $\psi_\nu^{(\beta)}$, where α and β stand for all numbers necessary to specify the first and second representations, and μ and ν stand for all numbers which differentiate among the different states within these representations. Then the basis tensors of the Kronecker product representation are given by $\psi_\mu^{(\alpha)}\psi_\nu^{(\beta)}$.

In general, these product basis tensors are not the basis tensors of an irreducible representation. However, the basis tensors of any irreducible representation contained in the product can be written as a linear combination of the product tensors. The coefficients in the sum are called *Clebsch–Gordan coefficients.* The irreducible tensors contained in the direct product may be written with a single index $\psi_m^{(j)}$, in which case they can be regarded as vectors. Since we shall have occasion to write them in both ways, we shall usually just call them basis functions. Likewise, the product tensors can be written as a linear combination of the irreducible functions. The coefficients in this sum are also called Clebsch–Gordan coefficients.

Let us illustrate with $SU(2)$ as an example. In this case, each of the symbols α, β, μ, and ν stands for a single number. Then we can write

$$\psi_m^{(j)} = \sum_{\mu\nu} (\alpha\beta\mu\nu \mid \alpha\beta j m)\psi_\mu^{(\alpha)}\psi_\nu^{(\beta)} \tag{8.1}$$

where the coefficients $(\alpha\beta\mu\nu \mid \alpha\beta jm)$ are the $SU(2)$ Clebsch–Gordan coefficients, often called Wigner coefficients. It is apparant from this expression that $\psi_m^{(j)}$ depends on α and β, since these indices are not summed over. However, it is customary to suppress these indices on $\psi_m^{(j)}$. Likewise, the product basis functions can be written as linear combinations of the basis functions of the irreducible representations:

$$\psi_\mu^{(\alpha)}\psi_\nu^{(\beta)} = \sum_{jm} (\alpha\beta jm \mid \alpha\beta\mu\nu)\psi_m^{(j)}. \tag{8.2}$$

The coefficients $(\alpha\beta jm \mid \alpha\beta\mu\nu)$ are also Clebsch–Gordan coefficients. It is seen that the indices $\alpha\beta$ are repeated in the symbol for a Clebsch–Gordan coefficient. Sometimes this redundancy is omitted, i.e., we write

$$(\alpha\beta jm \mid \alpha\beta\mu\nu) = (jm \mid \alpha\beta\mu\nu). \tag{8.3}$$

There is further redundancy in this expression, since the weights are additive, i.e., $m = \mu + \nu$. There are many other symbols for the Clebsch–Gordan coefficients.

A number of other relations can be proved to hold among these coefficients. For example, by taking the scalar product of Eq. (8.1) with $\psi_{\mu'}^{(\alpha)}\psi_{\nu'}^{(\beta)}$ we find, on using the notation of Eq. (8.3) and dropping the primes, that

$$(\psi_\mu^{(\alpha)}\psi_\nu^{(\beta)}, \psi_m^{(j)}) = (\alpha\beta\mu\nu \mid jm), \tag{8.4}$$

since the $\psi_\mu^{(\alpha)}\psi_\nu^{(\beta)}$ are orthonormal. Likewise, by taking the scalar product of Eq. (8.2) with $\psi_m^{(j)}$, we find

$$(\psi_m^{(j)}, \psi_\mu^{(\alpha)}\psi_\nu^{(\beta)}) = (jm \mid \alpha\beta\mu\nu). \tag{8.5}$$

Therefore, comparing Eq. (8.4) with Eq. (8.5), we find

$$(\alpha\beta\mu\nu \mid jm) = (jm \mid \alpha\beta\mu\nu)^*. \tag{8.6}$$

Without loss of generality, the phases of the basis functions can be chosen so that the Clebsch–Gordan coefficients are real. Then we obtain

$$(\alpha\beta\mu\nu \mid jm) = (jm \mid \alpha\beta\mu\nu). \tag{8.7}$$

We can also obtain summed relations among the coefficients. Putting the expression for $\psi_m^{(j)}$ of Eq. (8.1) into that for $\psi_\mu^{(\alpha)}\psi_\nu^{(\beta)}$ in Eq. (8.2), we obtain

$$\sum_{jm} (jm \mid \alpha\beta\mu\nu)(\alpha\beta\mu'\nu' \mid jm) = \delta_{\mu\mu'}\delta_{\nu\nu'}. \tag{8.8}$$

Similarly, substituting the expression for $\psi_\mu^{(\alpha)}\psi_\nu^{(\beta)}$ into that for $\psi_m^{(j)}$, we get

$$\sum_{\mu\nu} (\alpha\beta\mu\nu \mid jm)(j'm' \mid \alpha\beta\mu\nu) = \delta_{mm'}\delta_{jj'}. \tag{8.9}$$

The Clebsch–Gordan coefficients have many other interesting properties which we shall not discuss. See, for example, Wigner (1959), Hamermesh (1962), Rose (1957), and Edmonds (1957).

The expressions relating the basis tensors of product representations in terms of the basis functions of irreducible representations can be generalized to many other groups. We restrict ourselves to groups whose unitary representations can be decomposed into a direct sum of unitary irreducible representations. Furthermore, we consider the basis functions only of irreducible unitary representations and of representations formed from the Kronecker product of such representations. Then we can write an expression for the basis functions of a product representation in terms of the basis functions of irreducible representations in a form which looks very similar to the one for $SU(2)$. We write

$$\psi_\mu^{(\alpha)}\psi_\nu^{(\beta)} = \sum_{jm\gamma} (jm\gamma \mid \alpha\beta\mu\nu)\psi_m^{(j\gamma)}, \tag{8.10}$$

and

$$\psi_m^{(j\gamma)} = \sum_{\mu\nu} (\alpha\beta\mu\nu \mid jm\gamma)\psi_\mu^{(\alpha)}\psi_\nu^{(\beta)}. \tag{8.11}$$

There are two differences between these expressions and the analogous ones for $SU(2)$. The first difference is that, since the groups are not simply reducible, there is a sum over an additional index γ. This index distinguishes between irreducible representations with the same values of j and m.

The second difference is in the interpretation of the indices α, β, μ, ν, j, m. The upper index α on $\psi_\mu^{(\alpha)}$ again specifies the irreducible representation of which this function is a member, and the lower index μ specifies a particular basis function of that representation. But the indices α, μ can now each stand for a whole set of indices. For example, in $SU(n)$, α stands for $n-1$ integers $(p_1, p_2, \ldots, p_{n-1})$, and μ stands for a weight of $n-1$ dimensions, plus all other indices necessary to distinguish members of a representation. Furthermore, the expression $\sum_{\mu\nu}$ means a sum over all weights $\mu\nu$ appearing in the representation. A given weight may appear more than once in a given representation, and it is summed as often as it appears. Of course, one can still keep the interpretation that α is a single index by using a scheme of ordering the representations with a single number. Likewise, by ordering the basis functions within a representation, μ can retain its meaning as a single number.

In anology with the case of $SU(2)$ we can show in a straightforward way that the following relations hold among the Clebsch–Gordan coefficients:

$$\sum_{\mu\nu} (\alpha\beta\mu\nu \mid jm\gamma)(j'm'\gamma' \mid \alpha\beta\mu\nu) = \delta_{jj'}\,\delta_{mm'}\,\delta_{\gamma\gamma'}, \tag{8.12}$$

$$\sum_{jm\gamma} (jm\gamma \mid \alpha\beta\mu\nu)(\alpha\beta\mu'\nu' \mid jm\gamma) = \delta_{\mu\mu'}\,\delta_{\nu\nu'}, \tag{8.13}$$

$$(\alpha\beta\mu\nu \mid jm\gamma) = (jm\gamma \mid \alpha\beta\mu\nu), \tag{8.14}$$

where without loss of generality we have assumed that the Clebsch–Gordan coefficients are real.

For a group which is not simply reducible, the values of the Clebsch–Gordan coefficients are not completely specified by the group properties. Clebsch–Gordan coefficients depend on how the representations j are chosen whenever γ is greater than unity. We may make this choice on the basis of simplicity, but any choice is satisfactory from the standpoint of group theory.

We illustrate this point by comparing $SU(2)$ and $SU(3)$. In $SU(2)$, a reducible representation formed by the Kronecker product of two irreducible representations is specified by the numbers $\alpha\beta$ of the irreducible representations. The different basis tensors of a product representation $\alpha\beta$ are distinguished by the one-dimensional weights $\mu\nu$. Now consider the irreducible representations into which we decompose the product. These representations are characterized by the numbers $j\alpha\beta$ and the different basis functions of these representations are distinguished by the weight m. Therefore, four quantum numbers $\alpha\beta\mu\nu$ distinguish the basis tensors of various Kronecker product representations, and four quantum numbers $\alpha\beta jm$ distinguish basis tensors or irreducible representations obtained by decomposing Kronecker products.

Now let us consider $SU(3)$. We again specify a representation by α, but now α stands for two numbers, either $(p_1 p_2)$ or the quantum numbers (fg) of the Casimir operators. Alternatively, when no ambiguity arises, α can stand for the dimensionality of the representation. The different basis tensors of a representation are distinguished by three numbers: the two-dimensional weight μ and the eigenvalue of a Casimir operator of an $SU(2)$ subgroup for example, the isospin I. Therefore the basis tensors of a Kronecker product representation are fully specified by ten numbers $\alpha\beta\mu\nu I_1 I_2$ ($\alpha\beta\mu\nu$ each stands for two numbers). Now consider the irreducible representations of the direct sum. A given representation is specified by j and a weight by m. Again α and β are good quantum numbers, since they are not summed over. We also have the eigenvalue I of an $SU(2)$ subgroup to distinguish among tensors. Thus we have at our disposal only the nine quantum numbers $\alpha\beta jmI$ ($\alpha\beta jm$ each stands for two numbers) which are fully specified by group theory. In order to specify an irreducible tensor completely, we must add a tenth quantum number, denoted by γ, which must be defined in some convenient way without making reference to group properties.

We illustrate one way of defining γ with a specific example from $SU(3)$. The Kronecker product of two eight-dimensional representations has the Clebsch–Gordan series

$$\mathbf{8} \otimes \mathbf{8} = \mathbf{1} \oplus \mathbf{8}_s \oplus \mathbf{8}_a \oplus \mathbf{10} \oplus \overline{\mathbf{10}} \oplus \mathbf{27}. \tag{8.15}$$

In the direct sum, we have put a subscript "s" on one **8** and a subscript "a" on the other to indicate that the basis functions of the two eight-dimensional representations are formed from the symmetric and antisymmetric combinations, respectively of the product tensors. The Clebsch–Gordan coefficients will have these respective symmetries. It is customary, following Gell-Mann, to call the symmetric combination *D coupling* and the antisymmetric combination *F coupling*. However, the choice of symmetric and antisymmetric combinations is based on convenience, and is not dictated by group theory. We can take any linear combination of $\mathbf{8}_s$ and $\mathbf{8}_a$ to form two new representations with different Clebsch–Gordan coefficients.

The choice of symmetric and antisymmetric combinations is particularly convenient in this example. However, in other examples in which the Kronecker product representation is formed from two irreducible representations with different Young tableaux, symmetric and antisymmetric combinations are not possible, in general. Then the choice of Clebsch–Gordan coefficients is by no means obvious.

We encounter the problem of quantum numbers lying outside the group in $SU(2)$ when we consider the Kronecker product of *three* irreducible representations. For example, suppose we have three representations specified by j_1, j_2, and j_3. We can first decompose the Kronecker product of j_1 and j_2 to give an intermediate irreducible representation j'. Then we can decompose the Kronecker product of j' and j_3 to obtain an irreducible representation j. But alternatively, we can first decompose the Kronecker product of j_2 and j_3 to give an intermediate j'' and then decompose j'' and j_1 to give j. In general, we can take a linear combination of these possibilities, and the Clebsch–Gordan coefficients will depend on just what combination we choose. The choice may be dictated by our knowledge of the physics of the problem. For example, in atomic and nuclear physics, we usually couple the spins and orbital angular momenta of a many-particle system by either *LS* or *jj* coupling.

8.2 Raising and Lowering Operators

An expression for the Clebsch–Gordan coefficients of $SU(2)$ has been obtained in closed form by Wigner (see Wigner, 1959). However, we shall not use this method, but shall show how to obtain Clebsch–Gordan coefficients of simple compact groups by means of so-called lowering operators. To do this we first quote without proof several theorems about the basis functions of simple compact groups. In writing down these basis functions, we shall omit the superscripts specifying the representations to which the functions belong. Also, the subscripts will stand only for the weights, and we omit indices distinguishing between different functions with the same weight.

Consider two irreducible representations of a simple compact group. Let

ψ_μ be the basis vector belonging to the highest weight of the first representation and let ψ_ν be the basis vector belonging to the highest weight of the second representation. Then the basis tensor $\psi_\mu \psi_\nu$ of the Kronecker product representation is also a basis function of an irreducible representation of the group. Its weight is simple and is given by $\mathbf{M} = \boldsymbol{\mu} + \boldsymbol{\nu}$. Thus in the sum (8.11) there is only one Clebsch–Gordan coefficient different from zero and it is equal to one.

Once we have found one basis function of the irreducible representation, we can obtain other basis functions by using the nondiagonal standard operators E_α of the group. In fact we have the following

THEOREM. *If we operate on a basis function of an irreducible representation with a nondiagonal operator E_α, we obtain another basis function of the same irreducible representation or else get zero. Furthermore, by repeated operation on the basis function with the operators $E_\alpha, E_\beta \cdots E_\alpha E_\alpha, E_\alpha E_\beta \cdots E_\alpha E_\alpha E_\alpha \cdots$ we obtain a complete set of basis functions. They may not be orthogonal or normalized, however.*

We can also find the weight of the basis function $E_\alpha \psi_m$ if we know the weight of ψ_m from the following theorem:

THEOREM. *If we have a basis function ψ_m with weight* $\mathbf{m}$, i.e.,

$$\mathbf{H}\psi_m = \mathbf{m}\psi_m \tag{8.16}$$

then the basis function $E_\alpha \psi_m$ has weight $\mathbf{m}'$ given by

$$\mathbf{m}' = \mathbf{m} + \boldsymbol{\rho}(\alpha). \tag{8.17}$$

We prove the theorem as follows.

Proof. From Eq. (5.62) we have

$$[\mathbf{H}, E_\alpha] = \boldsymbol{\rho}(\alpha)E_\alpha, \tag{8.18}$$

where $\boldsymbol{\rho}(\alpha)$ is a nonvanishing root. Then

$$\mathbf{H}E_\alpha \psi_m = E_\alpha \mathbf{H}\psi_m + \boldsymbol{\rho}(\alpha)E_\alpha \psi_m. \tag{8.19}$$

Then, using Eq. (8.16) we obtain

$$\mathbf{H}E_\alpha \psi_m = E_\alpha \mathbf{m}\psi_m + \boldsymbol{\rho}(\alpha)E_\alpha \psi_m,$$

or

$$\mathbf{H}E_\alpha \psi_m = [\mathbf{m} + \boldsymbol{\rho}(\alpha)]E_\alpha \psi_m = \mathbf{m}'E_\alpha \psi_m, \tag{8.20}$$

which proves the theorem.

If the weight $\mathbf{m} + \boldsymbol{\rho}(\alpha)$ is higher than the weight $\mathbf{m}$, then E_α is said to be a *raising operator*; if $\mathbf{m} + \boldsymbol{\rho}(\alpha)$ is lower than $\mathbf{m}$, E_α is a *lowering operator*. Since $\boldsymbol{\rho}(-\alpha) = -\boldsymbol{\rho}(\alpha)$, if E_α is a raising operator, then $E_{-\alpha}$ is a lowering operator. Since our procedure to obtain the Clebsch–Gordan coefficients starts with the highest weight, we use lowering operators to obtain the other weights. However, this is a matter of convention. We can equally well start with the lowest weight and use raising operators. For $SU(3)$, all (3 or 6) weights at the corners of a weight diagram are equivalent and can serve as well as the highest weight for a starting point. In general, if we operate with a raising or lowering operator on a basis function, we do not obtain a normalized basis function. However, it is not too difficult to normalize the function. There is a theorem that helps in this regard. It says that if

$$E_\alpha \psi_m = 0, \tag{8.21}$$

then the function

$$[\boldsymbol{\rho}(\alpha) \cdot \mathbf{m}]^{1/2} E_{-\alpha} \psi_m, \tag{8.22}$$

is normalized.

We are now ready to give a prescription for decomposing the basis tensors of a representation which is the Kronecker product of two irreducible representations.

(1) We select from the basis tensors of the Kronecker product representation the one with the highest weight. This tensor is also a basis function of an irreducible representation contained in the decomposition of the Kronecker product.

(2) We apply the operators E_α, E_β, $E_\alpha E_\beta \cdots$ to the product basis tensor. By repeated application of the operators, we obtain a complete set of basis functions of the irreducible representation to which the product tensor belongs.

(3) The resulting functions will be orthogonal if they have different weights. This is clear because different eigenfunctions of the same hermitian operator with different eigenvalues are necessarily orthogonal. We can then normalize these basis functions. If two or more basis functions have the same weight, as can happen with any group of rank greater than one, we take orthogonal linear combinations of them. We have already given an example of how to make functions orthogonal to one another in Section 4.3 when we obtained the basis functions of the symmetric group S_3. This method is called the Schmidt orthogonalization procedure.

(4) In the subspace orthogonal to that generated from the highest weight we select the tensor with the highest weight. In general, such a tensor will be a linear combination of product tensors. This tensor will be a basis function of another irreducible representation. Furthermore, this function will have

the highest weight of the representation. We then repeat the process by operating on this function with the lowering operators $E_\alpha \cdots$. We repeat over and over until we have obtained the basis functions of all the irreducible representations contained in the direct sum of decomposition of the Kronecker product.

Thus, we obtain the irreducible functions as linear combinations of tensors of the Kronecker product. The coefficients obtained in this manner are the Clebsch–Gordan coefficients.

This procedure applies to any simple compact group. Therefore, we can apply it not only to the groups $SU(n)$, but also, for example, to the groups C_2 and G_2 discussed in Section 6.3. We can use this method to obtain the weight diagrams of C_2 and G_2 shown in Figs. 6.4 and 6.5, in addition to the Clebsch–Gordan coefficients of these groups. We leave this as an exercise for the interested reader.

8.3 Matrix Representation of the Algebra of $SU(n)$

In order to obtain the Clebsch–Gordan coefficients explicitly, we need to express the basis tensors $E_\alpha \psi_\mu \psi_\nu$, $E_\beta \psi_\mu \psi_\nu \cdots$ in terms of linear combinations of the product tensors. We do not need a matrix representation of the Lie algebra to do this, as we can work with the operator properties. For example, if we know that $E_\alpha^{(1)}$ is a lowering operator for $\psi_\mu^{(1)}$ and $E_\alpha^{(2)}$ is a lowering operator for $\psi_\nu^{(2)}$, then

$$E_\alpha = E_\alpha^{(1)} + E_\alpha^{(2)}, \tag{8.23}$$

is a lowering operator for $\psi_\mu^{(1)}\psi_\nu^{(2)}$. We can generalize further and state that if the product tensor $\psi_\mu^{(1)}\psi_\nu^{(2)} \cdots \psi_\lambda^{(k)}$ is also an irreducible tensor, then the lowering operator for this tensor is

$$E_\alpha = \sum_{i=1}^{k} E_\alpha^{(i)}. \tag{8.24}$$

However, it is helpful to have a matrix representation of the algebra so that our considerations are less abstract. We shall obtain the matrix generators of $SU(n)$. The lowest-dimensional nontrivial matrix representation of the algebra of $SU(n)$ is n-dimensional. Now consider a representation of the algebra of $SU(n)$ which operates on the basis functions of an N-dimensional irreducible representation of the group. The representation of the algebra may be by N-dimensional matrices if the basis functions are considered as vectors, or by a sum of n-dimensional matrices [Eq. (8.24)] if the basis functions are considered as tensors. We shall adopt the latter approach.

We have already written down in standard form the matrix generators of $SU(2)$ and $SU(3)$. These are given in Eqs. (6.7) and (6.22). It is easy to generalize to $SU(n)$, except for the complication of normalization. Therefore we introduce $n \times n$ matrix generators of $SU(n)$ in standard form except for normalization, which we shall choose in such a way as to be most convenient. We denote the diagonal matrices by h_i ($i = 1, 2, \ldots, n-1$) and the nondiagonal ones by e_{ij} ($i \neq j = 1, 2, \ldots, n$). The diagonal $n \times n$ matrices are defined as follows

$$(h_i)_{\mu\nu} = \begin{cases} \delta_{\mu\nu}, & \mu < i+1 \\ -i\delta_{\mu\nu}, & \mu < i+1 \\ 0, & \mu > i+1. \end{cases} \tag{8.25}$$

In Eq. (8.25), the letter i always stands for an integer, and not for $\sqrt{-1}$. The nondiagonal matrices e_{ij} are defined by

$$(e_{ij})_{\mu\nu} = \delta_{i\mu}\,\delta_{j\nu}. \tag{8.26}$$

It is easy to see that

$$e_{ij} = e_{ji}^{\dagger}. \tag{8.27}$$

If e_{ij} is a raising operator, then e_{ji} is a lowering operator. With our ordering of weights according to the last component rather than the first, e_{ij} is a raising operator if $j > i$.

As an example, for $SU(2)$ the generators are

$$h_1 = \begin{pmatrix} 1 & 0 \\ 0 & -1 \end{pmatrix}, \qquad e_{12} = \begin{pmatrix} 0 & 1 \\ 0 & 0 \end{pmatrix}, \qquad e_{21} = \begin{pmatrix} 0 & 0 \\ 1 & 0 \end{pmatrix}. \tag{8.28}$$

These are just the Pauli matrices: $h_1 = \sigma_3$, $e_{12} = \sigma_+$, $e_{21} = \sigma_-$. For $SU(3)$ the diagonal generators are

$$h_1 = \begin{pmatrix} 1 & 0 & 0 \\ 0 & -1 & 0 \\ 0 & 0 & 0 \end{pmatrix}, \qquad h_2 = \begin{pmatrix} 1 & 0 & 0 \\ 0 & 1 & 0 \\ 0 & 0 & -2 \end{pmatrix}. \tag{8.29}$$

The lowering operators are

$$e_{21} = \begin{pmatrix} 0 & 0 & 0 \\ 1 & 0 & 0 \\ 0 & 0 & 0 \end{pmatrix}, \qquad e_{31} = \begin{pmatrix} 0 & 0 & 0 \\ 0 & 0 & 0 \\ 1 & 0 & 0 \end{pmatrix}, \qquad e_{32} = \begin{pmatrix} 0 & 0 & 0 \\ 0 & 0 & 0 \\ 0 & 1 & 0 \end{pmatrix}. \tag{8.30}$$

The raising operators are obtained from Eq. (8.30) by using Eq. (8.27). The basis vectors u_i on which the operators h_i and e_{ij} act are the column vectors

$$(u_i)_j = \delta_{ij}, \qquad i, {}^{\cdot} = 1, 2, \ldots, n. \tag{8.31}$$

For $SU(2)$ these vectors are

$$u_1 = \begin{pmatrix} 1 \\ 0 \end{pmatrix}, \qquad u_2 = \begin{pmatrix} 0 \\ 1 \end{pmatrix}.$$

We do not need to write out the generators of $SU(n)$ explicitly. What is more relevant is a knowledge of how the generators act on the basis vectors u_j of the first fundamental representation. From Eqs. (8.25) and (8.31) we see that u_j is an eigenfunction of h_i

$$h_i u_j = h_{ij} u_j . \tag{8.32}$$

The eigenvalue h_{ij} is given by

$$h_{ij} = \begin{cases} 1, & i < j+1 \\ -i, & i = j+1 \\ 0, & i > j+1 \end{cases} \tag{8.33}$$

or in compact form

$$h_{ij} = \sum_{k=1}^{i} \delta_{kj} - i\delta_{i+1,j} . \tag{8.34}$$

From Eqs. (8.26) and (8.31) we have

$$e_{ij} u_k = u_i \delta_{jk} \tag{8.35}$$

From Eq. (8.35) we see that application of the raising or lowering operators to basis vectors of the first fundamental representation yields normalized basis vectors. This is not true in general for raising and lowering operators normalized in the standard way.

It is also useful to obtain expressions for the products of the generators. From the definitions (8.25) and (8.26) we obtain that

$$\begin{aligned} h_i e_{jk} &= h_{ij} e_{jk} \\ e_{jk} h_i &= h_{ik} e_{jk} \end{aligned} \tag{8.36}$$

where we have used Eq. (8.35). Similarly we obtain

$$e_{ij} e_{kl} = \delta_{jk} e_{il} , \qquad i \neq l. \tag{8.37}$$

The product $e_{ij} e_{ki}$ cannot be written as a linear combination of the generators, since its trace is different from zero. The product is given in terms of a projection operator P_i

$$e_{ij} e_{ki} = \delta_{jk} P_i , \tag{8.38}$$

where

$$(P_i)_{\mu\nu} = \delta_{i\mu} \delta_{\mu\nu} . \tag{8.39}$$

From Eqs. (8.36)–(8.39) we obtain the commutation relations of the generators. They are

$$[h_i , e_{jk}] = (h_{ij} - h_{ik}) e_{jk} , \tag{8.40}$$

$$[e_{ij} , e_{jl}] = e_{il} , \qquad i \neq l, \tag{8.41}$$

$$[e_{ij} , e_{ji}] = P_i - P_j . \tag{8.42}$$

The operators $P_i - P_j$ are traceless and therefore can be expressed as linear combinations of the h_i. We shall not do so, however, except for $SU(2)$ and $SU(3)$. Using Eqs. (8.40)–(8.42), we see that for $SU(2)$ we have only the following nonvanishing commutation relations

$$[h_1, e_{12}] = e_{12}, \qquad [h_1, e_{21}] = -e_{21}, \qquad [e_{12}, e_{21}] = h_1, \qquad (8.43)$$

since $P_1 - P_2 = h_1$. For $SU(3)$ the commutation relations of Eq. (8.42) can be expressed in terms of the h_i as follows:

$$[e_{12}, e_{21}] = h_1, \qquad [e_{13}, e_{31}] = \tfrac{1}{2}(h_1 + h_2), \qquad [e_{23}, e_{32}] = \tfrac{1}{2}(h_2 - h_1). \qquad (8.44)$$

The other commutation relations of $SU(3)$ are compactly given in Eqs. (8.40) and (8.41).

For convenience, we introduce still a bit more notation. We keep the convention that in writing a tensor as the product of ν vectors, each vector referring to a different particle, we write the vector of the first particle on the left, next the vector of the second particle etc. Then we can omit the integers specifying the particles. For example, a three-particle tensor is written

$$u_i(1)u_j(2)u_k(3) = u_i u_j u_k. \qquad (8.45)$$

The irreducible tensor with the highest weight of a product of ν single-particle vectors is clearly

$$u_1 u_1 \cdots u_1 \qquad (\nu \quad \text{times}). \qquad (8.46)$$

Also, we denote the raising or lowering operators which operate on product tensors by e_{ij}, where

$$\mathsf{e}_{ij} = \sum_{k=1}^{\nu} e_{ij}(k). \qquad (8.47)$$

Similarly, we denote the diagonal operators by

$$\mathsf{h}_i = \sum_{k=1}^{\nu} h_i(k). \qquad (8.48)$$

These operators are the generators of $SU(n)$ which act on the irreducible functions formed from linear combinations of the product tensors. The generators h_i and e_{ij} do not satisfy Eqs. (8.36)–(8.38) except in the case that $\nu = 1$. This is because for $\nu > 1$, these operators contain cross terms, for example $h_i(1)e_{jk}(2)$, which act on the state vectors of different particles. However, since any operator referring to one particle commutes with any operator referring to another particle, the h_i and e_{ij} do satisfy the commutation relations of Eqs. (8.40)–(8.42). In the case of $SU(3)$, we show in Fig. 8.1 the effect of operating on an irreducible function ψ_i with e_{ij}.

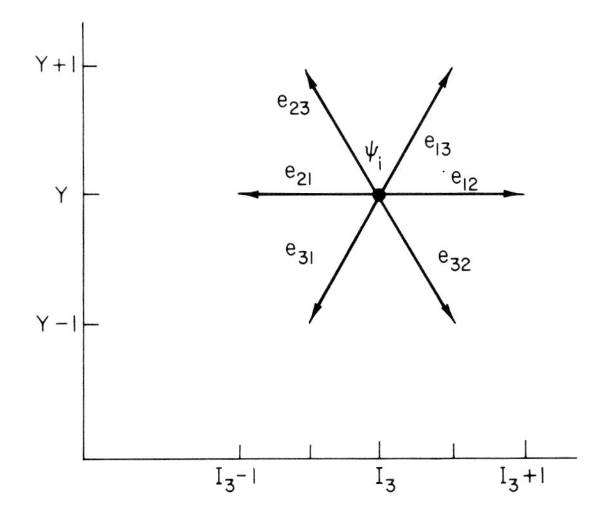

Fig. 8.1. Properties of the raising and lowering operators of $SU(3)$ when acting on a state ψ_i with weight $(I_3 Y)$.

We now have all the tools necessary to obtain the Clebsch–Gordan coefficients of $SU(n)$. We shall do this for $SU(2)$ and $SU(3)$ by the straightforward, if tedious, procedure of building up from the first fundamental representation. Such a procedure does not require knowledge of all the commutation relations of the generators. However, the procedure can be applied quite generally to $SU(n)$.

If we know all the commutation relations of the group, plus the expressions for the Casimir operators in terms of the generators, we can use more powerful methods to obtain the Clebsch–Gordan coefficients. However, since we have not given the Casimir operators of $SU(n)$, we shall not discuss this method.

8.4 Clebsch–Gordan Coefficients of $SU(2)$

We shall illustrate a procedure to obtain the Clebsch–Gordan coefficients of $SU(2)$ by a technique which does not make use of the Casimir operator J^2. We begin by decomposing the basis tensors of the Kronecker product of two fundamental representations. These four basis tensors are of the form $u_i u_j$ $(i, j = 1, 2)$. By a theorem quoted in Section 8.2 the basis tensor with the highest weight $u_1 u_1$ is already an irreducible function. This should be obvious, as there is only one way to obtain a function with the highest weight: namely, the way given in Eq. (8.46). Let us order the irreducible functions according to their weights, using a single subscript which denotes, *not* the weight, but the *order* of the weight. (This is to avoid fractions.) Then

$$\psi_1 = u_1 u_1. \tag{8.49}$$

Operating on ψ_1 with the lowering operator e_{21}, defined by Eq. (8.47), and, using Eq. (8.35), we obtain

$$\mathsf{e}_{21}\psi_1 = u_1 u_2 + u_2 u_1. \tag{8.50}$$

Normalizing, we obtain the irreducible function ψ_2 with the second highest weight

$$\psi_2 = (u_1 u_2 + u_2 u_1)/\sqrt{2}. \tag{8.51}$$

Again using e_{21}, we obtain

$$\mathsf{e}_{21}\psi_2 = \sqrt{2}\, u_2 u_2, \tag{8.52}$$

or normalizing,

$$\psi_3 = u_2 u_2. \tag{8.53}$$

If we operate on ψ_3 with e_{21} we obtain zero. Thus we have obtained the irreducible functions of a triplet representation. The coefficients multiplying the product tensors in Eqs. (8.49), (8.51), and (8.53) are $SU(2)$ Clebsch–Gordan coefficients.

From Eqs. (8.50) and (8.52), we see that for the triplet representation (which we now denote by a superscript),

$$\mathsf{e}_{21}\,\psi_i^{(3)} = \sqrt{2}\,\psi_{i+1}^{(3)}, \tag{8.54}$$

if $\psi_{i+1}^{(3)}$ is nonvanishing. Here the subscripts i and $i+1$ denote the order, not the weight.

The triplet function $\psi_2^{(3)}$ has the second-highest weight. Following the procedure outlined in Section 8.2, we next construct a function orthogonal to $\psi_2^{(3)}$ from the product tensors $u_1 u_2$ and $u_2 u_1$. Such a function is

$$\psi_1 = (u_1 u_2 - u_2 u_1)/\sqrt{2}. \tag{8.55}$$

If we operate on this function with e_{21}, we get zero. Thus ψ_1 is a singlet $\psi_1^{(1)}$, a fact we should know immediately, since there are only four product tensors $u_i u_j$.

The overall sign of $\psi_1^{(1)}$ is not obtained by this method. In fact, we must adopt a convention to determine it. The convention we use is that of Condon and Shortley (1935) which states that the coefficient

$$(\alpha\beta\alpha, j-\alpha \,|\, jj)$$

is positive. With this convention we have

$$(\alpha\beta\mu\nu \,|\, jm) = (-)^{j-\alpha-\beta}(\beta\alpha\nu\mu \,|\, jm) = (-)^{j-\alpha-\beta}(\alpha\beta, -\mu, -\nu \,|\, j, -m)$$
$$= (-1)^{\alpha-\mu}[(2j+1)/(2\beta+1)]^{1/2}(\alpha j\mu, -m \,|\, \beta, -\nu). \qquad (8.56)$$

The triplet and singlet irreducible functions can be obtained without the lowering operators by Young tableaux. Since the tableau $\boxed{}\boxed{}$ stands for a symmetric combination, we immediately have (after normalization)

$$\psi_1^{(3)} = \boxed{1}\boxed{1} = u_1 u_1,$$
$$\psi_2^{(3)} = \boxed{1}\boxed{2} = (u_1 u_2 + u_2 u_1)/\sqrt{2}, \qquad (8.57)$$
$$\psi_3^{(3)} = \boxed{2}\boxed{2} = u_2 u_2 .$$

Likewise, since

$$\begin{array}{c}\boxed{}\\ \boxed{}\end{array}$$

is antisymmetric, we have

$$\psi_1^{(1)} = \begin{array}{c}\boxed{1}\\ \boxed{2}\end{array} = \frac{1}{\sqrt{2}}(u_1 u_2 - u_2 u_1). \qquad (8.58)$$

We can extend these procedures to obtain any Clebsch–Gordan coefficients. For example, suppose we wish to obtain the irreducible functions contained in the Kronecker product of $\psi_i^{(3)} u_j$. Since the Kronecker product has the decomposition

$$\mathbf{3} \otimes \mathbf{2} = \mathbf{4} \oplus \mathbf{2},$$

we know that the irreducible functions contained in $\psi_i^{(3)} u_j$ belong to a quartet and a doublet. The product function with the highest weight must belong to the quartet:

$$\psi_1^{(4)} = \psi_1^{(3)} u_1. \qquad (8.59)$$

We operate on this function with e_{21}, using Eqs. (8.35) and (8.54). We then obtain, omitting the superscript (3)

$$\mathsf{e}_{21}\psi_2^{(4)} = \sqrt{2}\,\psi_2 u_1 + \psi_1 u_2 . \qquad (8.60)$$

Normalizing, we obtain

$$\psi_2^{(4)} = \sqrt{\tfrac{2}{3}}\,\psi_2 u_1 + \sqrt{\tfrac{1}{3}}\,\psi_1 u_2 . \qquad (8.61)$$

We can obtain the other members of this quartet state from Eqs. (8.60) and (8.61) either by Eq. (8.56) or by continued use of the lowering operators. The irreducible functions are

$$\begin{aligned} \psi_3^{(4)} &= \sqrt{\tfrac{2}{3}}\,\psi_2 u_2 + \sqrt{\tfrac{1}{3}}\,\psi_3 u_1, \\ \psi_4^{(4)} &= \psi_3^{(3)} u_2 . \end{aligned} \tag{8.62}$$

Alternatively, we can use Young tableaux. We have

$$\begin{aligned} \psi_1^{(4)} &= \boxed{1\,|\,1\,|\,1} = u_1 u_1 u_1, \\ \psi_2^{(4)} &= \boxed{1\,|\,1\,|\,2} = (u_1 u_1 u_2 + u_1 u_2 u_1 + u_2 u_1 u_1)/\sqrt{3}, \\ \psi_3^{(4)} &= \boxed{1\,|\,2\,|\,2} = (u_1 u_2 u_2 + u_2 u_1 u_2 + u_2 u_2 u_1)/\sqrt{3}, \\ \psi_4^{(4)} &= \boxed{2\,|\,2\,|\,2} = u_2 u_2 u_2 . \end{aligned} \tag{8.63}$$

Then by substituting Eqs. (8.49), (8.51), and (8.53) into Eqs. (8.63), we immediately get the normalized functions of Eqs. (8.59), (8.61), and (8.62).

We can obtain the other multiplet by constructing a function which is orthogonal to the function $\psi_2^{(4)}$, which has the second highest weight. It is

$$\psi_1^{(2)} = -\sqrt{\tfrac{1}{3}}\,\psi_2 u_1 + \sqrt{\tfrac{2}{3}}\,\psi_1 u_2 . \tag{8.64}$$

By using the lowering operator we find the other member of this doublet, which is

$$\psi_2^{(2)} = \sqrt{\tfrac{1}{3}}\,\psi_2 u_2 - \sqrt{\tfrac{2}{3}}\,\psi_3 u_1 . \tag{8.65}$$

Thus we have accomplished the decomposition of the product $\psi_i^{(3)} u_j$, obtaining all the relevant Clebsch–Gordan coefficients.

As a last example of this procedure, we shall obtain the decomposition of $\psi_i^{(3)} \psi_j^{(3)}$. The decomposition of the product is

$$\mathbf{3} \otimes \mathbf{3} = \mathbf{5} \oplus \mathbf{3} \oplus \mathbf{1}. \tag{8.66}$$

In this case we can obtain many of the Clebsch–Gordan coefficients by symmetry arguments. For example, we must have, omitting the superscript 3,

$$\begin{aligned} \psi_1^{(5)} &= \psi_1 \psi_1, \\ \psi_2^{(5)} &= (\psi_1 \psi_2 + \psi_2 \psi_1)/\sqrt{2}. \end{aligned} \tag{8.67}$$

Symmetry arguments are not sufficient to obtain $\psi_3^{(5)}$ as there exist two independent symmetric functions with the relevant weight. We therefore operate on $\psi_2^{(5)}$ with e_{21}, using Eq. (8.54). We obtain

$$e_{21}\psi_2^{(5)} = \psi_1\psi_3 + 2\psi_2\psi_2 + \psi_3\psi_1. \tag{8.68}$$

Normalizing, we obtain

$$\psi_3^{(5)} = (\psi_1\psi_3 + 2\psi_2\psi_2 + \psi_3\psi_1)/\sqrt{6}. \tag{8.69}$$

The remaining functions $\psi_4^{(5)}$ and $\psi_5^{(5)}$ are easily obtained either by symmetry arguments or by the lowering operator. We obtain $\psi_1^{(3)}$ by constructing a function orthogonal to $\psi_2^{(5)}$. It is

$$\psi_1^{(3)} = (\psi_1\psi_2 - \psi_2\psi_1)/\sqrt{2}. \tag{8.70}$$

Since this function is antisymmetric, the other members of the triplet must also be antisymmetric. We thus have

$$\begin{aligned} \psi_2^{(3)} &= (\psi_1\psi_3 - \psi_3\psi_1)/\sqrt{2} \\ \psi_3^{(3)} &= (\psi_2\psi_3 - \psi_3\psi_2)/\sqrt{2} \end{aligned} \tag{8.71}$$

The singlet function $\psi_1^{(1)}$ must be orthogonal to $\psi_3^{(5)}$ and $\psi_2^{(3)}$. Therefore it is given by

$$\psi_1^{(1)} = (\psi_1\psi_3 - \psi_2\psi_2 + \psi_3\psi_1)/\sqrt{3} \tag{8.72}$$

Thus, we have obtained all the Clebsch–Gordan coefficients relevant in decomposing the product $\psi_i^{(3)}\psi_j^{(3)}$.

It should be clear from these examples that in many cases we can obtain Clebsch–Gordan coefficients by considerations of symmetry. However, in general, we use the lowering operator e_{21}. We have seen that e_{21} operating on $\psi_m^{(j)}$ gives a constant multiplied by $\psi_{m-1}^{(j)}$. From our examples, it can be seen that this constant, which we denote by $c(jm)$, depends in general on j and m. In order to obtain the Clebsch–Gordan coefficients, we need to obtain an expression for the constant $c(jm)$ in terms of j and m.

We have described a procedure for obtaining the constants $c(jm)$ by building up irreducible functions from the fundamental irreducible vectors. This procedure is straightforward and can be generalized to any $SU(n)$.

Unfortunately, the method we have described can become rather tedious if the dimensionality of the multiplets is large. There exist a number of short-cut methods which make use of the Casimir operator J^2. These are described in many places, for example, Rose (1957); Edmonds (1957), and Gottfried (1966). We shall not reproduce these methods. A number of $SU(2)$ Clebsch–Gordan coefficients are given in Table 8.1 with the phases of Condon and Shortley (1935). In Table 8.2 are given some spherical harmonics.

TABLE 8.1

SOME $SU(2)$ CLEBSCH-GORDAN COEFFICIENTS WITH CONDON AND SHORTLEY (1935) PHASES. THE FORMAT FOLLOWS THAT OF THE PARTICLE DATA GROUP (1969).

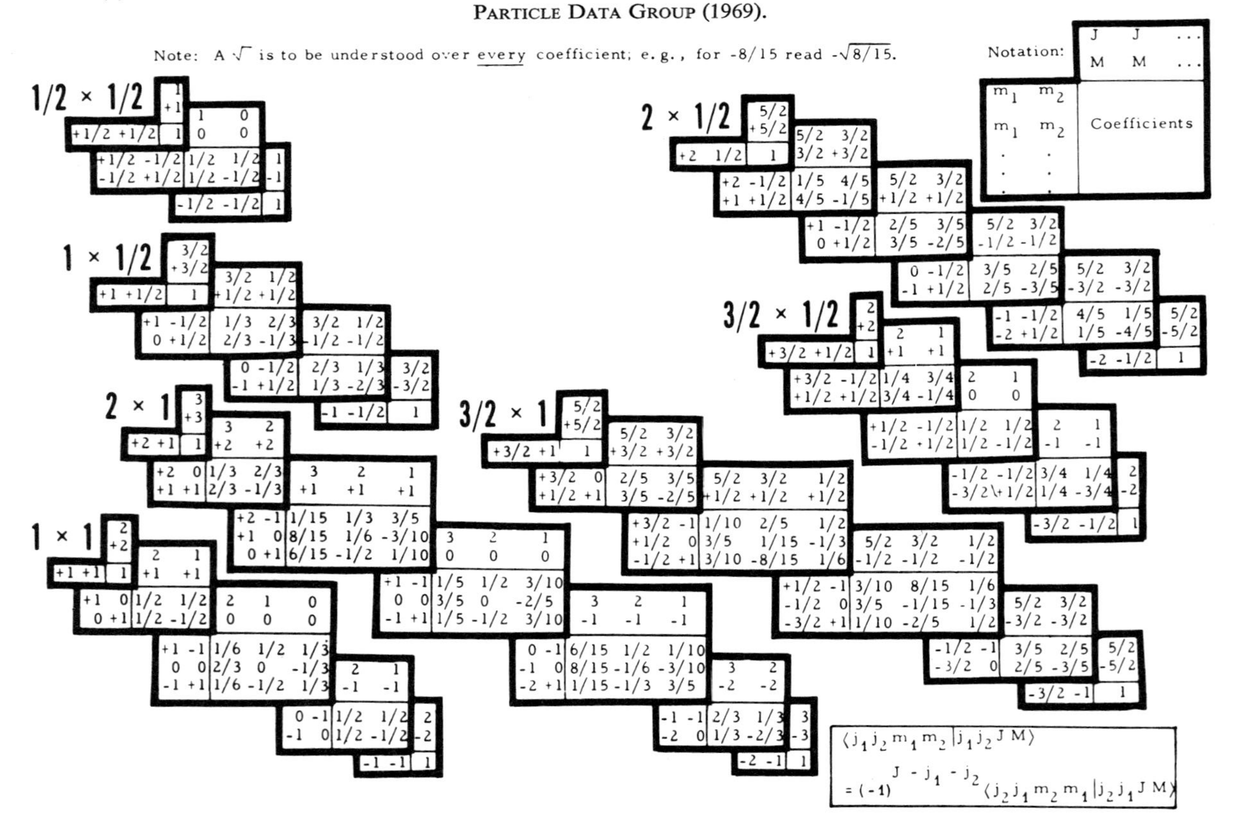

Note: A $\sqrt{\ }$ is to be understood over every coefficient; e.g., for -8/15 read $-\sqrt{8/15}$.

Notation:

m_1	m_2	J	J	...
		M	M	...
m_1	m_2	Coefficients		
.	.			

1/2 × 1/2

m_1	m_2	J=1, M=+1
+1/2	+1/2	1

m_1	m_2	J=1, M=0	J=0, M=0
+1/2	-1/2	1/2	1/2
-1/2	+1/2	1/2	-1/2

m_1	m_2	J=1, M=-1
-1/2	-1/2	1

1 × 1/2

m_1	m_2	3/2, +3/2
+1	+1/2	1

m_1	m_2	3/2, +1/2	1/2, +1/2
+1	-1/2	1/3	2/3
0	+1/2	2/3	-1/3

m_1	m_2	3/2, -1/2	1/2, -1/2
0	-1/2	2/3	1/3
-1	+1/2	1/3	-2/3

m_1	m_2	3/2, -3/2
-1	-1/2	1

2 × 1/2

m_1	m_2	5/2, +5/2
+2	1/2	1

m_1	m_2	5/2, +3/2	3/2, +3/2
+2	-1/2	1/5	4/5
+1	+1/2	4/5	-1/5

m_1	m_2	5/2, +1/2	3/2, +1/2
+1	-1/2	2/5	3/5
0	+1/2	3/5	-2/5

m_1	m_2	5/2, -1/2	3/2, -1/2
0	-1/2	3/5	2/5
-1	+1/2	2/5	-3/5

m_1	m_2	5/2, -3/2	3/2, -3/2
-1	-1/2	4/5	1/5
-2	+1/2	1/5	-4/5

m_1	m_2	5/2, -5/2
-2	-1/2	1

3/2 × 1/2

m_1	m_2	2, +2
+3/2	+1/2	1

m_1	m_2	2, +1	1, +1
+3/2	-1/2	1/4	3/4
+1/2	+1/2	3/4	-1/4

m_1	m_2	2, 0	1, 0
+1/2	-1/2	1/2	1/2
-1/2	+1/2	1/2	-1/2

m_1	m_2	2, -1	1, -1
-1/2	-1/2	3/4	1/4
-3/2	+1/2	1/4	-3/4

m_1	m_2	2, -2
-3/2	-1/2	1

2 × 1

m_1	m_2	3, +3
+2	+1	1

m_1	m_2	3, +2	2, +2
+2	0	1/3	2/3
+1	+1	2/3	-1/3

m_1	m_2	3, +1	2, +1	1, +1
+2	-1	1/15	1/3	3/5
+1	0	8/15	1/6	-3/10
0	+1	6/15	-1/2	1/10

m_1	m_2	3, 0	2, 0	1, 0
+1	-1	1/5	1/2	3/10
0	0	3/5	0	-2/5
-1	+1	1/5	-1/2	3/10

m_1	m_2	3, -1	2, -1	1, -1
0	-1	6/15	1/2	1/10
-1	0	8/15	-1/6	-3/10
-2	+1	1/15	-1/3	3/5

m_1	m_2	3, -2	2, -2
-1	-1	2/3	1/3
-2	0	1/3	-2/3

m_1	m_2	3, -3
-2	-1	1

1 × 1

m_1	m_2	2, +2
+1	+1	1

m_1	m_2	2, +1	1, +1
+1	0	1/2	1/2
0	+1	1/2	-1/2

m_1	m_2	2, 0	1, 0	0, 0
+1	-1	1/6	1/2	1/3
0	0	2/3	0	-1/3
-1	+1	1/6	-1/2	1/3

m_1	m_2	2, -1	1, -1
0	-1	1/2	1/2
-1	0	1/2	-1/2

m_1	m_2	2, -2
-1	-1	1

3/2 × 1

m_1	m_2	5/2, +5/2
+3/2	+1	1

m_1	m_2	5/2, +3/2	3/2, +3/2
+3/2	0	2/5	3/5
+1/2	+1	3/5	-2/5

m_1	m_2	5/2, +1/2	3/2, +1/2	1/2, +1/2
+3/2	-1	1/10	2/5	1/2
+1/2	0	3/5	1/15	-1/3
-1/2	+1	3/10	-8/15	1/6

m_1	m_2	5/2, -1/2	3/2, -1/2	1/2, -1/2
+1/2	-1	3/10	8/15	1/6
-1/2	0	3/5	-1/15	-1/3
-3/2	+1	1/10	-2/5	1/2

m_1	m_2	5/2, -3/2	3/2, -3/2
-1/2	-1	3/5	2/5
-3/2	0	2/5	-3/5

m_1	m_2	5/2, -5/2
-3/2	-1	1

$$\langle j_1 j_2 m_1 m_2 | j_1 j_2 J M\rangle = (-1)^{J - j_1 - j_2} \langle j_2 j_1 m_2 m_1 | j_2 j_1 J M\rangle$$

TABLE 8-2.

SOME SPHERICAL HARMONICS WITH THE PHASES OF CONDON AND SHORTLEY (1935).

$Y_0^0 = \sqrt{\frac{1}{4\pi}}$	
$Y_1^0 = \sqrt{\frac{3}{4\pi}} \cos\theta;$	$Y_1^1 = -\sqrt{\frac{3}{8\pi}} \sin\theta \, e^{i\phi}$
$Y_2^0 = \sqrt{\frac{5}{4\pi}} \left(\frac{3}{2}\cos^2\theta - \frac{1}{2}\right);$	$Y_2^1 = -\sqrt{\frac{15}{8\pi}} \sin\theta \cos\theta \, e^{i\phi}$
$Y_2^2 = \frac{1}{4}\sqrt{\frac{15}{2\pi}} \sin^2\theta e^{2i\phi}$	
$Y_3^0 = \sqrt{\frac{7}{4\pi}} \left(\frac{5}{2}\cos^3\theta - \frac{3}{2}\cos\theta\right);$	$Y_3^1 = -\frac{1}{4}\sqrt{\frac{21}{4\pi}} \sin\theta \, (5\cos^2\theta - 1) e^{i\phi}$
$Y_3^2 = \frac{1}{4}\sqrt{\frac{105}{2\pi}} \sin^2\theta \cos\theta \, e^{2i\phi};$	$Y_3^3 = -\frac{1}{4}\sqrt{\frac{35}{4\pi}} \sin^3\theta \, e^{3i\phi}$
$(Y_j^m)^* = (-1)^m \, Y_j^{-m}$	

8.5 Clebsch–Gordan Coefficients of $SU(3)$

The procedure outlined in the last section can be used with only a few modifications which are necessary because $SU(3)$ is not a simply reducible group. The basis functions of the first fundamental representation of $SU(3)$ are u_1, u_2 and u_3 with values of (I_3, Y) given by $(\frac{1}{2}, \frac{1}{3})$, $(-\frac{1}{2}, \frac{1}{3})$, $(0, -\frac{2}{3})$, respectively. (Alternatively, we can specify the weights shown in Fig. 6.6.)

As in the case of $SU(2)$, we shall distinguish the functions belonging to a given representation by a single subscript, which denotes the *order* of the function. We shall adopt a convention which orders the states by the values of Y, I, and I_3, respectively.

Let us consider the Clebsch–Gordan coefficients which arise in reducing the Kronecker product $\mathbf{3} \otimes \mathbf{3}$. The basis tensors of the Kronecker product representation are the nine two-particle states

$$u_i u_j, \qquad i, j = 1, 2, 3. \tag{8.73}$$

We know that the product representation can be composed into a symmetric representation with Young tableau □□ and an antisymmetric representation with Young tableau

$$\begin{array}{c} \square \\ \square \end{array}$$

By writing down symmetric and antisymmetric combinations, we immediately get the relevant Clebsch–Gordan coefficients, except for possible sign ambiguities which must be decided by convention. In the symmetric case the normalized irreducible functions are (suppressing a superscript which denotes the dimensionality of the representation):

$$
\begin{aligned}
\boxed{1\,|\,1} &= u_1 u_1 = \psi_1, \\
\boxed{1\,|\,2} &= \frac{1}{\sqrt{2}}(u_1 u_2 + u_2 u_1) = \psi_2, \\
\boxed{2\,|\,2} &= u_2 u_2 = \psi_3, \\
\boxed{1\,|\,3} &= \frac{1}{\sqrt{2}}(u_1 u_3 + u_3 u_1) = \psi_4, \\
\boxed{2\,|\,3} &= \frac{1}{\sqrt{2}}(u_2 u_3 + u_3 u_2) = \psi_5, \\
\boxed{3\,|\,3} &= u_3 u_3 = \psi_6 .
\end{aligned}
\tag{8.74}
$$

The Clebsch–Gordan coefficients are just the coefficients in these sums. In the antisymmetric case, the irreducible functions are

$$
\begin{aligned}
\begin{array}{|c|}\hline 1 \\ \hline 2 \\ \hline \end{array} &= \frac{1}{\sqrt{2}}(u_1 u_2 - u_2 u_1) = \bar{u}_3, \\
\begin{array}{|c|}\hline 1 \\ \hline 3 \\ \hline \end{array} &= \frac{1}{\sqrt{2}}(u_3 u_1 - u_1 u_3) = \bar{u}_2, \\
\begin{array}{|c|}\hline 2 \\ \hline 3 \\ \hline \end{array} &= \frac{1}{\sqrt{2}}(u_2 u_3 - u_3 u_2) = \bar{u}_1,
\end{aligned}
\tag{8.75}
$$

with a particular sign convention which we shall discuss shortly.

We have introduced the notation $\bar{u}_1$, $\bar{u}_2$, and $\bar{u}_3$ for the basis functions of the antisymmetric representation because these functions also serve as basis vectors for the second fundamental representation of $SU(3)$. These states can be interpreted as the ones which denote the antiparticles of the states u_1, u_2, and u_3. This can be seen first by noting that $\bar{u}_i$ has a weight which is opposite in sign to the weight of u_i:

$$\mathsf{h}_i \bar{u}_j = -h_{ij} \bar{u}_j . \tag{8.76}$$

Furthermore, if we operate on $\bar{u}_k$ with e_{ij} we find, using the definitions of Eq. (8.75), that

$$\mathsf{e}_{ij}\bar{u}_k = -\bar{u}_j\,\delta_{ik}. \tag{8.77}$$

Thus, comparing (8.76) and (8.77) with Eqs. (8.32) and (8.35) we see that the generators of the $\bar{\mathbf{3}}$ representation (denoted by $\bar{h}_i$ and $\bar{e}_{ij}$) satisfy

$$\bar{h}_i = -h_i, \qquad \bar{e}_{ij} = -e_{ji}, \tag{8.78}$$

in agreement with Eq. (6.16).

Although the sign convention of Eq. (8.75) is satisfactory, another convention is used by de Swart (1963). To conform to de Swart's sign convention, we introduce the vectors v_1, v_2, and v_3 which are ordered according to Y, I, and I_3:

$$v_3 = -\bar{u}_1, \qquad v_2 = \bar{u}_2, \qquad v_1 = \bar{u}_3. \tag{8.79}$$

If we operate on the states v_1, v_2, and v_3 with the lowering operators, we obtain

$$\mathsf{e}_{31}v_1 = v_3, \qquad \mathsf{e}_{21}v_2 = v_3, \qquad \mathsf{e}_{32}\,v_1 = -v_2. \tag{8.80}$$

Thus, we see that by adopting the sign convention of Eq. (8.79), we can force e_{31} and e_{21} to have only positive matrix elements. The matrix elements of e_{32} are then determined by the $SU(3)$ commutation relations. In order to deal with matrix elements with only positive signs, we shall not use the operator e_{32}, but only the lowering operators e_{31} and e_{21} (and their corresponding raising operators e_{13} and e_{12}).

We are now ready to construct the Clebsch–Gordan coefficients of other irreducible representations. Let us consider the decomposition

$$\bar{\mathbf{3}} \otimes \mathbf{3} = \mathbf{8} \oplus \mathbf{1}. \tag{8.81}$$

By symmetry, the six states at the periphery of the octet weight diagram can be obtained at once. They are (suppressing a superscript on $\psi_i^{(8)}$)

$$\begin{aligned} \psi_1 &= v_1 u_1, & \psi_2 &= v_1 u_2, & \psi_3 &= v_2 u_1, \\ \psi_5 &= v_3 u_2, & \psi_7 &= v_2 u_3, & \psi_8 &= v_3 u_3. \end{aligned} \tag{8.82}$$

The signs are correct, as can be checked by operating with the e_{ij}. The states of Eq. (8.82) are all eigenstates of isospin. We can obtain another eigenstate of isospin by operating on ψ_3 with e_{21}:

$$\mathsf{e}_{21}\psi_3 = v_3 u_1 + v_2 u_2.$$

Normalizing, we get

$$\psi_4 = (v_3 u_1 + v_2 u_2)/\sqrt{2}. \tag{8.83}$$

The functions ψ_3, ψ_4, and ψ_5 are an isospin triplet with $Y = 0$, ψ_1, ψ_2 are a doublet with $Y = 1$, and ψ_7, ψ_8 are a doublet with $Y = -1$.

Thus, we have obtained seven functions of the octet. The remaining function ψ_6 has the same weight as ψ_4, but differs from ψ_4 in that it has $I = 0$ rather than $I = 1$. To obtain ψ_6, we first obtain a function φ by operating on ψ_1 with the lowering operator e_{31}. We get

$$\varphi = \mathsf{e}_{31}\psi_1 = v_1 u_3 + v_3 u_1.$$

However, φ is not orthogonal to ψ_4. Furthermore, φ is not an eigenstate of the isospin I, but is an eigenstate of another $SU(2)$ subgroup of $SU(3)$. In general, we obtain an eigenstate of I when we operate on an I-eigenstate with e_{21}, but not when we operate with e_{31}. We get an I-eigenstate by operating with e_{31} only if we reach a simple weight as a result of the operation. We must now obtain a function orthogonal to ψ_4 by the Schmidt procedure. We let

$$\psi_6 = c(\varphi - a\psi_4),$$

where c is a normalization constant and a is a parameter chosen so that $(\psi_6, \psi_4) = 0$. We obtain

$$\psi_6 = (v_2 u_2 - v_3 u_1 - 2v_1 u_3)/\sqrt{6}, \tag{8.84}$$

where the overall sign has been chosen to conform to the convention of de Swart (1963). We shall say more about this sign convention later in this section when we discuss the so-called isoscalar factors. In Eqs. (8.82)–(8.84) we have exhibited the Clebsch–Gordan coefficients which arise in constructing an octet from $\bar{\mathbf{3}} \otimes \mathbf{3}$.

We obtain a singlet by constructing a state function which is orthogonal to Eqs. (8.83) and (8.84). It is

$$\psi = (v_1 u_3 - v_3 u_1 + v_2 u_2)/\sqrt{3}. \tag{8.85}$$

As another example, let us calculate the Clebsch–Gordan coefficients relevant to the decomposition

$$\mathbf{6} \otimes \mathbf{3} = \mathbf{10} \oplus \mathbf{8}. \tag{8.86}$$

It is easiest to obtain the Clebsch–Gordan coefficients for the 10 by using Young tableaux. In terms of the one-particle states u_i, the **10** is completely symmetric and the states $\psi_i^{(10)}$ of the **10** can be written down by inspection. They are:

$$
\begin{aligned}
&\boxed{1\,|\,1\,|\,1} = u_1 u_1 u_1 = \psi_1^{(10)},\\
&\boxed{1\,|\,1\,|\,2} = (u_1 u_1 u_2 + u_1 u_2 u_1 + u_2 u_1 u_1)/\sqrt{3} = \psi_2^{(10)},\\
&\boxed{1\,|\,2\,|\,2} = (u_1 u_2 u_2 + u_2 u_1 u_2 + u_2 u_2 u_1)/\sqrt{3} = \psi_3^{(10)},\\
&\boxed{2\,|\,2\,|\,2} = u_2 u_2 u_2 = \psi_4^{(10)},\\
&\boxed{1\,|\,1\,|\,3} = (u_1 u_1 u_3 + u_1 u_3 u_1 + u_3 u_1 u_1)/\sqrt{3} = \psi_5^{(10)},\\
&\boxed{1\,|\,2\,|\,3} = (u_1 u_2 u_3 + u_1 u_3 u_2 + u_2 u_1 u_3\\
&\qquad\qquad + u_2 u_3 u_1 + u_3 u_1 u_2 + u_3 u_2 u_1)/\sqrt{6} = \psi_6^{(10)},\\
&\boxed{2\,|\,2\,|\,3} = (u_2 u_2 u_3 + u_2 u_3 u_2 + u_3 u_2 u_2)/\sqrt{3} = \psi_7^{(10)},\\
&\boxed{1\,|\,3\,|\,3} = (u_1 u_3 u_3 + u_3 u_1 u_3 + u_3 u_3 u_1)/\sqrt{3} = \psi_8^{(10)},\\
&\boxed{2\,|\,3\,|\,3} = (u_2 u_3 u_3 + u_3 u_2 u_3 + u_3 u_3 u_2)/\sqrt{3} = \psi_9^{(10)},\\
&\boxed{3\,|\,3\,|\,3} = u_3 u_3 u_3 = \psi_{10}^{(10)}.
\end{aligned}
\tag{8.87}
$$

Next, we substitute the states of Eq. (8.74) for the first two particles of Eq. (8.87), obtaining

$$
\begin{aligned}
\psi_1^{(10)} &= \psi_1 u_1,\\
\psi_2^{(10)} &= (\psi_1 u_2 + \sqrt{2}\,\psi_2 u_1)/\sqrt{3},\\
\psi_3^{(10)} &= (\sqrt{2}\,\psi_2 u_2 + \psi_3 u_1)/\sqrt{3},\\
\psi_4^{(10)} &= \psi_3 u_2,\\
\psi_5^{(10)} &= (\psi_1 u_3 + \sqrt{2}\,\psi_4 u_1)/\sqrt{3},\\
\psi_6^{(10)} &= (\psi_2 u_3 + \psi_4 u_2 + \psi_5 u_1)/\sqrt{3},\\
\psi_7^{(10)} &= (\psi_3 u_3 + \sqrt{2}\,\psi_5 u_2)/\sqrt{3},\\
\psi_8^{(10)} &= (\sqrt{2}\,\psi_4 u_3 + \psi_6 u_1)/\sqrt{3},\\
\psi_9^{(10)} &= (\sqrt{2}\,\psi_5 u_3 + \psi_6 u_2)/\sqrt{3},\\
\psi_{10}^{(10)} &= \psi_6 u_3.
\end{aligned}
\tag{8.88}
$$

We obtain the highest member of the **8** by constructing a function $\psi_1^{(8)}$ orthogonal to $\psi_2^{(10)}$. It is given by

$$\psi_1^{(8)} = (\sqrt{2}\,\psi_1 u_2 - \psi_2 u_1)/\sqrt{3},$$

with de Swart's sign convention. By symmetry we can obtain the other members of the octet which are at the corners of the octet weight diagram.

However, such a procedure does not give us the overall signs of the octet functions relative to $\psi_1^{(8)}$. Therefore, we use the lowering operators e_{21} and e_{31} and the raising operator e_{12} to obtain all the remaining octet functions from $\psi_1^{(8)}$. The functions are

$$\begin{aligned}
\psi_1^{(8)} &= (\sqrt{2}\,\psi_1 u_2 - \psi_2 u_1)/\sqrt{3}, \\
\psi_2^{(8)} &= (\psi_2 u_2 - \sqrt{2}\,\psi_3 u_1)/\sqrt{3}, \\
\psi_3^{(8)} &= (\psi_4 u_1 - \sqrt{2}\,\psi_1 u_3)/\sqrt{3}, \\
\psi_4^{(8)} &= (\psi_4 u_2 + \psi_5 u_1 - 2\psi_2 u_3)/\sqrt{6}, \\
\psi_5^{(8)} &= (\sqrt{2}\,\psi_3 u_3 - \psi_5 u_2)/\sqrt{3}, \\
\psi_6^{(8)} &= (\psi_4 u_2 - \psi_5 u_1)/\sqrt{2}, \\
\psi_7^{(8)} &= (\sqrt{2}\,\psi_6 u_1 - \psi_4 u_3)/\sqrt{3}, \\
\psi_8^{(8)} &= (\sqrt{2}\,\psi_6 u_2 - \psi_5 u_3)/\sqrt{3}.
\end{aligned} \tag{8.89}$$

In obtaining $\psi_6^{(8)}$, we have used the Schmidt orthogonalization procedure.

It is instructive to compare the Clebsch–Gordan coefficients of $SU(2)$ and $SU(3)$. If we do so, we see that some of them are the same in both cases. For example, the coefficients of Eq. (8.63) are the same as those in the first four equations of (8.87). Such coefficients do not need to be recalculated for $SU(3)$ after they have been obtained for $SU(2)$.

In order to take advantage of tables of $SU(2)$ Clebsch–Gordan coefficients, it is useful to introduce the concept of isoscalar factors (Edmonds, 1962). We begin with the Kronecker product tensors $\psi_\mu^{(\alpha)}\psi_\nu^{(\beta)}$ where μ and ν each stands for three numbers: the isospin, the third component of isospin, and the hypercharge, as these are the quantities which distinguish different members of a multiplet.

We now use a somewhat less compact notation. We let

$$\mu = yii_3, \qquad \nu = y'i'i_3', \qquad m = YII_3, \qquad j = N, \tag{8.90}$$

where i, i' and I stand for the isospin, y, y' and Y for the hypercharge, i_3, i_3' and I_3 for the third component of the isospin, and N for the dimensionality of the $SU(3)$ representation. We next form linear combinations of the product tensors $\psi_{yii_3}^{(\alpha)}\psi_{y'i'i_3'}^{(\beta)}$, which correspond to eigenstates of the total isospin by summing over i_3 and i_3'. Denoting these isospin eigenstates by $\chi_{yiy'i'}^{(I)}$, we have

$$\chi_{yiy'i'}^{(I)} = \sum_{i_3 i_3'} (ii' i_3 i_3' \mid II_3)\psi_{yii_3}^{(\alpha)}\psi_{y'i'i_3'}^{(\beta)}, \tag{8.91}$$

where the coefficients $(ii'i_3 i_3' \mid II_3)$ are $SU(2)$ Clebsch–Gordan coefficients. We then obtain the eigenstates $\psi_{YII_3}^{(N\gamma)}$ by taking linear combinations of the $\chi_{yiy'i'}^{(I)}$ as follows:

$$\psi_{YII_3}^{(N\gamma)} = \sum_{yiy'i'} (\alpha\beta yiy'i' \mid NYI\gamma)\chi_{yiy'i'}^{(I)}. \qquad (8.92)$$

The coefficients $(\alpha\beta yiy'i' \mid NIY\gamma)$ are called *isoscalar factors* because they depend on the magnitude of the isospin but not on the value of its third component. If we substitute Eq. (8.91) in Eq. (8.92) we find that an $SU(3)$ Clebsch–Gordan coefficient is given by

$$(\alpha\beta\mu\nu \mid Nm\gamma) = (\alpha\beta yiy'i' \mid NYI\gamma)(ii'i_3 i_3' \mid II_3), \qquad (8.93)$$

where $\mu\nu m$ are given by Eq. (8.90). Some tables of isoscalar factors are given by de Swart (1963). Following de Swart, we adopt the sign convention that the isoscalar factor with the largest value of i is positive. If this leaves an ambiguity, we take the isoscalar factor with the largest value of i' to be positive. This convention may not be sufficient to determine all the signs of the $SU(3)$ Clebsch–Gordan coefficients, but it is adequate for the ones we consider here. With this convention, we have

$$(\alpha\beta\mu\nu \mid Nm\gamma) = x(\beta\alpha\nu\mu \mid Nm\gamma) = y(\bar{\alpha}\bar{\beta} - \mu - \nu \mid \bar{N} - m\gamma) \qquad (8.94)$$

where x and y are either plus or minus one. These signs are given in Table 8.3

TABLE 8.3

SOME SIGN FACTORS x AND y TO BE USED IN EQ. (8.94) RELATING CERTAIN $SU(3)$ CLEBSCH–COEFFICIENTS

α	β	N	x	y
8	8	27	1	1
		10	−1	1
		$\overline{10}$	−1	1
		8_s	1	1
		8_a	−1	−1
		1	1	1
8	10	35	1	1
		27	−1	1
		10	−1	−1
		8	1	−1

for some selected Clebsch–Gordan coefficients. More extensive tables are given by de Swart (1963). If $m = YII_3$, then by the notation $-m$, we mean $-Y, I, -I_3$.

Let us obtain the isoscalar factors which arise in constructing an **8** from $\bar{\mathbf{3}} \otimes \mathbf{3}$. The $SU(3)$ Clebsch–Gordan coefficients are given in Eqs. (8.82)–(8.84). The isoscalar factors of Eq. (8.82) are obviously all unity, since the $SU(3)$ Clebsch–Gordan coefficients are unity. The function ψ_4 of Eq. (8.83) is constructed from two isospin doublets. Therefore, the Clebsch–Gordan coefficient $1/\sqrt{2}$ is an $SU(2)$ coefficient, and the isoscalar factor is unity. The function ψ_6 of Eq. (8.84) is an isosinglet. It is composed partly from the isosinglet

$$\chi_1 = v_1 u_3,$$

which is composed of two isosinglets, and partly from the isosinglet

$$\varphi_1 = (1/\sqrt{2})(v_2 u_2 - v_3 u_1),$$

which is made from two isodoublets. Thus we can write Eq. (8.84) as

$$\psi_6 = (\varphi_1 - \sqrt{2}\,\chi_1)/\sqrt{3}.$$

The coefficients $1/\sqrt{3}$ and $-\sqrt{2/3}$ are the isoscalar factors.

As another example, consider the $\psi_i^{(10)}$ of Eq. (8.88). The functions $\psi_i^{(10)}$ $(i = 1, \ldots, 4)$, form an isospin quartet. Since the isoscalar factor of $\psi_1^{(10)}$ is obviously 1, all the isoscalar factors of the quartet are unity, and the coefficients are Clebsch–Gordan coefficients of $SU(2)$. The function $\psi_5^{(10)}$ is composed partly of the isospin triplet state $\chi_3 = \psi_1 u_3$ which is made of an isospin triplet and singlet and partly of $\varphi_3 = \psi_4 u_1$ which is made of two isospin doublets. We have

$$\psi_4^{(10)} = (\chi_3 + \sqrt{2}\,\varphi_3)/\sqrt{3}.$$

Therefore isoscalar factors of the triplet $\psi_i^{(10)}$ $(i = 4, 5, 6)$ are $1/\sqrt{3}$ and $\sqrt{2/3}$. The other isoscalar factors of this $SU(3)$ multiplet are obtained in an analogous manner.

Some isoscalar factors taken from de Swart (1963) are given in Tables 8.4 and 8.5. These two tables use the symbols I_1, I_2, Y_1, Y_2, μ_γ instead of i, i', y, y', N, γ as in the text. For more extensive tables, see de Swart (1963). In Tables 8.6 and 8.7 are given some $SU(3)$ Clebsch–Gordan coefficients taken from McNamee and Chilton (1964). See their work for more extensive tables.

TABLE 8.4

ISOSCALAR FACTORS FOR $8 \otimes 8$ (AFTER DE SWART, 1963).

$Y = 2 \quad I = 1$

$I_1, Y_1; I_2, Y_2$	27	μ_γ
$\frac{1}{2}, 1; \frac{1}{2}, 1$	1	

$Y = 2 \quad I = 0$

$I_1, Y_1; I_2, Y_2$	10^*	μ_γ
$\frac{1}{2}, 1; \frac{1}{2}, 1$	-1	

$Y = 1 \quad I = \frac{3}{2}$

$I_1, Y_1; I_2, Y_2$	27	10	μ_γ
$\frac{1}{2}, 1; 1, 0$	$\sqrt{2}/2$	$-\sqrt{2}/2$	
$1, 0; \frac{1}{2}, 1$	$\sqrt{2}/2$	$\sqrt{2}/2$	

$Y = 0 \quad I = 0$

$I_1, Y_1; I_2, Y_2$	27	8_1	1	8_2	μ_γ
$\frac{1}{2}, 1; \frac{1}{2}, -1$	$\sqrt{15}/10$	$\sqrt{10}/10$	$1/2$	$\sqrt{2}/2$	
$\frac{1}{2}, -1; \frac{1}{2}, 1$	$-\sqrt{15}/10$	$-\sqrt{10}/10$	$-1/2$	$\sqrt{2}/2$	
$1, 0; 1, 0$	$-\sqrt{10}/20$	$-\sqrt{15}/5$	$\sqrt{6}/4$	0	
$0, 0; 0, 0$	$3\sqrt{30}/20$	$-\sqrt{5}/5$	$-\sqrt{2}/4$	0	

$Y = 1 \quad I = \frac{1}{2}$

$I_1, Y_1; I_2, Y_2$	27	8_1	8_2	10^*	μ_γ
$\frac{1}{2}, 1; 1, 0$	$\sqrt{5}/10$	$3\sqrt{5}/10$	$1/2$	$-1/2$	
$1, 0; \frac{1}{2}, 1$	$-\sqrt{5}/10$	$-3\sqrt{5}/10$	$1/2$	$-1/2$	
$\frac{1}{2}, 1; 0, 0$	$3\sqrt{5}/10$	$-\sqrt{5}/10$	$1/2$	$1/2$	
$0, 0; \frac{1}{2}, 1$	$3\sqrt{5}/10$	$-\sqrt{5}/10$	$-1/2$	$-1/2$	

$Y = -1 \quad I = \frac{3}{2}$

$I_1, Y_1; I_2, Y_2$	27	10^*	μ_γ
$\frac{1}{2}, -1; 1, 0$	$\sqrt{2}/2$	$-\sqrt{2}/2$	
$1, 0; \frac{1}{2}, -1$	$\sqrt{2}/2$	$\sqrt{2}/2$	

$Y = 0 \quad I = 2$

$I_1, Y_1; I_2, Y_2$	27	μ_γ
$1, 0; 1, 0$	1	

$Y = -1 \quad I = \frac{1}{2}$

$I_1, Y_1; I_2, Y_2$	27	8_1	8_2	10	μ_γ
$\frac{1}{2}, -1; 1, 0$	$-\sqrt{5}/10$	$-3\sqrt{5}/10$	$1/2$	$1/2$	
$1, 0; \frac{1}{2}, -1$	$\sqrt{5}/10$	$3\sqrt{5}/10$	$1/2$	$1/2$	
$\frac{1}{2}, -1; 0, 0$	$3\sqrt{5}/10$	$-\sqrt{5}/10$	$-1/2$	$1/2$	
$0, 0; \frac{1}{2}, -1$	$3\sqrt{5}/10$	$-\sqrt{5}/10$	$1/2$	$-1/2$	

$Y = 0 \quad I = 1$

$I_1, Y_1; I_2, Y_2$	27	8_1	8_2	10	10^*	μ_γ
$\frac{1}{2}, 1; \frac{1}{2}, -1$	$\sqrt{5}/5$	$-\sqrt{30}/10$	$\sqrt{6}/6$	$-\sqrt{6}/6$	$\sqrt{6}/6$	
$\frac{1}{2}, -1; \frac{1}{2}, 1$	$\sqrt{5}/5$	$-\sqrt{30}/10$	$-\sqrt{6}/6$	$\sqrt{6}/6$	$-\sqrt{6}/6$	
$1, 0; 1, 0$	0	0	$\sqrt{6}/3$	$\sqrt{6}/6$	$-\sqrt{6}/6$	
$1, 0; 0, 0$	$\sqrt{30}/10$	$\sqrt{5}/5$	0	$1/2$	$1/2$	
$0, 0; 1, 0$	$\sqrt{30}/10$	$\sqrt{5}/5$	0	$-1/2$	$-1/2$	

$Y = -2 \quad I = 1$

$I_1, Y_1; I_2, Y_2$	27	μ_γ
$\frac{1}{2}, -1; \frac{1}{2}, -1$	1	

$Y = -2 \quad I = 0$

$I_1, Y_1; I_2, Y_2$	10	μ_γ
$\frac{1}{2}, -1; \frac{1}{2}, -1$	1	

TABLE 8.5

ISOSCALAR FACTORS FOR $\mathbf{8} \otimes \mathbf{10}$ (AFTER DE SWART, 1963).

$Y = 2 \quad I = 2$

$I_1, Y_1; I_2, Y_2$	35	μ_γ
$\frac{1}{2}$, 1; $\frac{3}{2}$, 1	1	

$Y = 2 \quad I = 1$

$I_1, Y_1; I_2, Y_2$	27	μ_γ
$\frac{1}{2}$, 1; $\frac{3}{2}$, 1	-1	

$Y = 1 \quad I = \frac{5}{2}$

$I_1, Y_1; I_2, Y_2$	35	μ_γ
1, 0; $\frac{3}{2}$, 1	1	

$Y = 1 \quad I = \frac{3}{2}$

$I_1, Y_1; I_2, Y_2$	35	27	10	μ_γ
1, 0; $\frac{3}{2}$, 1	$-1/4$	$-\sqrt{5}/4$	$\sqrt{10}/4$	
0, 0; $\frac{3}{2}$, 1	$\sqrt{5}/4$	$-3/4$	$-\sqrt{2}/4$	
$\frac{1}{2}$, 1; 1, 0	$\sqrt{10}/4$	$\sqrt{2}/4$	$1/2$	

$Y = 0 \quad I = 0$

$I_1, Y_1; I_2, Y_2$	27	8	μ_γ
1, 0; 1, 0	$\sqrt{10}/5$	$-\sqrt{15}/5$	
$\frac{1}{2}$, 1; $\frac{1}{2}$, -1	$-\sqrt{15}/5$	$-\sqrt{10}/5$	

$Y = -1 \quad I = \frac{3}{2}$

$I_1, Y_1; I_2, Y_2$	35	27	μ_γ
1, 0; $\frac{1}{2}$, -1	$\sqrt{2}/2$	$\sqrt{2}/2$	
$\frac{1}{2}$, -1; 1, 0	$\sqrt{2}/2$	$-\sqrt{2}/2$	

$Y = -1 \quad I = \frac{1}{2}$

$I_1, Y_1; I_2, Y_2$	35	27	10	8	μ_γ
1, 0; $\frac{1}{2}$, -1	$-1/4$	$-7\sqrt{5}/20$	$\sqrt{2}/4$	$\sqrt{5}/5$	
0, 0; $\frac{1}{2}$, -1	$3/4$	$-3\sqrt{5}/20$	$\sqrt{2}/4$	$-\sqrt{5}/5$	
$\frac{1}{2}$, 1; 0, -2	$\sqrt{2}/4$	$3\sqrt{10}/20$	$1/2$	$\sqrt{10}/5$	
$\frac{1}{2}$, -1; 1, 0	$-1/2$	$\sqrt{5}/10$	$\sqrt{2}/2$	$-\sqrt{5}/5$	

$Y = 1 \quad I = \frac{1}{2}$

$I_1, Y_1; I_2, Y_2$	27	8	μ_γ
1, 0; $\frac{3}{2}$, 1	$\sqrt{5}/5$	$-2\sqrt{5}/5$	
$\frac{1}{2}$, 1; 1, 0	$-2\sqrt{5}/5$	$-\sqrt{5}/5$	

$Y = 0 \quad I = 2$

$I_1, Y_1; I_2, Y_2$	35	27	μ_γ
1, 0; 1, 0	$\sqrt{3}/2$	$1/2$	
$\frac{1}{2}$, -1; $\frac{3}{2}$, 1	$1/2$	$-\sqrt{3}/2$	

$Y = 0 \quad I = 1$

$I_1, Y_1; I_2, Y_2$	35	27	10	8	μ_γ
1, 0; 1, 0	$-\sqrt{3}/6$	$-3\sqrt{5}/10$	$\sqrt{3}/3$	$\sqrt{30}/15$	
0, 0; 1, 0	$\sqrt{2}/2$	$-\sqrt{30}/10$	0	$-\sqrt{5}/5$	
$\frac{1}{2}$, 1; $\frac{1}{2}$, -1	$\sqrt{3}/3$	$\sqrt{5}/5$	$\sqrt{3}/3$	$\sqrt{30}/15$	
$\frac{1}{2}$, -1; $\frac{3}{2}$, 1	$-\sqrt{3}/6$	$\sqrt{5}/10$	$\sqrt{3}/3$	$-2\sqrt{30}/15$	

$Y = -2 \quad I = 1$

$I_1, Y_1; I_2, Y_2$	35	27	μ_γ
1, 0; 0, -2	$1/2$	$\sqrt{3}/2$	
$\frac{1}{2}$, -1; $\frac{1}{2}$, -1	$\sqrt{3}/2$	$-1/2$	

$Y = -2 \quad I = 0$

$I_1, Y_1; I_2, Y_2$	35	10	μ_γ
0, 0; 0, -2	$\sqrt{2}/2$	$\sqrt{2}/2$	
$\frac{1}{2}$, -1; $\frac{1}{2}$, -1	$-\sqrt{2}/2$	$\sqrt{2}/2$	

$Y = -3 \quad I = \frac{1}{2}$

$I_1, Y_1; I_2, Y_2$	35	μ_γ
$\frac{1}{2}$, -1; 0, -2	1	

$SU(3)$ CLEBSCH–GORDAN COEFFICIENTS FOR $8 \otimes 8$ (AFTER MCNAMEE AND CHILTON, 1964).

Form: $8 \otimes 8$

	N
	Y
	I
	I_3
$y\ i\ i_3$ $y'\ i'\ i_3'$	

N			27	10
Y			1	1
I			$\frac{3}{2}$	$\frac{3}{2}$
I_3			$\frac{3}{2}$	$\frac{3}{2}$
$1\ \frac{1}{2}\ \frac{1}{2}$	$0\ 1\ 1$		$\sqrt{\frac{1}{2}}$	$-\sqrt{\frac{1}{2}}$
$0\ 1\ 1$	$1\ \frac{1}{2}\ \frac{1}{2}$		$\sqrt{\frac{1}{2}}$	$\sqrt{\frac{1}{2}}$

N		27	10	27	$\overline{10}$	8	8′
Y		1	1	1	1	1	1
I		$\frac{3}{2}$	$\frac{3}{2}$	$\frac{1}{2}$	$\frac{1}{2}$	$\frac{1}{2}$	$\frac{1}{2}$
I_3		$\frac{1}{2}$	$\frac{1}{2}$	$\frac{1}{2}$	$\frac{1}{2}$	$\frac{1}{2}$	$\frac{1}{2}$
$1\ \frac{1}{2}\ \frac{1}{2}$	$0\ 1\ 0$	$\sqrt{\frac{1}{3}}$	$-\sqrt{\frac{1}{3}}$	$\sqrt{\frac{1}{60}}$	$-\sqrt{\frac{1}{12}}$	$\sqrt{\frac{3}{20}}$	$\sqrt{\frac{1}{12}}$
$1\ \frac{1}{2}\ \frac{1}{2}$	$0\ 0\ 0$	0	0	$\sqrt{\frac{9}{20}}$	$\sqrt{\frac{1}{4}}$	$-\sqrt{\frac{1}{20}}$	$\sqrt{\frac{1}{4}}$
$1\ \frac{1}{2}\ -\frac{1}{2}$	$0\ 1\ 1$	$\sqrt{\frac{1}{6}}$	$-\sqrt{\frac{1}{6}}$	$-\sqrt{\frac{1}{30}}$	$\sqrt{\frac{1}{6}}$	$-\sqrt{\frac{3}{10}}$	$-\sqrt{\frac{1}{6}}$
$0\ 1\ 1$	$1\ \frac{1}{2}\ -\frac{1}{2}$	$\sqrt{\frac{1}{6}}$	$\sqrt{\frac{1}{6}}$	$-\sqrt{\frac{1}{30}}$	$-\sqrt{\frac{1}{6}}$	$-\sqrt{\frac{3}{10}}$	$\sqrt{\frac{1}{6}}$
$0\ 1\ 0$	$1\ \frac{1}{2}\ \frac{1}{2}$	$\sqrt{\frac{1}{3}}$	$\sqrt{\frac{1}{3}}$	$\sqrt{\frac{1}{60}}$	$\sqrt{\frac{1}{12}}$	$\sqrt{\frac{3}{20}}$	$-\sqrt{\frac{1}{12}}$
$0\ 0\ 0$	$1\ \frac{1}{2}\ \frac{1}{2}$	0	0	$\sqrt{\frac{9}{20}}$	$-\sqrt{\frac{1}{4}}$	$-\sqrt{\frac{1}{20}}$	$-\sqrt{\frac{1}{4}}$

N		27	10	27	$\overline{10}$	8	8′
Y		1	1	1	1	1	1
I		$\frac{3}{2}$	$\frac{3}{2}$	$\frac{1}{2}$	$\frac{1}{2}$	$\frac{1}{2}$	$\frac{1}{2}$
I_3		$-\frac{1}{2}$	$-\frac{1}{2}$	$-\frac{1}{2}$	$-\frac{1}{2}$	$-\frac{1}{2}$	$-\frac{1}{2}$
$1\ \frac{1}{2}\ \frac{1}{2}$	$0\ 1\ -1$	$\sqrt{\frac{1}{6}}$	$-\sqrt{\frac{1}{6}}$	$\sqrt{\frac{1}{30}}$	$-\sqrt{\frac{1}{6}}$	$\sqrt{\frac{3}{10}}$	$\sqrt{\frac{1}{6}}$
$1\ \frac{1}{2}\ -\frac{1}{2}$	$0\ 1\ 0$	$\sqrt{\frac{1}{3}}$	$-\sqrt{\frac{1}{3}}$	$-\sqrt{\frac{1}{60}}$	$\sqrt{\frac{1}{12}}$	$-\sqrt{\frac{3}{20}}$	$-\sqrt{\frac{1}{12}}$
$1\ \frac{1}{2}\ -\frac{1}{2}$	$0\ 0\ 0$	0	0	$\sqrt{\frac{9}{20}}$	$\sqrt{\frac{1}{4}}$	$-\sqrt{\frac{1}{20}}$	$\sqrt{\frac{1}{4}}$
$0\ 1\ 0$	$1\ \frac{1}{2}\ -\frac{1}{2}$	$\sqrt{\frac{1}{3}}$	$\sqrt{\frac{1}{3}}$	$-\sqrt{\frac{1}{60}}$	$-\sqrt{\frac{1}{12}}$	$-\sqrt{\frac{3}{20}}$	$\sqrt{\frac{1}{12}}$
$0\ 1\ -1$	$1\ \frac{1}{2}\ \frac{1}{2}$	$\sqrt{\frac{1}{6}}$	$\sqrt{\frac{1}{6}}$	$\sqrt{\frac{1}{30}}$	$\sqrt{\frac{1}{6}}$	$\sqrt{\frac{3}{10}}$	$-\sqrt{\frac{1}{6}}$
$0\ 0\ 0$	$1\ \frac{1}{2}\ -\frac{1}{2}$	0	0	$\sqrt{\frac{9}{20}}$	$-\sqrt{\frac{1}{4}}$	$-\sqrt{\frac{1}{20}}$	$-\sqrt{\frac{1}{4}}$

N		27	10
Y		1	1
I		$\frac{3}{2}$	$\frac{3}{2}$
I_3		$-\frac{3}{2}$	$-\frac{3}{2}$
$1\ \frac{1}{2}\ -\frac{1}{2}$	$0\ 1\ -1$	$\sqrt{\frac{1}{2}}$	$-\sqrt{\frac{1}{2}}$
$0\ 1\ -1$	$1\ \frac{1}{2}\ -\frac{1}{2}$	$\sqrt{\frac{1}{2}}$	$\sqrt{\frac{1}{2}}$

N		27
Y		2
I		1
I_3		1
$1\ \frac{1}{2}\ \frac{1}{2}$	$1\ \frac{1}{2}\ \frac{1}{2}$	1

N		27	$\overline{10}$
Y		2	2
I		1	0
I_3		0	0
$1\ \frac{1}{2}\ \frac{1}{2}$	$1\ \frac{1}{2}\ -\frac{1}{2}$	$\sqrt{\frac{1}{2}}$	$-\sqrt{\frac{1}{2}}$
$1\ \frac{1}{2}\ -\frac{1}{2}$	$1\ \frac{1}{2}\ \frac{1}{2}$	$\sqrt{\frac{1}{2}}$	$\sqrt{\frac{1}{2}}$

N		27
Y		2
I		1
I_3		-1
$1\ \frac{1}{2}\ -\frac{1}{2}$	$1\ \frac{1}{2}\ -\frac{1}{2}$	1

TABLE 8.6—(*continued*)

$8 \otimes 8$

	27 0 2 2
$0\ 1\ 1 \quad 0\ 1\ 1$	1

	27 0 2 1	27 0 1 1	10 0 1 1	$\overline{10}$ 0 1 1	8 0 1 1	8' 0 1 1
$1\ \frac{1}{2}\ \frac{1}{2} \quad -1\ \frac{1}{2}\ \frac{1}{2}$	0	$\sqrt{\frac{1}{5}}$	$-\sqrt{\frac{1}{6}}$	$\sqrt{\frac{1}{6}}$	$-\sqrt{\frac{3}{10}}$	$\sqrt{\frac{1}{6}}$
$0\ 1\ 1 \quad 0\ 1\ 0$	$\sqrt{\frac{1}{2}}$	0	$\sqrt{\frac{1}{12}}$	$-\sqrt{\frac{1}{12}}$	0	$\sqrt{\frac{1}{3}}$
$0\ 1\ 1 \quad 0\ 0\ 0$	0	$\sqrt{\frac{3}{10}}$	$\sqrt{\frac{1}{4}}$	$\sqrt{\frac{1}{4}}$	$\sqrt{\frac{1}{5}}$	0
$0\ 1\ 0 \quad 0\ 1\ 1$	$\sqrt{\frac{1}{2}}$	0	$-\sqrt{\frac{1}{12}}$	$\sqrt{\frac{1}{12}}$	0	$-\sqrt{\frac{1}{3}}$
$0\ 0\ 0 \quad 0\ 1\ 1$	0	$\sqrt{\frac{3}{10}}$	$-\sqrt{\frac{1}{4}}$	$-\sqrt{\frac{1}{4}}$	$\sqrt{\frac{1}{5}}$	0
$-1\ \frac{1}{2}\ \frac{1}{2} \quad 1\ \frac{1}{2}\ \frac{1}{2}$	0	$\sqrt{\frac{1}{5}}$	$\sqrt{\frac{1}{6}}$	$-\sqrt{\frac{1}{6}}$	$-\sqrt{\frac{3}{10}}$	$-\sqrt{\frac{1}{6}}$

	27 0 2 -1	27 0 1 -1	10 0 1 -1	$\overline{10}$ 0 1 -1	8 0 1 -1	8' 0 1 -1
$1\ \frac{1}{2}\ -\frac{1}{2} \quad -1\ \frac{1}{2}\ -\frac{1}{2}$	0	$\sqrt{\frac{1}{5}}$	$-\sqrt{\frac{1}{6}}$	$\sqrt{\frac{1}{6}}$	$-\sqrt{\frac{3}{10}}$	$\sqrt{\frac{1}{6}}$
$0\ 1\ 0 \quad 0\ 1\ -1$	$\sqrt{\frac{1}{2}}$	0	$\sqrt{\frac{1}{12}}$	$-\sqrt{\frac{1}{12}}$	0	$\sqrt{\frac{1}{3}}$
$0\ 1\ -1 \quad 0\ 1\ 0$	$\sqrt{\frac{1}{2}}$	0	$-\sqrt{\frac{1}{12}}$	$\sqrt{\frac{1}{12}}$	0	$-\sqrt{\frac{1}{3}}$
$0\ 1\ -1 \quad 0\ 0\ 0$	0	$\sqrt{\frac{3}{10}}$	$\sqrt{\frac{1}{4}}$	$\sqrt{\frac{1}{4}}$	$\sqrt{\frac{1}{5}}$	0
$0\ 0\ 0 \quad 0\ 1\ -1$	0	$\sqrt{\frac{3}{10}}$	$-\sqrt{\frac{1}{4}}$	$-\sqrt{\frac{1}{4}}$	$\sqrt{\frac{1}{5}}$	0
$-1\ \frac{1}{2}\ -\frac{1}{2} \quad 1\ \frac{1}{2}\ -\frac{1}{2}$	0	$\sqrt{\frac{1}{5}}$	$\sqrt{\frac{1}{6}}$	$-\sqrt{\frac{1}{6}}$	$-\sqrt{\frac{3}{10}}$	$-\sqrt{\frac{1}{6}}$

	27 0 2 -2
$0\ 1\ -1 \quad 0\ 1\ -1$	1

	27 0 2 0	27 0 1 0	10 0 1 0	$\overline{10}$ 0 1 0	8 0 1 0	8' 0 1 0	27 0 0 0	8 0 0 0	8' 0 0 0	1 0 0 0
$1\ \frac{1}{2}\ \frac{1}{2} \quad -1\ \frac{1}{2}\ -\frac{1}{2}$	0	$\sqrt{\frac{1}{10}}$	$-\sqrt{\frac{1}{12}}$	$\sqrt{\frac{1}{12}}$	$-\sqrt{\frac{3}{20}}$	$\sqrt{\frac{1}{12}}$	$\sqrt{\frac{3}{40}}$	$\sqrt{\frac{1}{20}}$	$\sqrt{\frac{1}{4}}$	$\sqrt{\frac{1}{8}}$
$1\ \frac{1}{2}\ -\frac{1}{2} \quad -1\ \frac{1}{2}\ \frac{1}{2}$	0	$\sqrt{\frac{1}{10}}$	$-\sqrt{\frac{1}{12}}$	$\sqrt{\frac{1}{12}}$	$-\sqrt{\frac{3}{20}}$	$\sqrt{\frac{1}{12}}$	$-\sqrt{\frac{3}{40}}$	$-\sqrt{\frac{1}{20}}$	$-\sqrt{\frac{1}{4}}$	$-\sqrt{\frac{1}{8}}$
$0\ 1\ 1 \quad 0\ 1\ -1$	$\sqrt{\frac{1}{6}}$	0	$\sqrt{\frac{1}{12}}$	$-\sqrt{\frac{1}{12}}$	0	$\sqrt{\frac{1}{3}}$	$-\sqrt{\frac{1}{120}}$	$-\sqrt{\frac{1}{5}}$	0	$\sqrt{\frac{1}{8}}$
$0\ 1\ 0 \quad 0\ 1\ 0$	$\sqrt{\frac{2}{3}}$	0	0	0	0	0	$\sqrt{\frac{1}{120}}$	$\sqrt{\frac{1}{5}}$	0	$-\sqrt{\frac{1}{8}}$
$0\ 1\ 0 \quad 0\ 0\ 0$	0	$\sqrt{\frac{3}{10}}$	$\sqrt{\frac{1}{4}}$	$\sqrt{\frac{1}{4}}$	$\sqrt{\frac{1}{5}}$	0	0	0	0	0
$0\ 1\ -1 \quad 0\ 1\ 1$	$\sqrt{\frac{1}{6}}$	0	$-\sqrt{\frac{1}{12}}$	$\sqrt{\frac{1}{12}}$	0	$-\sqrt{\frac{1}{3}}$	$-\sqrt{\frac{1}{120}}$	$-\sqrt{\frac{1}{5}}$	0	$\sqrt{\frac{1}{8}}$
$0\ 0\ 0 \quad 0\ 1\ 0$	0	$\sqrt{\frac{3}{10}}$	$-\sqrt{\frac{1}{4}}$	$-\sqrt{\frac{1}{4}}$	$\sqrt{\frac{1}{5}}$	0	0	0	0	0
$0\ 0\ 0 \quad 0\ 0\ 0$	0	0	0	0	0	0	$\sqrt{\frac{27}{40}}$	$-\sqrt{\frac{1}{5}}$	0	$-\sqrt{\frac{1}{8}}$
$-1\ \frac{1}{2}\ \frac{1}{2} \quad 1\ \frac{1}{2}\ -\frac{1}{2}$	0	$\sqrt{\frac{1}{10}}$	$\sqrt{\frac{1}{12}}$	$-\sqrt{\frac{1}{12}}$	$-\sqrt{\frac{3}{20}}$	$-\sqrt{\frac{1}{12}}$	$-\sqrt{\frac{3}{40}}$	$-\sqrt{\frac{1}{20}}$	$\sqrt{\frac{1}{4}}$	$-\sqrt{\frac{1}{8}}$
$-1\ \frac{1}{2}\ -\frac{1}{2} \quad 1\ \frac{1}{2}\ \frac{1}{2}$	0	$\sqrt{\frac{1}{10}}$	$\sqrt{\frac{1}{12}}$	$-\sqrt{\frac{1}{12}}$	$-\sqrt{\frac{3}{20}}$	$-\sqrt{\frac{1}{12}}$	$\sqrt{\frac{3}{40}}$	$\sqrt{\frac{1}{20}}$	$-\sqrt{\frac{1}{4}}$	$\sqrt{\frac{1}{8}}$

TABLE 8.6—(*continued*)

$\underset{\sim}{8} \otimes \underset{\sim}{8}$

	$\underset{\sim}{27}$ -1 $\frac{3}{2}$ $\frac{3}{2}$	$\overline{\underset{\sim}{10}}$ -1 $\frac{3}{2}$ $\frac{3}{2}$
$0\ 1\ 1 \quad -1\ \frac{1}{2}\ \frac{1}{2}$	$\sqrt{\frac{1}{2}}$	$\sqrt{\frac{1}{2}}$
$-1\ \frac{1}{2}\ \frac{1}{2} \quad 0\ 1\ 1$	$\sqrt{\frac{1}{2}}$	$-\sqrt{\frac{1}{2}}$

	$\underset{\sim}{27}$ -1 $\frac{3}{2}$ $\frac{1}{2}$	$\overline{\underset{\sim}{10}}$ -1 $\frac{3}{2}$ $\frac{1}{2}$	$\underset{\sim}{27}$ -1 $\frac{1}{2}$ $\frac{1}{2}$	$\underset{\sim}{10}$ -1 $\frac{1}{2}$ $\frac{1}{2}$	$\underset{\sim}{8}$ -1 $\frac{1}{2}$ $\frac{1}{2}$	$\underset{\sim}{8}'$ -1 $\frac{1}{2}$ $\frac{1}{2}$
$0\ 1\ 1 \quad -1\ \frac{1}{2}\ -\frac{1}{2}$	$\sqrt{\frac{1}{6}}$	$\sqrt{\frac{1}{6}}$	$\sqrt{\frac{1}{30}}$	$\sqrt{\frac{1}{6}}$	$\sqrt{\frac{3}{10}}$	$\sqrt{\frac{1}{6}}$
$0\ 1\ 0 \quad -1\ \frac{1}{2}\ \frac{1}{2}$	$\sqrt{\frac{1}{3}}$	$\sqrt{\frac{1}{3}}$	$-\sqrt{\frac{1}{60}}$	$-\sqrt{\frac{1}{12}}$	$-\sqrt{\frac{3}{20}}$	$-\sqrt{\frac{1}{12}}$
$0\ 0\ 0 \quad -1\ \frac{1}{2}\ \frac{1}{2}$	0	0	$\sqrt{\frac{9}{20}}$	$-\sqrt{\frac{1}{4}}$	$-\sqrt{\frac{1}{20}}$	$\sqrt{\frac{1}{4}}$
$-1\ \frac{1}{2}\ \frac{1}{2} \quad 0\ 1\ 0$	$\sqrt{\frac{1}{3}}$	$-\sqrt{\frac{1}{3}}$	$-\sqrt{\frac{1}{60}}$	$\sqrt{\frac{1}{12}}$	$-\sqrt{\frac{3}{20}}$	$\sqrt{\frac{1}{12}}$
$-1\ \frac{1}{2}\ \frac{1}{2} \quad 0\ 0\ 0$	0	0	$\sqrt{\frac{9}{20}}$	$\sqrt{\frac{1}{4}}$	$-\sqrt{\frac{1}{20}}$	$-\sqrt{\frac{1}{4}}$
$-1\ \frac{1}{2}\ -\frac{1}{2} \quad 0\ 1\ 1$	$\sqrt{\frac{1}{6}}$	$-\sqrt{\frac{1}{6}}$	$\sqrt{\frac{1}{30}}$	$-\sqrt{\frac{1}{6}}$	$\sqrt{\frac{3}{10}}$	$-\sqrt{\frac{1}{6}}$

	$\underset{\sim}{27}$ -1 $\frac{3}{2}$ $-\frac{1}{2}$	$\overline{\underset{\sim}{10}}$ -1 $\frac{3}{2}$ $-\frac{1}{2}$	$\underset{\sim}{27}$ -1 $\frac{1}{2}$ $-\frac{1}{2}$	$\underset{\sim}{10}$ -1 $\frac{1}{2}$ $-\frac{1}{2}$	$\underset{\sim}{8}$ -1 $\frac{1}{2}$ $-\frac{1}{2}$	$\underset{\sim}{8}$ -1 $\frac{1}{2}$ $-\frac{1}{2}$
$0\ 1\ 0 \quad -1\ \frac{1}{2}\ -\frac{1}{2}$	$\sqrt{\frac{1}{3}}$	$\sqrt{\frac{1}{3}}$	$\sqrt{\frac{1}{60}}$	$\sqrt{\frac{1}{12}}$	$\sqrt{\frac{3}{20}}$	$\sqrt{\frac{1}{12}}$
$0\ 1\ -1 \quad -1\ \frac{1}{2}\ \frac{1}{2}$	$\sqrt{\frac{1}{6}}$	$\sqrt{\frac{1}{6}}$	$-\sqrt{\frac{1}{30}}$	$-\sqrt{\frac{1}{6}}$	$-\sqrt{\frac{3}{10}}$	$-\sqrt{\frac{1}{6}}$
$0\ 0\ 0 \quad -1\ \frac{1}{2}\ -\frac{1}{2}$	0	0	$\sqrt{\frac{9}{20}}$	$-\sqrt{\frac{1}{4}}$	$-\sqrt{\frac{1}{20}}$	$\sqrt{\frac{1}{4}}$
$-1\ \frac{1}{2}\ \frac{1}{2} \quad 0\ 1\ -1$	$\sqrt{\frac{1}{6}}$	$-\sqrt{\frac{1}{6}}$	$-\sqrt{\frac{1}{30}}$	$\sqrt{\frac{1}{6}}$	$-\sqrt{\frac{3}{10}}$	$\sqrt{\frac{1}{6}}$
$-1\ \frac{1}{2}\ -\frac{1}{2} \quad 0\ 1\ 0$	$\sqrt{\frac{1}{3}}$	$-\sqrt{\frac{1}{3}}$	$\sqrt{\frac{1}{60}}$	$-\sqrt{\frac{1}{12}}$	$\sqrt{\frac{3}{20}}$	$-\sqrt{\frac{1}{12}}$
$-1\ \frac{1}{2}\ -\frac{1}{2} \quad 0\ 0\ 0$	0	0	$\sqrt{\frac{9}{20}}$	$\sqrt{\frac{1}{4}}$	$-\sqrt{\frac{1}{20}}$	$-\sqrt{\frac{1}{4}}$

	$\underset{\sim}{27}$ -1 $\frac{3}{2}$ $-\frac{3}{2}$	$\overline{\underset{\sim}{10}}$ -1 $\frac{3}{2}$ $-\frac{3}{2}$
$0\ 1\ -1 \quad -1\ \frac{1}{2}\ -\frac{1}{2}$	$\sqrt{\frac{1}{2}}$	$\sqrt{\frac{1}{2}}$
$-1\ \frac{1}{2}\ -\frac{1}{2} \quad 0\ 1\ -1$	$\sqrt{\frac{1}{2}}$	$-\sqrt{\frac{1}{2}}$

	$\underset{\sim}{27}$ -2 1 1
$-1\ \frac{1}{2}\ \frac{1}{2} \quad -1\ \frac{1}{2}\ \frac{1}{2}$	1

	$\underset{\sim}{27}$ -2 1 0	$\underset{\sim}{10}$ -2 0 0
$-1\ \frac{1}{2}\ \frac{1}{2} \quad -1\ \frac{1}{2}\ -\frac{1}{2}$	$\sqrt{\frac{1}{2}}$	$\sqrt{\frac{1}{2}}$
$-1\ \frac{1}{2}\ -\frac{1}{2} \quad -1\ \frac{1}{2}\ \frac{1}{2}$	$\sqrt{\frac{1}{2}}$	$-\sqrt{\frac{1}{2}}$

	$\underset{\sim}{27}$ -2 1 -1
$-1\ \frac{1}{2}\ -\frac{1}{2} \quad -1\ \frac{1}{2}\ -\frac{1}{2}$	1

TABLE 8.7

$SU(3)$ CLEBSCH-GORDAN COEFFICIENTS FOR $\mathbf{8} \otimes \mathbf{10}$ (AFTER MCNAMEE AND CHILTON, 1964).

Form: $\mathbf{8} \otimes \mathbf{10}$

$y\ i\ i_3 \quad y'\ i'\ i_3'$	N Y I I_3

$y\ i\ i_3 \quad y'\ i'\ i_3'$	$\underset{\sim}{35}$ 1 $\frac{5}{2}$ $\frac{5}{2}$
0 1 1 1 $\frac{3}{2}$ $\frac{3}{2}$	1

$y\ i\ i_3 \quad y'\ i'\ i_3'$	$\underset{\sim}{35}$ 1 $\frac{5}{2}$ $\frac{3}{2}$	$\underset{\sim}{35}$ 1 $\frac{3}{2}$ $\frac{3}{2}$	$\underset{\sim}{27}$ 1 $\frac{3}{2}$ $\frac{3}{2}$	$\underset{\sim}{10}$ 1 $\frac{3}{2}$ $\frac{3}{2}$
1 $\frac{1}{2}$ $\frac{1}{2}$ 0 1 1	0	$\sqrt{\frac{5}{8}}$	$\sqrt{\frac{1}{8}}$	$\sqrt{\frac{1}{4}}$
0 1 1 1 $\frac{3}{2}$ $\frac{1}{2}$	$\sqrt{\frac{3}{5}}$	$-\sqrt{\frac{1}{40}}$	$-\sqrt{\frac{1}{8}}$	$\sqrt{\frac{1}{4}}$
0 1 0 1 $\frac{3}{2}$ $\frac{3}{2}$	$\sqrt{\frac{2}{5}}$	$\sqrt{\frac{3}{80}}$	$\sqrt{\frac{3}{16}}$	$-\sqrt{\frac{3}{8}}$
0 0 0 1 $\frac{3}{2}$ $\frac{3}{2}$	0	$\sqrt{\frac{5}{16}}$	$-\sqrt{\frac{9}{16}}$	$-\sqrt{\frac{1}{8}}$

$y\ i\ i_3 \quad y'\ i'\ i_3'$	$\underset{\sim}{35}$ 1 $\frac{5}{2}$ $\frac{1}{2}$	$\underset{\sim}{35}$ 1 $\frac{3}{2}$ $\frac{1}{2}$	$\underset{\sim}{27}$ 1 $\frac{3}{2}$ $\frac{1}{2}$	$\underset{\sim}{10}$ 1 $\frac{3}{2}$ $\frac{1}{2}$	$\underset{\sim}{27}$ 1 $\frac{1}{2}$ $\frac{1}{2}$	$\underset{\sim}{8}$ 1 $\frac{1}{2}$ $\frac{1}{2}$
1 $\frac{1}{2}$ $\frac{1}{2}$ 0 1 0	0	$\sqrt{\frac{5}{12}}$	$\sqrt{\frac{1}{12}}$	$\sqrt{\frac{1}{6}}$	$-\sqrt{\frac{4}{15}}$	$-\sqrt{\frac{1}{15}}$
1 $\frac{1}{2}$ $-\frac{1}{2}$ 0 1 1	0	$\sqrt{\frac{5}{24}}$	$\sqrt{\frac{1}{24}}$	$\sqrt{\frac{1}{12}}$	$\sqrt{\frac{8}{15}}$	$\sqrt{\frac{2}{15}}$
0 1 1 1 $\frac{3}{2}$ $-\frac{1}{2}$	$\sqrt{\frac{3}{10}}$	$-\sqrt{\frac{1}{30}}$	$-\sqrt{\frac{1}{6}}$	$\sqrt{\frac{1}{3}}$	$\sqrt{\frac{1}{30}}$	$-\sqrt{\frac{2}{15}}$
0 1 0 1 $\frac{3}{2}$ $\frac{1}{2}$	$\sqrt{\frac{3}{5}}$	$\sqrt{\frac{1}{240}}$	$\sqrt{\frac{1}{48}}$	$-\sqrt{\frac{1}{24}}$	$-\sqrt{\frac{1}{15}}$	$\sqrt{\frac{4}{15}}$
0 1 -1 1 $\frac{3}{2}$ $\frac{3}{2}$	$\sqrt{\frac{1}{10}}$	$\sqrt{\frac{1}{40}}$	$\sqrt{\frac{1}{8}}$	$-\sqrt{\frac{1}{4}}$	$\sqrt{\frac{1}{10}}$	$-\sqrt{\frac{2}{5}}$
0 0 0 1 $\frac{3}{2}$ $\frac{1}{2}$	0	$\sqrt{\frac{5}{16}}$	$-\sqrt{\frac{9}{16}}$	$-\sqrt{\frac{1}{8}}$	0	0

TABLE 8.7—(continued)

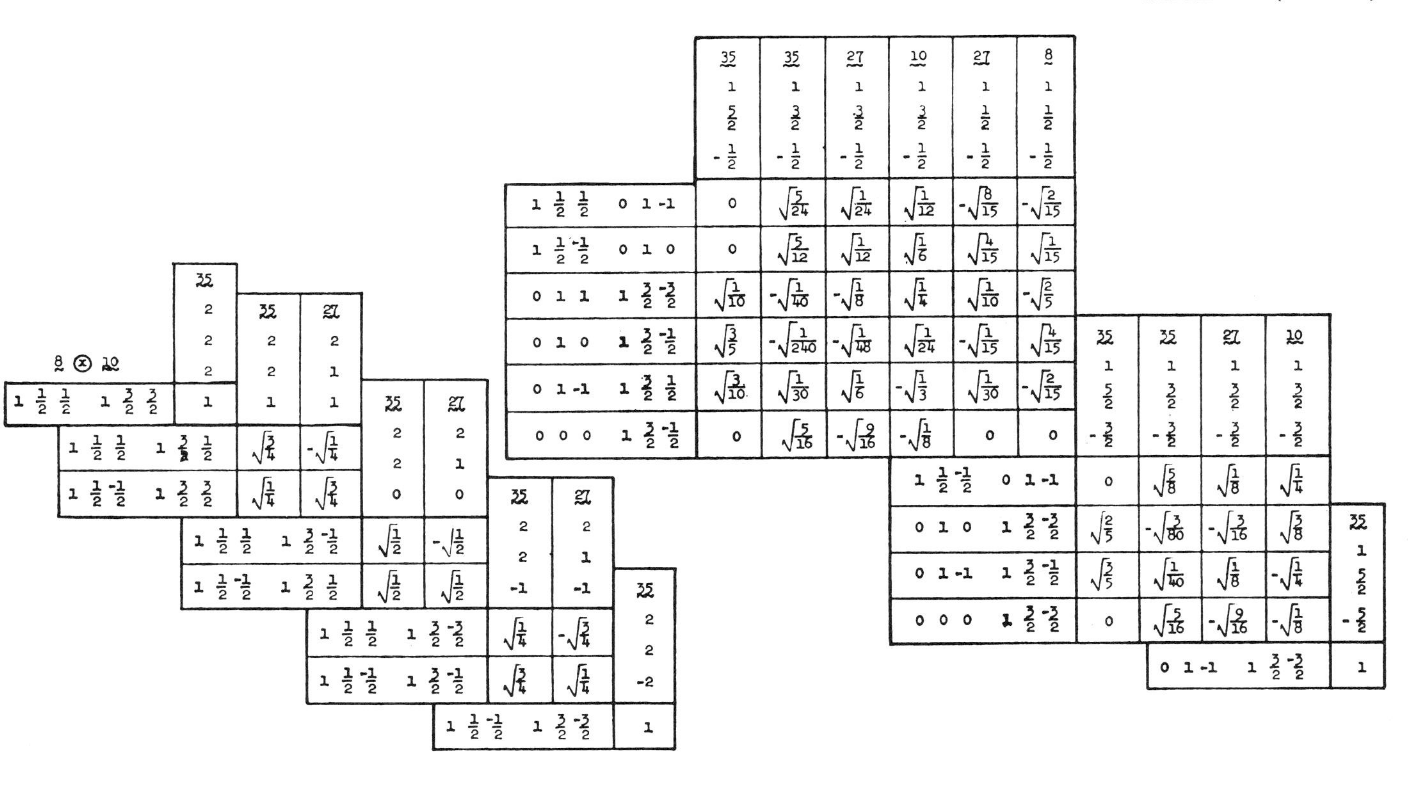

$8 \otimes 10$

	$35\ \ 2\ \ 2\ \ 2$
$1\ \frac{1}{2}\ \frac{1}{2}\quad 1\ \frac{3}{2}\ \frac{3}{2}$	1

	$35\ \ 2\ \ 2\ \ 1$	$27\ \ 2\ \ 1\ \ 1$
$1\ \frac{1}{2}\ \frac{1}{2}\quad 1\ \frac{3}{2}\ \frac{1}{2}$	$\sqrt{\frac{3}{4}}$	$-\sqrt{\frac{1}{4}}$
$1\ \frac{1}{2}\ -\frac{1}{2}\quad 1\ \frac{3}{2}\ \frac{3}{2}$	$\sqrt{\frac{1}{4}}$	$\sqrt{\frac{3}{4}}$

	$35\ \ 2\ \ 2\ \ 0$	$27\ \ 2\ \ 1\ \ 0$
$1\ \frac{1}{2}\ \frac{1}{2}\quad 1\ \frac{3}{2}\ -\frac{1}{2}$	$\sqrt{\frac{1}{2}}$	$-\sqrt{\frac{1}{2}}$
$1\ \frac{1}{2}\ -\frac{1}{2}\quad 1\ \frac{3}{2}\ \frac{1}{2}$	$\sqrt{\frac{1}{2}}$	$\sqrt{\frac{1}{2}}$

	$35\ \ 2\ \ 2\ \ -1$	$27\ \ 2\ \ 1\ \ -1$
$1\ \frac{1}{2}\ \frac{1}{2}\quad 1\ \frac{3}{2}\ -\frac{3}{2}$	$\sqrt{\frac{1}{4}}$	$-\sqrt{\frac{3}{4}}$
$1\ \frac{1}{2}\ -\frac{1}{2}\quad 1\ \frac{3}{2}\ -\frac{1}{2}$	$\sqrt{\frac{3}{4}}$	$\sqrt{\frac{1}{4}}$

	$35\ \ 2\ \ 2\ \ -2$
$1\ \frac{1}{2}\ -\frac{1}{2}\quad 1\ \frac{3}{2}\ -\frac{3}{2}$	1

	$35\ \ 1\ \ \frac{5}{2}\ \ -\frac{1}{2}$	$35\ \ 1\ \ \frac{3}{2}\ \ -\frac{1}{2}$	$27\ \ 1\ \ \frac{3}{2}\ \ -\frac{1}{2}$	$10\ \ 1\ \ \frac{3}{2}\ \ -\frac{1}{2}$	$27\ \ 1\ \ \frac{1}{2}\ \ -\frac{1}{2}$	$8\ \ 1\ \ \frac{1}{2}\ \ -\frac{1}{2}$
$1\ \frac{1}{2}\ \frac{1}{2}\quad 0\ 1\ -1$	0	$\sqrt{\frac{5}{24}}$	$\sqrt{\frac{1}{24}}$	$\sqrt{\frac{1}{12}}$	$-\sqrt{\frac{8}{15}}$	$-\sqrt{\frac{2}{15}}$
$1\ \frac{1}{2}\ -\frac{1}{2}\quad 0\ 1\ 0$	0	$\sqrt{\frac{5}{12}}$	$\sqrt{\frac{1}{12}}$	$\sqrt{\frac{1}{6}}$	$\sqrt{\frac{4}{15}}$	$\sqrt{\frac{1}{15}}$
$0\ 1\ 1\quad 1\ \frac{3}{2}\ -\frac{3}{2}$	$\sqrt{\frac{1}{10}}$	$-\sqrt{\frac{1}{40}}$	$-\sqrt{\frac{1}{8}}$	$\sqrt{\frac{1}{4}}$	$\sqrt{\frac{1}{10}}$	$-\sqrt{\frac{2}{5}}$
$0\ 1\ 0\quad 1\ \frac{3}{2}\ -\frac{1}{2}$	$\sqrt{\frac{3}{5}}$	$-\sqrt{\frac{1}{240}}$	$-\sqrt{\frac{1}{48}}$	$\sqrt{\frac{1}{24}}$	$-\sqrt{\frac{1}{15}}$	$\sqrt{\frac{4}{15}}$
$0\ 1\ -1\quad 1\ \frac{3}{2}\ \frac{1}{2}$	$\sqrt{\frac{3}{10}}$	$\sqrt{\frac{1}{30}}$	$\sqrt{\frac{1}{6}}$	$-\sqrt{\frac{1}{3}}$	$\sqrt{\frac{1}{30}}$	$-\sqrt{\frac{2}{15}}$
$0\ 0\ 0\quad 1\ \frac{3}{2}\ -\frac{1}{2}$	0	$\sqrt{\frac{5}{16}}$	$-\sqrt{\frac{9}{16}}$	$-\sqrt{\frac{1}{8}}$	0	0

	$35\ \ 1\ \ \frac{5}{2}\ \ -\frac{3}{2}$	$35\ \ 1\ \ \frac{3}{2}\ \ -\frac{3}{2}$	$27\ \ 1\ \ \frac{3}{2}\ \ -\frac{3}{2}$	$10\ \ 1\ \ \frac{3}{2}\ \ -\frac{3}{2}$
$1\ \frac{1}{2}\ -\frac{1}{2}\quad 0\ 1\ -1$	0	$\sqrt{\frac{5}{8}}$	$\sqrt{\frac{1}{8}}$	$\sqrt{\frac{1}{4}}$
$0\ 1\ 0\quad 1\ \frac{3}{2}\ -\frac{3}{2}$	$\sqrt{\frac{2}{5}}$	$-\sqrt{\frac{3}{80}}$	$-\sqrt{\frac{3}{16}}$	$\sqrt{\frac{3}{8}}$
$0\ 1\ -1\quad 1\ \frac{3}{2}\ -\frac{1}{2}$	$\sqrt{\frac{3}{5}}$	$\sqrt{\frac{1}{40}}$	$\sqrt{\frac{1}{8}}$	$-\sqrt{\frac{1}{4}}$
$0\ 0\ 0\quad 1\ \frac{3}{2}\ -\frac{3}{2}$	0	$\sqrt{\frac{5}{16}}$	$-\sqrt{\frac{9}{16}}$	$-\sqrt{\frac{1}{8}}$

	$35\ \ 1\ \ \frac{5}{2}\ \ -\frac{5}{2}$
$0\ 1\ -1\quad 1\ \frac{3}{2}\ -\frac{3}{2}$	1

TABLE 8.7—(continued)

$8 \otimes 10$	35 0 2 2	27 0 2 2
0 1 1 0 1 1	$\sqrt{\frac{3}{4}}$	$\sqrt{\frac{1}{4}}$
$-1\ \frac{1}{2}\ \frac{1}{2}\quad 1\ \frac{3}{2}\ \frac{3}{2}$	$\sqrt{\frac{1}{4}}$	$-\sqrt{\frac{3}{4}}$

	35 0 2 1	27 0 2 1	35 0 1 1	27 0 1 1	10 0 1 1	8 0 1 1
$1\ \frac{1}{2}\ \frac{1}{2}\quad -1\ \frac{1}{2}\ \frac{1}{2}$	0	0	$\sqrt{\frac{1}{3}}$	$\sqrt{\frac{1}{5}}$	$\sqrt{\frac{1}{3}}$	$\sqrt{\frac{2}{15}}$
0 1 1 0 1 0	$\sqrt{\frac{3}{8}}$	$\sqrt{\frac{1}{8}}$	$-\sqrt{\frac{1}{24}}$	$-\sqrt{\frac{9}{40}}$	$\sqrt{\frac{1}{6}}$	$\sqrt{\frac{1}{15}}$
0 1 0 0 1 1	$\sqrt{\frac{3}{8}}$	$\sqrt{\frac{1}{8}}$	$\sqrt{\frac{1}{24}}$	$\sqrt{\frac{9}{40}}$	$-\sqrt{\frac{1}{6}}$	$-\sqrt{\frac{1}{15}}$
0 0 0 0 1 1	0	0	$\sqrt{\frac{1}{2}}$	$-\sqrt{\frac{3}{10}}$	0	$-\sqrt{\frac{1}{5}}$
$-1\ \frac{1}{2}\ \frac{1}{2}\quad 1\ \frac{3}{2}\ \frac{1}{2}$	$\sqrt{\frac{3}{16}}$	$-\sqrt{\frac{9}{16}}$	$-\sqrt{\frac{1}{48}}$	$\sqrt{\frac{1}{80}}$	$\sqrt{\frac{1}{12}}$	$-\sqrt{\frac{2}{15}}$
$-1\ \frac{1}{2}\ -\frac{1}{2}\quad 1\ \frac{3}{2}\ \frac{3}{2}$	$\sqrt{\frac{1}{16}}$	$-\sqrt{\frac{3}{16}}$	$\sqrt{\frac{1}{16}}$	$-\sqrt{\frac{3}{80}}$	$-\sqrt{\frac{1}{4}}$	$\sqrt{\frac{2}{5}}$

	35 0 2 0	27 0 2 0	35 0 1 0	27 0 1 0	10 0 1 0	8 0 1 0	27 0 0 0	8 0 0 0
$1\ \frac{1}{2}\ \frac{1}{2}\quad -1\ \frac{1}{2}\ -\frac{1}{2}$	0	0	$\sqrt{\frac{1}{6}}$	$\sqrt{\frac{1}{10}}$	$\sqrt{\frac{1}{6}}$	$\sqrt{\frac{1}{15}}$	$-\sqrt{\frac{3}{10}}$	$-\sqrt{\frac{1}{5}}$
$1\ \frac{1}{2}\ -\frac{1}{2}\quad -1\ \frac{1}{2}\ \frac{1}{2}$	0	0	$\sqrt{\frac{1}{6}}$	$\sqrt{\frac{1}{10}}$	$\sqrt{\frac{1}{6}}$	$\sqrt{\frac{1}{15}}$	$\sqrt{\frac{3}{10}}$	$\sqrt{\frac{1}{5}}$
0 1 1 0 1 -1	$\sqrt{\frac{1}{8}}$	$\sqrt{\frac{1}{24}}$	$-\sqrt{\frac{1}{24}}$	$-\sqrt{\frac{9}{40}}$	$\sqrt{\frac{1}{6}}$	$\sqrt{\frac{1}{15}}$	$\sqrt{\frac{2}{15}}$	$-\sqrt{\frac{1}{5}}$
0 1 0 0 1 0	$\sqrt{\frac{1}{2}}$	$\sqrt{\frac{1}{6}}$	0	0	0	0	$-\sqrt{\frac{2}{15}}$	$\sqrt{\frac{1}{5}}$
0 1 -1 0 1 1	$\sqrt{\frac{1}{8}}$	$\sqrt{\frac{1}{24}}$	$\sqrt{\frac{1}{24}}$	$\sqrt{\frac{9}{40}}$	$-\sqrt{\frac{1}{6}}$	$-\sqrt{\frac{1}{15}}$	$\sqrt{\frac{2}{15}}$	$-\sqrt{\frac{1}{5}}$
0 0 0 0 1 0	0	0	$\sqrt{\frac{1}{2}}$	$-\sqrt{\frac{3}{10}}$	0	$-\sqrt{\frac{1}{5}}$	0	0
$-1\ \frac{1}{2}\ \frac{1}{2}\quad 1\ \frac{3}{2}\ -\frac{1}{2}$	$\sqrt{\frac{1}{8}}$	$-\sqrt{\frac{3}{8}}$	$-\sqrt{\frac{1}{24}}$	$\sqrt{\frac{1}{40}}$	$\sqrt{\frac{1}{6}}$	$-\sqrt{\frac{4}{15}}$	0	0
$-1\ \frac{1}{2}\ -\frac{1}{2}\quad 1\ \frac{3}{2}\ \frac{1}{2}$	$\sqrt{\frac{1}{8}}$	$-\sqrt{\frac{3}{8}}$	$\sqrt{\frac{1}{24}}$	$-\sqrt{\frac{1}{40}}$	$-\sqrt{\frac{1}{6}}$	$\sqrt{\frac{4}{15}}$	0	0

	35 0 2 -1	27 0 2 -1	35 0 1 -1	27 0 1 -1	10 0 1 -1	8 0 1 -1
$1\ \frac{1}{2}\ -\frac{1}{2}\quad -1\ \frac{1}{2}\ -\frac{1}{2}$	0	0	$\sqrt{\frac{1}{3}}$	$\sqrt{\frac{1}{5}}$	$\sqrt{\frac{1}{3}}$	$\sqrt{\frac{2}{15}}$
0 1 0 0 1 -1	$\sqrt{\frac{3}{8}}$	$\sqrt{\frac{1}{8}}$	$-\sqrt{\frac{1}{24}}$	$-\sqrt{\frac{9}{40}}$	$\sqrt{\frac{1}{6}}$	$\sqrt{\frac{1}{15}}$
0 1 -1 0 1 0	$\sqrt{\frac{3}{8}}$	$\sqrt{\frac{1}{8}}$	$\sqrt{\frac{1}{24}}$	$\sqrt{\frac{9}{40}}$	$-\sqrt{\frac{1}{6}}$	$-\sqrt{\frac{1}{15}}$
0 0 0 0 1 -1	0	0	$\sqrt{\frac{1}{2}}$	$-\sqrt{\frac{3}{10}}$	0	$-\sqrt{\frac{1}{5}}$
$-1\ \frac{1}{2}\ \frac{1}{2}\quad 1\ \frac{3}{2}\ -\frac{3}{2}$	$\sqrt{\frac{1}{16}}$	$-\sqrt{\frac{3}{16}}$	$-\sqrt{\frac{1}{16}}$	$\sqrt{\frac{3}{80}}$	$\sqrt{\frac{1}{4}}$	$-\sqrt{\frac{2}{5}}$
$-1\ \frac{1}{2}\ -\frac{1}{2}\quad 1\ \frac{3}{2}\ -\frac{1}{2}$	$\sqrt{\frac{3}{16}}$	$-\sqrt{\frac{9}{16}}$	$\sqrt{\frac{1}{48}}$	$-\sqrt{\frac{1}{80}}$	$-\sqrt{\frac{1}{12}}$	$\sqrt{\frac{2}{15}}$

	35 0 2 -2	27 0 2 -2
0 1 -1 0 1 -1	$\sqrt{\frac{3}{4}}$	$\sqrt{\frac{1}{4}}$
$-1\ \frac{3}{2}\ -\frac{1}{2}\quad 1\ \frac{3}{2}\ -\frac{3}{2}$	$\sqrt{\frac{1}{4}}$	$-\sqrt{\frac{3}{4}}$

TABLE 8.7—*(continued)*

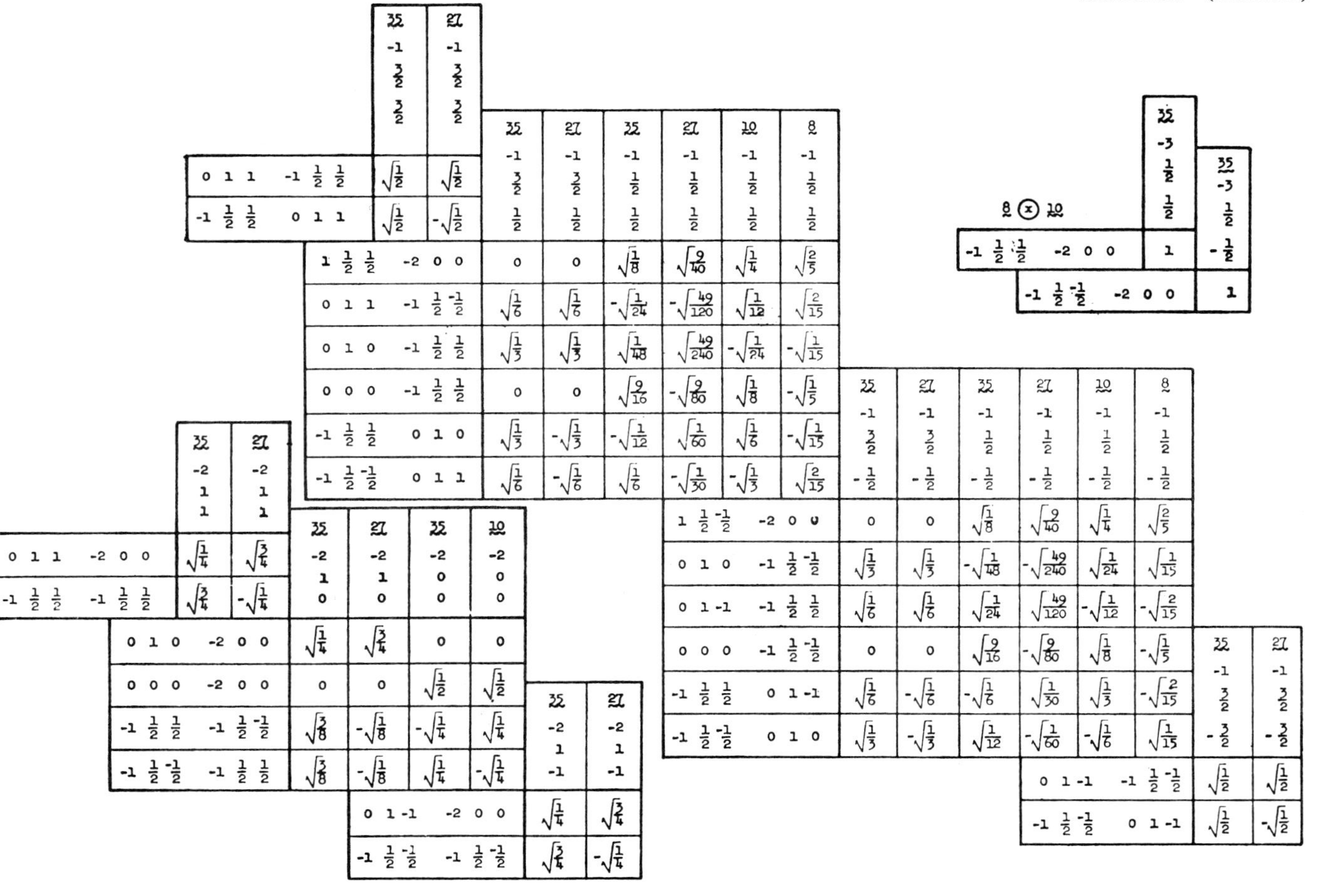

$8 \otimes 10$

	35, -3, $\frac{1}{2}$, $\frac{1}{2}$	35, -3, $\frac{1}{2}$, $-\frac{1}{2}$
-1 $\frac{1}{2}$ $\frac{1}{2}$ -2 0 0	1	
-1 $\frac{1}{2}$ $-\frac{1}{2}$ -2 0 0		1

	35, -1, $\frac{3}{2}$, $\frac{3}{2}$	27, -1, $\frac{3}{2}$, $\frac{3}{2}$
0 1 1 -1 $\frac{1}{2}$ $\frac{1}{2}$	$\sqrt{\frac{1}{2}}$	$\sqrt{\frac{1}{2}}$
-1 $\frac{1}{2}$ $\frac{1}{2}$ 0 1 1	$\sqrt{\frac{1}{2}}$	$-\sqrt{\frac{1}{2}}$

	35, -1, $\frac{3}{2}$, $\frac{1}{2}$	27, -1, $\frac{3}{2}$, $\frac{1}{2}$	35, -1, $\frac{1}{2}$, $\frac{1}{2}$	27, -1, $\frac{1}{2}$, $\frac{1}{2}$	10, -1, $\frac{1}{2}$, $\frac{1}{2}$	8, -1, $\frac{1}{2}$, $\frac{1}{2}$
1 $\frac{1}{2}$ $\frac{1}{2}$ -2 0 0	0	0	$\sqrt{\frac{1}{8}}$	$\sqrt{\frac{9}{40}}$	$\sqrt{\frac{1}{4}}$	$\sqrt{\frac{2}{5}}$
0 1 1 -1 $\frac{1}{2}$ $-\frac{1}{2}$	$\sqrt{\frac{1}{6}}$	$\sqrt{\frac{1}{6}}$	$-\sqrt{\frac{1}{24}}$	$-\sqrt{\frac{49}{120}}$	$\sqrt{\frac{1}{12}}$	$\sqrt{\frac{2}{15}}$
0 1 0 -1 $\frac{1}{2}$ $\frac{1}{2}$	$\sqrt{\frac{1}{3}}$	$\sqrt{\frac{1}{3}}$	$\sqrt{\frac{1}{48}}$	$\sqrt{\frac{49}{240}}$	$-\sqrt{\frac{1}{24}}$	$-\sqrt{\frac{1}{15}}$
0 0 0 -1 $\frac{1}{2}$ $\frac{1}{2}$	0	0	$\sqrt{\frac{9}{16}}$	$-\sqrt{\frac{9}{80}}$	$\sqrt{\frac{1}{8}}$	$-\sqrt{\frac{1}{5}}$
-1 $\frac{1}{2}$ $\frac{1}{2}$ 0 1 0	$\sqrt{\frac{1}{3}}$	$-\sqrt{\frac{1}{3}}$	$-\sqrt{\frac{1}{12}}$	$\sqrt{\frac{1}{60}}$	$\sqrt{\frac{1}{6}}$	$-\sqrt{\frac{1}{15}}$
-1 $\frac{1}{2}$ $-\frac{1}{2}$ 0 1 1	$\sqrt{\frac{1}{6}}$	$-\sqrt{\frac{1}{6}}$	$\sqrt{\frac{1}{6}}$	$-\sqrt{\frac{1}{30}}$	$-\sqrt{\frac{1}{3}}$	$\sqrt{\frac{2}{15}}$

	35, -1, $\frac{3}{2}$, $-\frac{1}{2}$	27, -1, $\frac{3}{2}$, $-\frac{1}{2}$	35, -1, $\frac{1}{2}$, $-\frac{1}{2}$	27, -1, $\frac{1}{2}$, $-\frac{1}{2}$	10, -1, $\frac{1}{2}$, $-\frac{1}{2}$	8, -1, $\frac{1}{2}$, $-\frac{1}{2}$
1 $\frac{1}{2}$ $-\frac{1}{2}$ -2 0 0	0	0	$\sqrt{\frac{1}{8}}$	$\sqrt{\frac{9}{40}}$	$\sqrt{\frac{1}{4}}$	$\sqrt{\frac{2}{5}}$
0 1 0 -1 $\frac{1}{2}$ $-\frac{1}{2}$	$\sqrt{\frac{1}{3}}$	$\sqrt{\frac{1}{3}}$	$-\sqrt{\frac{1}{48}}$	$-\sqrt{\frac{49}{240}}$	$\sqrt{\frac{1}{24}}$	$\sqrt{\frac{1}{15}}$
0 1 -1 -1 $\frac{1}{2}$ $\frac{1}{2}$	$\sqrt{\frac{1}{6}}$	$\sqrt{\frac{1}{6}}$	$\sqrt{\frac{1}{24}}$	$\sqrt{\frac{49}{120}}$	$-\sqrt{\frac{1}{12}}$	$-\sqrt{\frac{2}{15}}$
0 0 0 -1 $\frac{1}{2}$ $-\frac{1}{2}$	0	0	$\sqrt{\frac{9}{16}}$	$-\sqrt{\frac{9}{80}}$	$\sqrt{\frac{1}{8}}$	$-\sqrt{\frac{1}{5}}$
-1 $\frac{1}{2}$ $\frac{1}{2}$ 0 1 -1	$\sqrt{\frac{1}{6}}$	$-\sqrt{\frac{1}{6}}$	$-\sqrt{\frac{1}{6}}$	$\sqrt{\frac{1}{30}}$	$\sqrt{\frac{1}{3}}$	$-\sqrt{\frac{2}{15}}$
-1 $\frac{1}{2}$ $-\frac{1}{2}$ 0 1 0	$\sqrt{\frac{1}{3}}$	$-\sqrt{\frac{1}{3}}$	$\sqrt{\frac{1}{12}}$	$-\sqrt{\frac{1}{60}}$	$-\sqrt{\frac{1}{6}}$	$\sqrt{\frac{1}{15}}$

	35, -1, $\frac{3}{2}$, $-\frac{3}{2}$	27, -1, $\frac{3}{2}$, $-\frac{3}{2}$
0 1 -1 -1 $\frac{1}{2}$ $-\frac{1}{2}$	$\sqrt{\frac{1}{2}}$	$\sqrt{\frac{1}{2}}$
-1 $\frac{1}{2}$ $-\frac{1}{2}$ 0 1 -1	$\sqrt{\frac{1}{2}}$	$-\sqrt{\frac{1}{2}}$

	35, -2, 1, 1	27, -2, 1, 1
0 1 1 -2 0 0	$\sqrt{\frac{1}{4}}$	$\sqrt{\frac{3}{4}}$
-1 $\frac{1}{2}$ $\frac{1}{2}$ -1 $\frac{1}{2}$ $\frac{1}{2}$	$\sqrt{\frac{3}{4}}$	$-\sqrt{\frac{1}{4}}$

	35, -2, 1, 0	27, -2, 1, 0	35, -2, 0, 0	10, -2, 0, 0
0 1 0 -2 0 0	$\sqrt{\frac{1}{4}}$	$\sqrt{\frac{3}{4}}$	0	0
0 0 0 -2 0 0	0	0	$\sqrt{\frac{1}{2}}$	$\sqrt{\frac{1}{2}}$
-1 $\frac{1}{2}$ $\frac{1}{2}$ -1 $\frac{1}{2}$ $-\frac{1}{2}$	$\sqrt{\frac{3}{8}}$	$-\sqrt{\frac{1}{8}}$	$-\sqrt{\frac{1}{4}}$	$\sqrt{\frac{1}{4}}$
-1 $\frac{1}{2}$ $-\frac{1}{2}$ -1 $\frac{1}{2}$ $\frac{1}{2}$	$\sqrt{\frac{3}{8}}$	$-\sqrt{\frac{1}{8}}$	$\sqrt{\frac{1}{4}}$	$-\sqrt{\frac{1}{4}}$

	35, -2, 1, -1	27, -2, 1, -1
0 1 -1 -2 0 0	$\sqrt{\frac{1}{4}}$	$\sqrt{\frac{3}{4}}$
-1 $\frac{1}{2}$ $-\frac{1}{2}$ -1 $\frac{1}{2}$ $-\frac{1}{2}$	$\sqrt{\frac{3}{4}}$	$-\sqrt{\frac{1}{4}}$

Several further remarks about these tables are in order. In Tables 8.4 and 8.5, the multiplets $\mathbf{8}_s$ and$\mathbf{8}_a$ are called 8_1 and 8_2, respectively, while in Tables 8.6 and 8.7, they are called 8 and 8′. Furthermore, Tables 8.5 and 8.7 give isoscalar factors and $SU(3)$ Clebsch–Gordan coefficients, respectively, for $\mathbf{8} \otimes \mathbf{10}$, whereas in much work dealing with meson–baryon scattering, the coefficients for $\mathbf{8} \otimes \overline{\mathbf{10}}$ are used. Also, in making use of the dynamical principle of crossing symmetry, it is useful to reverse the order of the states. The appropriate sign changes for reversal of states can be found by using Eq. (8.56) for the $SU(2)$ Clebsch–Gordan coefficients and Eq. (8.94) together with Table 8.3 for the $SU(3)$ Clebsch–Gordan coefficients. If we use Eq. (8.93), we can also find the appropriate sign changes for the isoscalar factors.

8.6 Wigner–Eckart Theorem

In Section 6.6 we discussed irreducible tensor operators $T_m^{(j)}$ for a simple compact Lie group. Let us consider the matrix element of $T_m^{(j)}$ between two states $\psi_\mu^{(\alpha)}$ and $\psi_\nu^{(\beta)}$ which are basis vectors of two irreducible representations of the group. The Wigner–Eckart theorem states that such a matrix element $(\psi_\mu^{(\alpha)}, T_m^{(j)}\psi_\nu^{(\beta)})$ can be written

$$(\psi_\mu^{(\alpha)}, T_m^{(j)}\, \psi_\nu^{(\beta)}) = \sum_\gamma (\alpha\beta\mu\nu \,|\, jm\gamma)(\alpha \,\|\, T_m^{(j)} \,\|\, \beta)_\gamma, \tag{8.95}$$

where $(\alpha\beta\mu\nu \,|\, jm\gamma)$ is a Clebsch–Gordan coefficient of the group and the $(\alpha \,\|\, T_m^{(j)} \,\|\, \beta)_\gamma$ are independent of the quantum numbers μ and ν which distinguish different basis vectors of the representations α and β. The quantities $(\alpha \,\|\, T_m^{(j)} \,\|\, \beta)_\gamma$ are known as *reduced matrix elements*. In Eq. (8.95), the sum over γ includes as many terms as the irreducible representation j is contained in the decomposition of the kronecker product $\alpha \otimes \beta$. Since $SU(2)$ is simply reducible, the right-hand side of Eq. (8.95) consists of only a single term in this case.

The proof of the theorem for the rotation group is given in Wigner (1959, p. 244). The method given by Wigner can be generalized in a straightforward manner to a simple compact Lie group, although if the group is not simply reducible there is a certain arbitrariness in how the γ are chosen and how the states are labeled. The values of the Clebsch–Gordan coefficients depend of course on this arbitrary choice. The proof of the Wigner–Eckart theorem for $SU(3)$ is sketched by de Swart (1963).

Let us consider the matrix elements of a tensor between two states belonging to the same irreducible representation of a group. If the tensor T_μ is of rank one, or an irreducible *vector*, then the matrix element is proportional to the matrix element of the irreducible vector X_μ whose components are the generators of the group. For $SU(2)$, we have

$$(\psi_{m'}^{(j)}, T_\mu \psi_m^{(j')}) = C(j)(\psi_{m'}^{(j)}, J_\mu \psi_m^{(j')}), \tag{8.96}$$

where $C(j)$ is a constant of proportionality and we have put $X_\mu = J_\mu$, the components of the angular momentum. For $SU(3)$, on the other hand, there are two independent irreducible vectors, the generators F_μ, and the symmetric combination

$$D_\mu = \tfrac{2}{3} \sum_{\alpha\beta} d_{\mu\alpha\beta} F_\alpha F_\beta .$$

Then the matrix element of T_μ is given by

$$(\psi_{m'}^{(j)}, T_\mu \psi_m^{(j')}) = C_F(j)(\psi_{m'}^{(j)}, F_\mu \psi_m^{(j')}) + C_D(j)(\psi_{m'}^{(j)}, D_\mu \psi_m^{(j')}). \tag{8.97}$$

The index γ has been replaced by the symbols F and D in this case. The fact that there are two matrix elements on the right hand side of Eq. (8.97) is related to the fact that in reducing the Kronecker product of any representation $D^{(\alpha)}$ with an **8**, the representation $D^{(\alpha)}$ appears twice in the direct sum. An exception is that if $D^{(\alpha)}$ has a rectangular Young diagram, then $D^{(\alpha)}$ appears only once in the direct sum. The ratio of C_F to C_D is called the F/D ratio. By the Wigner–Eckart theorem, equations similar to (8.96) and (8.97) can be written for the reduced matrix elements.

CHAPTER 9

THE EIGHTFOLD WAY

9.1 $SU(3)$ and Hadrons

Many of the hadrons (strongly interacting particles) which have been observed so far can be conveniently classified as members of $SU(3)$ multiplets. The unclassified hadrons are also believed to belong to such multiplets, but not all members have been seen experimentally. Since additional hadrons continue to be discovered, it seems reasonable to assume that many of the known states that are now unclassified will later be found to fit in the $SU(3)$ scheme.

The $SU(3)$ multiplets that are presently known exhibit a number of characteristics which merit comment. First of all, not all members of a multiplet have the same mass, a fact which shows that the symmetry is at best approximate. However, the mass differences appear to have certain regularities, which we shall discuss later in this chapter.

A second characteristic of the known multiplets is that they all have zero triality. The lowest-dimensional representation of $SU(3)$ with zero triality (except for the trivial one-dimensional representation) has eight dimensions. In fact, all zero-triality representations of $SU(3)$ can be obtained from the reduction of Kronecker products of eight-dimensional representations. Thus, the **8**, rather than the **3**, is basic to this approach, as first postulated by Gell-Mann (1961) and Ne'eman (1961). This idea led Gell-Mann to name the symmetry scheme "the eightfold way."

As we have previously pointed out, the Clebsch–Gordan series for the Kronecker product of two **8**'s is

$$\mathbf{8} \otimes \mathbf{8} = \mathbf{1} \oplus \mathbf{8} \oplus \mathbf{8} \oplus \mathbf{10} \oplus \overline{\mathbf{10}} \oplus \mathbf{27}. \tag{9.1}$$

We can then take the Kronecker product of an **8** with a **10** to obtain

$$\mathbf{8} \otimes \mathbf{10} = \mathbf{8} \oplus \mathbf{10} \oplus \mathbf{27} \oplus \mathbf{35}. \tag{9.2}$$

In similar fashion, we can build up all the zero-triality representations.

If we restrict ourselves to hadronic states with baryon number $0 \leq B \leq 1$, then all states definitely established experimentally thus far belong to multiplets of 1, 8, or 10 states, that is to singlets, octets, or decuplets.[1] There is some inconclusive evidence for the existence of states not belonging to singlets, octets, or decuplets—the so-called *exotic* states. If exotic states exist, they are evidently coupled relatively weakly to the more commonly observed hadrons. The reason that only singlets, octets, and decuplets with $|B| \leq 1$ appear as strongly-coupled states is a question that cannot be answered on the basis of the octet model alone. Not all representations of a symmetry group are required to be realized by states in nature.

Suppose we consider an $SU(3)$ multiplet which is not a singlet. In general, the state vectors of these particles have symmetry properties under transformations of coordinates other than their $SU(3)$ indices. For example, the individual states of an $SU(3)$ multiplet may be subjected to a gauge transformation corresponding to the baryon number, to an inhomogeneous Lorentz transformation, or a parity transformation. If we neglect all symmetry breaking, we may consider as the overall symmetry group of the multiplet the direct-product group of the individual symmetry groups. Then $SU(3)$ is an invariant subgroup of this larger direct-product group.

Now transformations within the $SU(3)$ subgroup do not change any of the quantum numbers of the larger group which are not part of the $SU(3)$ subgroup. These quantum numbers then remain invariant operators for transformations within the $SU(3)$ subgroup. We conclude that all members of an $SU(3)$ multiplet should have the same values of all quantum numbers outside the subgroup.

[1] There is perhaps an inconsistent nomenclature for the multiplets. Multiplets of 3, 4, . . . , 8, members should be designated "triplets, quadruplets, . . . , octuplets" if the similarity of the members is emphasized rather than "trios, quartets, . . . , octets." However, in the old days of atomic spectroscopy, after naming multiplets of three states "triplets," physicists named multiplets of 4, 5, "quartets, quintets." Gell-Mann continued this spectroscopic designation by naming the 8 an "octet" and the 10 a "decimet." However, others (especially Glashow and Rosenfeld, 1963) have tried to switch to the other notation. The term "octet" has stuck, even though there are a few "octuplets" in the literature. The terms "decimet" and "decuplet" are both used, but we shall use "decuplet."

his picture runs into difficulty when we consider symmetry breaking. The square of the mass of a particle is an invariant of the Poincaré group. Therefore, when we make an $SU(3)$ transformation, we should not change the mass of a particle. But we know that the states of an $SU(3)$ multiplet are not degenerate in mass. Thus, it appears that we should not consider a symmetry group which is a direct product of an approximate $SU(3)$ and the Poincaré group.

This difficulty persists at the $SU(2)$ level, since isospin is also an approximate symmetry. It is instructive to compare the case of isospin with ordinary spin. The ordinary spin group is a subgroup of the Poincaré group, whereas the isospin group is not. If we consider the two spin states of an electron in an external magnetic field, these states lose their degeneracy. But the magnetic field breaks the Poincaré group (rotational invariance is lost), so that we do not expect mass degeneracy. On the other hand, the electromagnetic interaction which is always present breaks the mass degeneracy between proton and neutron without breaking the Poincaré group.

We have seen that the Hamiltonian describing the elementary particles cannot be invariant under the direct-product group $SU(2) \times P$, where P is the Poincaré group. Under such circumstances we may ask whether there exists a group G, containing both $SU(3)$ and P but not as a direct product, such that the masses of the members of an $SU(3)$ multiplet may differ from one another. However, if G is a finite Lie group, this possibility is ruled out, as has been discussed by many authors under various assumptions. McGlinn (1964) has given one of the early proofs that no such G exists. O'Raifeartaigh (1968) has given a proof with quite weak assumptions that an internal symmetry group and the Poincaré cannot be embedded in a finite Lie group in such a way as to break the mass degeneracy of a multiplet.

We shall assume that the Hamiltonian which describes the hadrons consists of two parts, one of which H_0 is invariant under $SU(3)$ and another H' which is not. Then we can regard the eigenstates of H_0 as multiplets of the direct product of $SU(3)$ and the Poincaré group. The symmetry-breaking term H' must be of such a form as to break the mass degeneracy, but to leave other quantum numbers outside $SU(3)$, such as baryon number, spin, and parity, invariant for the entire multiplet.

9.2 Baryon Multiplets

Any state with baryon number $B = 1$ we call a baryon. Baryons have half-integral spin, and all except the proton are unstable. If a baryon has hypercharge $Y \neq 1$, it is called a hyperon. Sometimes a quantity S (the strangeness) is defined by

$$S = Y - B. \tag{9.3}$$

Any hadron with $S \neq 0$ is sometimes called a strange particle. We denote a baryon by a symbol which depends on the value of its isospin I and hypercharge Y according to the scheme of Table 9.1. The electric charge of a

TABLE 9.1

NOMENCLATURE FOR BARYONS[a]

Symbol	Isospin I	Hypercharge Y
N	$\frac{1}{2}$	1
Λ	0	0
Σ	1	0
Ξ	$\frac{1}{2}$	-1
Δ	$\frac{3}{2}$	1
Ω	0	-2

[a] An asterisk on the symbol for a particle denotes an excited state. If the mass of the state is given in parentheses following the symbol, the asterisk is omitted.

baryon is denoted by a superscript except in the case of the nucleon N, where we use p (for proton) instead of N^+, and n (for neutron) instead of N^0. We sometimes give the spin and parity J^P in parentheses and sometimes also the mass in MeV. (We let $c = 1$.) Thus the symbol $N(939, \frac{1}{2}^+)$ means the nucleon has $I = \frac{1}{2}$, $Y = 1$, $J^P = \frac{1}{2}^+$, and a mass (averaged over the different charge states) of 939 MeV. The symbol unmodified denotes the state of lowest mass with the values of I and Y specified by Table 9.1. If we do not give the mass of an excited state, we put an asterisk on the symbol. For example, the symbol Σ means $\Sigma(1193)$; Σ^* means any higher-mass baryon with $I = 1$, $Y = 0$. Baryons with $Y = 0$ are usually given the symbol Y^* in the literature, but we shall not use this notation.

A bar on the symbol for a particle denotes the antiparticle. If a particle has quantum numbers J, B, I, I_3, Y, then the antiparticle has quantum numbers $J, -B, I, -I_3, -Y$. A baryon and its antiparticle have opposite parity, while a meson and its antiparticle have the same parity. A further discussion of the notation used in the classification of hadrons is given by Chew *et al.* (1964). A discussion of the quantum numbers of the hadrons and how they are determined is given by Lichtenberg (1965).

The baryon of lowest mass is the proton p. This particle is a member of the nucleon isospin doublet N, which also includes the neutron n. The N

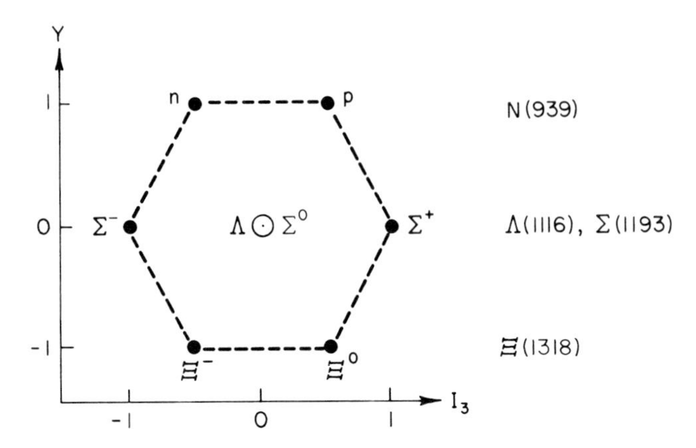

Fig. 9.1. Weight diagram of the baryon octet with spin and parity $J^P = \frac{1}{2}^+$. The average mass in MeV of each isospin multiplet is also given.

belongs in a baryon octet together with another isospin doublet (the Ξ hyperon), an isospin singlet (the Λ hyperon), and an isospin triplet (the Σ hyperon). These states are shown on a weight diagram in Fig. 9.1, with their masses in MeV in parentheses, neglecting the electromagnetic mass splitting within isospin multiplets. The hypercharge is plotted on the vertical scale in this section, but to preserve the 120° symmetry, the vertical scale is contracted by a factor $\frac{1}{2}\sqrt{3}$. The masses of all members of the octet are given in Table 9.2. Also given are their lifetimes (mean lives), and principal decay models. These values are adapted from the compilation of the Particle Data Group (1969) and from their subsequent compilation.

TABLE 9.2

SOME PROPERTIES OF THE BARYON OCTET

Symbol	Mass (MeV)	Mean life (sec)	Principal decay modes
p	938.26	Stable	—
n	939.55	932 ± 14	$pe^- \nu$
Λ	1115.6 ± 0.1	$(2.51 \pm 0.03) \times 10^{-10}$	$p\pi^-$ 65% $n\pi^0$ 35%
Σ^+	1189.4 ± 0.2	$(0.80 \pm 0.01) \times 10^{-10}$	$p\pi^0$ 52% $n\pi^+$ 48%
Σ^0	1192.5 ± 0.1	$<10^{-14}$	$\Lambda\gamma$
Σ^-	1197.3 ± 0.1	$(1.49 \pm 0.03) \times 10^{-10}$	$n\pi^-$
Ξ^0	1314.7 ± 0.7	$(3.0 \pm 0.2) + 10^{-10}$	$\Lambda\pi^0$
Ξ^-	1321.3 ± 0.2	$(1.66 \pm 0.04) \times 10^{-10}$	$\Lambda\pi^-$

According to our assumptions of the previous section, an $SU(3)$ multiplet should have the same values of all quantum numbers which refer to other symmetry groups, except for the mass. Let us see how this agrees with experiment for the baryon octet. All members of this octet have $B = 1$, as they should have. Likewise each baryon has spin measured to be $J = \frac{1}{2}$. The parity of the Ξ has not been definitively measured, but the others all have positive parity. (The parity of the p, n, and Λ can be defined to be positive, but then the measurement of the parity of the Σ is a confirmation of the assignment $P = +$. See, for example, Lichtenberg, 1965.)

There is another octet consisting of the antiparticles of the baryon octet. In general, for every $SU(3)$ multiplet of baryons, there will exist an $SU(3)$ multiplet of antibaryons belonging to the conjugate weight diagram. In the case of an octet, the $SU(3)$ weight diagram is self-conjugate. However, the octet of antibaryons is distinguished from the octet of baryons by the opposite baryon number and opposite intrinsic parity.

Another baryon multiplet which has been well established experimentally is the baryon decuplet, consisting of an isospin quartet Δ, triplet Σ^* (or Y^*), doublet Ξ^*, and singlet Ω. The members of the decuplet are shown on the weight diagram of Fig. 9.2. In Table 9.3 are given the masses of the members of the decuplet. Also given are the full widths of the states at half maximum (except in the case of the Ω where the lifetime is given) and the principal decay modes.

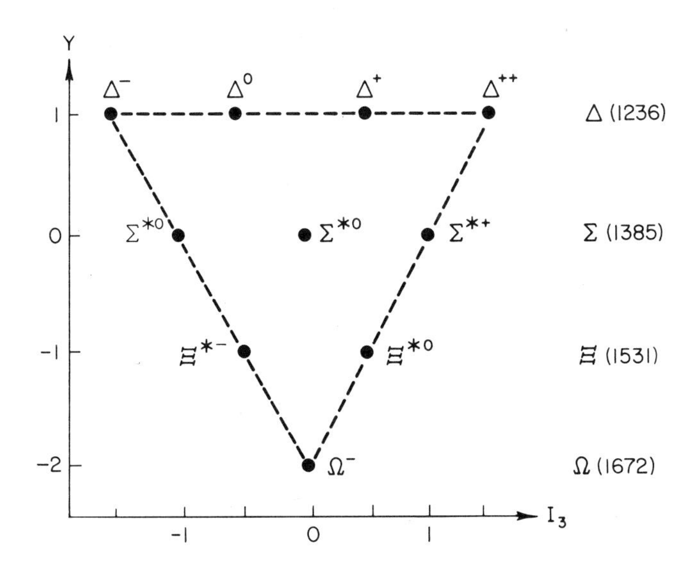

Fig. 9.2. Weight diagram of the baryon decuplet with $J^P = \frac{3}{2}^+$. The Σ^* is often called Y^*.

TABLE 9.3

SOME PROPERTIES OF THE BARYON DECUPLET

Symbol	Mass (MeV)	Full width (MeV) or mean life (sec)	Principal decay modes
Δ^{++}	1236 ± 1	120 ± 2 MeV	$N\pi$
Δ^{+}	?		
Δ^{0}	1237 ± 2		
Δ^{-}	1244 ± 8		
Σ^{*+}	1382 ± 1	36 ± 3 MeV	$\Lambda\pi$ $(90 \pm 3)\%$ $\Sigma\pi$ $(10 \pm 3)\%$
Σ^{*0}	?		
Σ^{*-}	1386 ± 2		
Ξ^{*0}	1529 ± 2	7 ± 2 MeV	$\Xi\pi$
Ξ^{*-}	1534 ± 2		
Ω^{-}	1672 ± 1	$(1.3 \pm 0.4) \times 10^{-10}$ sec	$\Xi^{0}\pi^{-}$ $\Xi^{-}\pi^{0}$ ΛK^{-}

If all these particles are correctly classified as members of the same $SU(3)$ multiplet, they should all have the same baryon number, spin, and parity (but not the same mass). All of them have $B = 1$. The spins and parities of the Δ, Σ^*, and Ξ^* have been measured and found to be $J^P = \frac{3}{2}^+$. The spin and parity of the Ω has not been measured, and it can be regarded as a prediction of the classification scheme that the Ω also has $J^P = \frac{3}{2}^+$. An interesting property of the baryon decuplet is that the Δ, Σ^*, and Ξ^* decay strongly, while the Ω decays only weakly. The Ω lives about 10^{13} times as long as a Δ. Thus, stability against decay by strong interactions is not a good criterion to use in classification of the hadrons.

The octet and decuplet are the two best established baryon $SU(3)$ multiplets. There also appear to be $SU(3)$ singlets among the baryons. The weight diagram corresponding to an $SU(3)$ singlet consists merely of a point with $Y = I_3 = 0$. Many of the baryon multiplets are experimentally incomplete. This is because it is easiest experimentally to establish evidence for N^* and Δ^* baryons, somewhat more difficult to produce Λ^* and Σ^* states, still harder to produce Ξ^* states, and extremely difficult to produce Ω^* baryons. The reason for this is, of course, the symmetry breaking. Because of the symmetry-breaking interaction, protons are the only stable target particles with $B = 1$ and pions are the most readily available projectiles with $B = 0$. Thus, states with hypercharge $Y = 1$ are most accessible to experiment. Since K^- mesons are also copiously produced in accelerators, states with $Y = 0$ are also fairly accessible to experiment. A further discussion of experimental problems in obtaining evidence for multiplets is beyond the scope of this book. In Table 9.4 we list some possible baryon multiplets, some of which are incomplete, adapted from the Particle Data Group (1969). The assignments to $SU(3)$ multiplets are only tentative. Presumably, when an N^* state is discovered it belongs to an octet and when a Δ^* is discovered it belongs to a decuplet. A Λ^* can belong to a singlet or octet, while a Σ^* and Ξ^* can belong to an octet or decuplet. Many baryon states have been discovered which are not listed in Table 9.4. For further information, see, for example, the compilation of the Particle Data Group (1969).

TABLE 9.4

TENTATIVE ASSIGNMENTS OF BARYON STATES TO $SU(3)$ SINGLETS, OCTETS, AND DECUPLETS[a]

$SU(3)$ Multiplicity	Spin and parity	Symbols and masses
1	$\frac{1}{2}^-$	$\Lambda(1405)$
8	$\frac{1}{2}^-$	$N(1535)$, $\Lambda(1670)$, ?, ?
1	$\frac{3}{2}^-$	$\Lambda(1520)$
8	$\frac{3}{2}^-$	$N(1520)$, $\Lambda(1690)$, $\Sigma(1670)$, ?
8	$\frac{5}{2}^+$	$N(1690)$, $\Lambda(1815)$, $\Sigma(1915)$, $\Xi(2030)$
10	$\frac{7}{2}^+$	$\Delta(1950)$, $\Sigma(2030)$, ?, ?

[a] Some of these multiplets are experimentally incomplete. Also, not all the quantum numbers have been verified. A question mark means that a state has not been seen.

We had remarked in Section 9.1 that all prominently observed hadron multiplets with $|B| \leq 1$ appear to belong to multiplets of 1, 8, or 10 dimensions. However, multiplets of dimensionality greater than 10 may exist for baryon number $B > 1$. An argument why this should be so is as follows: The **1**, **8**, and **10** representations are restricted to have $Y \leq 1$, as can be seen from the weight diagrams of these multiplets. In general, the larger the highest value of a weight in a multiplet (i.e., the highest value of Y), the larger is the dimensionality of the multiplet. Since a nucleon has $Y = 1$, a nucleus consisting of ν nucleons has $Y = \nu$. Therefore, if nuclei such as uranium with $Y = 238$ are members of $SU(3)$ multiplets, the multiplicity will be very large indeed. Unfortunately, at the present time it is not experimentally feasible to test whether heavy nuclei do in fact belong to $SU(3)$ multiplets.

Oakes (1963) has looked at the predictions of $SU(3)$ for baryon number $B = 2$. If $SU(3)$ is a good symmetry, it is plausible that a low-mass bound state of two baryons should belong in a multiplet contained in the Kronecker product of $\mathbf{8} \otimes \mathbf{8}$. (An alternative possibility is that such a state belongs to a multiplet contained in the Kronecker product of $\mathbf{8} \times \mathbf{10}$.) Consider the deuteron as an example. This two-baryon state has $I = 0$, $Y = 2$. Such a state is contained in $\mathbf{8} \otimes \mathbf{8}$, but not in $\mathbf{8} \otimes \mathbf{10}$. In Eq. (9.1) we have given the Clebsch–Gordan series for the decomposition of $\mathbf{8} \otimes \mathbf{8}$. Using the methods of Chapter 6 or 7, it is easy to obtain the (IY) content of the multiplets contained in $\mathbf{8} \otimes \mathbf{8}$. We find that a state with $(I, Y) = (0, 2)$ is contained only of the $\overline{\mathbf{10}}$ of the Clebsch–Gordan series. Thus, if a deuteron belongs to an $SU(3)$ multiplet, it must belong to a $\overline{\mathbf{10}}$. Because of symmetry-breaking, the other members of the $\overline{\mathbf{10}}$, if they exist, are expected to be unstable. There is some inconclusive experimental evidence in favor of a $p - \Lambda$ resonance, but that is all.

Symmetry-breaking effects are expected to be especially important in states with more than one baryon. This is because the binding energy $\sim$8 MeV per nucleon is small compared to typical symmetry-breaking energies of more than 100 MeV. The question of whether $SU(3)$ is useful for systems with $B \geq 2$ has not yet been answered.

9.3 Meson Multiplets

A hadron with $B = 0$ is called a meson. All mesons have integral spin and are unstable. The meson multiplets of $SU(3)$ differ from the baryon multiplets in several important ways. In the first place, thus far all prominently observed mesons belong only to one of two multiplets, either a singlet or an octet.

A second difference between baryons and mesons is that whereas, for any multiplet of baryons there is a distinct (conjugate) multiplet of antibaryons, for those mesons observed thus far, particles and antiparticles are contained in the same $SU(3)$ multiplet. We see how this can come about as follows. Both the octet and singlet are self-conjugate multiplets, and therefore for every state in a multiplet with a given value of I_3 and Y, there is also a state with the values of these quantum numbers given by $-I_3$, $-Y$. (If $I_3 = Y = 0$, there is only one state.)

Furthermore, for mesons, unlike baryons, the particle and antiparticle have the same values of all quantum numbers outside the group, namely spin, parity, and baryon number. (The baryon number, in general, is opposite for particle and antiparticle, but mesons have $B = 0$.) Therefore, for every state of a multiplet, there exists a state in the same multiplet (either the same state or another one) with just the quantum numbers of the antiparticle. That this state is, in fact, the antiparticle is an experimental observation.

A third difference between mesons and baryons concerns the manner of $SU(3)$ breaking, although this difference may be quantitative rather than qualitative. Suppose there exist two $SU(3)$ multiplets with the same values of spin, parity, and baryon number. Further, suppose that one of the members of the first $SU(3)$ multiplet has the same values of I, I_3, and Y as a member of the second $SU(3)$ multiplet. Then the $SU(3)$ symmetry-breaking interaction can "mix" the two multiplets. For example, consider an $SU(3)$ singlet, which has $I = Y = 0$. Also one member of an $SU(3)$ octet also has $I = Y = 0$. Then it is possible that for the symmetry-breaking interaction to cause the two *physical* states to be linear combinations of the $SU(3)$ singlet and $SU(3)$ octet. We call this $SU(3)$ mixing.

It appears that $SU(3)$ mixing occurs to a greater degree in the meson states than in the baryon octet and decuplet of lowest mass, as we shall see in Section 9.6. Furthermore, every well-established $SU(3)$ meson octet appears to be accompanied by an $SU(3)$ singlet of the same spin and parity. Since this singlet appears to mix with the octet, it is sometimes more convenient to classify all members of singlet and octet together as a nonet. Such a nonet also belongs to a *reducible* representation of $SU(3)$. However, a nonet does belong to an irreducible representation of a larger group, the group $U(3) \times U(3)$.

The nomenclature for the mesons is much more disorganized than that for baryons, as a large number of the states are given individual symbols. A scheme to label the mesons by the values of I and Y has been proposed, but has not become popular. This scheme is given in Table 9.5. In much of the literature, however, the symbols π, η, and η' are reserved for mesons with $J^P = 0^-$. We shall use the symbols of Table 9.5, but shall also give on occasion the more common symbols for mesons. Again the unmodified

TABLE 9.5

NOMENCLATURE FOR THE MESONS[a]

Symbol	I	Y	$SU(3)$ multiplicity
π	1	0	8
K	$\frac{1}{2}$	1	8
η	0	0	Mixed 8 and 1, mostly 8
η'	0	0	Mixed 8 and 1, mostly 1

[a] An asterisk on the symbol for a particle denotes an excited state. If the mass is given in parentheses following the symbol, the asterisk is omitted. A subscript "n" is sometimes used on a symbol to denote normal (or natural) parity $[P = (-1)^J]$, and a subscript "a" to denote abnormal (or unnatural) parity $[P = -(-1)^J]$.

symbol means the state of lowest mass with the values of I and Y, while an asterisk indicates an excited state. The spin, parity, and mass are sometimes given in parentheses; if the mass is given, the asterisk is omitted. The neutral states with $Y = 0$ are eigenstates of the charge conjugation operator. Thus, these states have a definite C parity, which we also give. For example, the π^0 has $J^{PC} = 0^{-+}$. Thus, we denote a pion by $\pi(138, 0^{-+})$.

Sometimes, instead of the C parity, G parity is used to characterize a meson. For mesons with $Y = 0$, G is given by $G = C(-1)^I$. Although C is a good quantum number for the neutral number of a $Y = 0$ meson multiplet, G is a good quantum number for all members of such a multiplet. For example, a neutral pion has $J^{PC} = 0^{-+}$, but all pions have $J^{PG} = 0^{--}$. For this reason G is sometimes a more convenient quantum number than C, especially since G appears to be conserved in strong interactions. However, we shall use only C, which appears to be conserved in both strong and electromagnetic interactions.

The meson multiplet of lowest average mass and the first to be known experimentally is the pseudoscalar[2] meson octet, with spin and parity $J^P = 0^-$. The weight diagram of this octet is shown in Fig. 9.3. The π and η^0 mesons are in the center of the weight diagram and are their own antiparticlcs. Since for these states, particle and antiparticle are indistinguishable and therefore

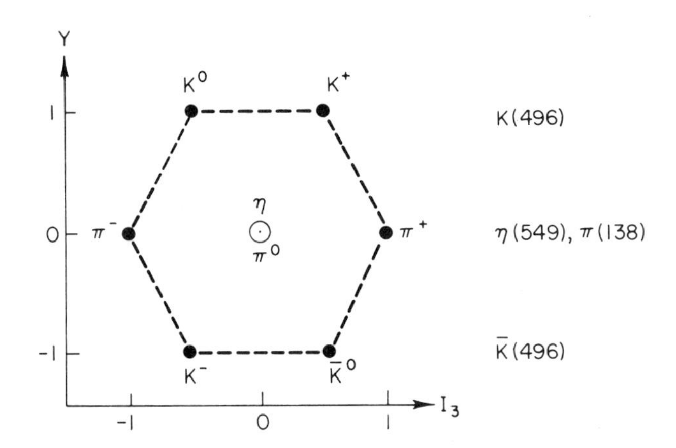

Fig. 9.3. Weight diagram of pseudoscalar meson octet. The average mass in MeV of each isospin multiplet is also given.

[2] A scalar particle has spin and parity $J^P = 0^+$, a pseudoscalar has $J^P = 0^-$, a vector $J^P = 1^-$, and an axial vector $J^P = 1^+$. By a tensor particle we usually mean a particle with $J^P = 2^+$ and by a pseudotensor $J^P = 2^-$. A particle with spin and parity related by $P = (-1)^J$ is sometimes said to have normal or natural parity; if $P = -(-1)^J$, then the parity is said to be abnormal or unnatural.

contained in the same multiplet, all the antiparticles must be in the same $SU(3)$ multiplet as the particles. The weight of an antiparticle of a given state is obtained by reflection the weight of the particle through the origin.

Accompanying the pseudoscalar octet is a pseudoscalar singlet, the η'. The η' is its own antiparticle. In Table 9.6 the masses, lifetimes, and principal decay modes of the pseudoscalar octet plus singlet, or nonet, are given.

TABLE 9.6

SOME PROPERTIES OF THE PSEUDOSCALAR MESON OCTET PLUS SINGLET[a] WITH $J^{PC} = 0^{-+}$

$SU(3)$ Assignment	Symbol	Mass (MeV)	Mean life (sec) or width	Principal decay modes
8	π^+	139.58 ± 0.02	$(2.60 \pm 0.01) \times 10^{-8}$	$\mu\nu$
	π^0	134.97 ± 0.02	$(0.9 \pm 0.2) \times 10^{-16}$	$\gamma\gamma$
	$K^\pm$	493.8 ± 0.1	$(1.23 \pm 0.01) \times 10^{-8}$	$\mu\nu$ $(63.8 \pm 0.3)\%$
				$\pi^\pm\pi^0$ $(20.9 \pm 0.3)\%$
				$\pi^\pm\pi^+\pi^-$ $(5.6 \pm 0.1)\%$
				$\pi^\pm\pi^0\pi^0$ $(1.7 \pm 0.1)\%$
				$\mu\pi^0\nu$ $(3.2 \pm 0.1)\%$
				$e\pi^0\nu$ $(4.8 \pm 0.1)\%$
	K^0, $\bar{K}^0$	497.8 ± 0.2	50% K_L, 50% K_S	
	(K_S)		$(0.86 \pm 0.01) \times 10^{-10}$	$\pi^+\pi^-$ $(68 \pm 1)\%$
				$\pi^0\pi^0$ $(32 \pm 1)\%$
	(K_L)		$(5.3 \pm 0.2) \times 10^{-8}$	$3\pi^0$ $(22 \pm 1)\%$
				$\pi^+\pi^-\pi^0$ $(13 \pm 1)\%$
				$\pi\mu\nu$ $(27 \pm 1)\%$
				$\pi e\nu$ $(39 \pm 1)\%$
	η	548.8 ± 0.6	$(2.6 \pm 0.6) \times 10^{-3}$ MeV	$\gamma\gamma$ $(38 \pm 2)\%$
				$\pi^0\gamma\gamma$ $(2 \pm 3)\%$
				$3\pi^0$ $(31 \pm 3)\%$
				$\pi^+\pi^-\pi^0$ $(23 \pm 1)\%$
				$\pi^+\pi^-\gamma$ $(5.4 \pm 0.5)\%$
1	η'	958 or 1422		

[a] The C-parity quantum number refers only to neutral mesons with $Y = 0$.

The vector mesons are more conveniently classified as a nonet than as an octet plus singlet, because the evidence indicates that there is significant mixing between the octet and singlet. We shall discuss this $SU(3)$ mixing in more detail in Section 9.6. Some properties of the vector nonet are given in Table 9.7.

TABLE 9.7

SOME PROPERTIES OF THE VECTOR MESON NONET[a] WITH $J^{PC} = 1^{--}$

Symbol	Common symbol	Mass (MeV)	Width (MeV)	Principal decay modes
$\pi^{*\pm}$	$\rho^{\pm}$	765 ± 10	125 ± 20	$\pi\pi$
π^{*0}	ρ^{0}	765 ± 10	125 ± 20	$\pi\pi$
$K^{*\pm}$	$K^{*\pm}$	892 ± 1	50 ± 1	$K\pi$
K^{*0}	K^{*0}	898 ± 4		$K\pi$
η^{*}	φ	1019.5 ± 0.6	3.7 ± 0.6	K^+K^- $(46 \pm 3)\%$ $K_L K_S$ $(36 \pm 3)\%$ $\pi^+\pi^-\pi^0$ $(18 \pm 5)\%$
$\eta^{*\prime}$	ω	783.7 ± 0.4	12 ± 1	$\pi^+\pi^-\pi^0$ $(87 \pm 4)\%$ $\pi^0\gamma$ $(9 \pm 2)\%$

[a] The C-parity quantum number refers only to neutral mesons with $Y = 0$.

A tensor meson nonet is also well established experimentally. Some properties of the members are given in Table 9.8. In Table 9.9 are given the masses of other meson nonets for which there is some experimental evidence.

TABLE 9.8

SOME PROPERTIES OF THE TENSOR MESON NONET[a] WITH $J^{PC} = 2^{++}$

Symbol	Common symbol	Mass (MeV)	Width (MeV)	Principal decay modes
π^{*}	$A2$[b]	1320 ± 5	21 ± 4	$\rho\pi$, $K\bar{K}$, $\eta\pi$
K^{*}	K^{*}	1422 ± 4	96 ± 7	$K\pi$ $(49 \pm 3)\%$ $K^{*}\pi$ $(36 \pm 3)\%$ $K\rho$ $(8 \pm 4)\%$ $K\omega$ $(4 \pm 1)\%$ $K\eta$ $(2 \pm 1)\%$
η^{*}	f'	1514 ± 5	73 ± 23	$K\bar{K}$ $(72 \pm 12)\%$ $\eta\pi\pi$ $(18 \pm 10)\%$ $K\bar{K}\pi$ $(10 \pm 10)\%$
$\eta^{*\prime}$	f	1264 ± 10	151 ± 25	$\pi\pi$

[a] The C-parity quantum number refers only to neutral mesons with $Y = 0$.
[b] There is some evidence that the $A2$ resonance is split into two peaks. We give only the properties of the higher-energy peak.

TABLE 9.9

TENTATIVE ASSIGNMENTS OF MESONS TO NONETS[a]

Spin and parity	C-Parity of neutral, $Y=0$ mesons	Symbols and masses
1^+	+	$\pi(1070)$, $K(1240)$, $\eta(1285)$, $\eta(?)$
1^+	−	$\pi(1235)$, $K(1330)$, $\eta(?)$ $\eta'(?)$
0^+	+	$\pi(962)$, $K(?)$, $\eta(1060)$ $\eta'(?)$

[a] Not all the quantum numbers have been verified. A question mark means a state has not yet been observed.

9.4 *U*-Spin

Thus far, we have discussed extensively the isospin subgroup of $SU(3)$, but we have not considered the possible physical significance of the other two $SU(2)$ subgroups. In Fig. 9.4 is shown the baryon octet, with the weights of three $SU(2)$ subgroups labeled as I_3, U_3, and V_3. Use of all three $SU(2)$ subgroups has been proposed by Meshkov *et al.* (1963), who have named them *I*-spin,[3] *U*-spin, and *V*-spin. These authors have discussed the three subgroups especially *U*-spin, in some detail. (see also Lipkin, 1965.)

If a weight is simple, the corresponding state vector is an eigenstate of *I*-spin, *U*-spin, and *V*-spin. However, if a weight is not simple, the eigenstates of *I*-spin having this weight are not, in general, eigenstates of *U*-spin or *V*-spin. The states Λ and Σ^0 are eigenstates of *I*-spin, but not of *U*-spin or *V*-spin. All other members of the baryon octet are eigenstates of I, U, and V. The generator e_{21} is a lowering operator for *I*-spin, e_{32} for U spin, and e_{31} for V spin.

It can be seen from the masses given in Fig. 9.4 that *U*-spin and *V*-spin are more badly broken than *I*-spin. For example, the different members of the *I*-spin multiplets differ in mass by at most 8 MeV, while different members of a *U*- or *V*-spin multiplet differ in mass by as much as several hundred MeV. Of course, since even *I*-spin is not exact, the physical states Λ and Σ° may not correspond to pure eigenstates of *I*-spin, but there may be a small amount of mixing between $I = 0$ and $I = 1$. Nevertheless, it is a convenient approximation for most purposes to regard the Σ^0 and Λ as eigenstates of *I*-spin.

[3] In this section only, we shall refer to isospin as "*I*-spin."

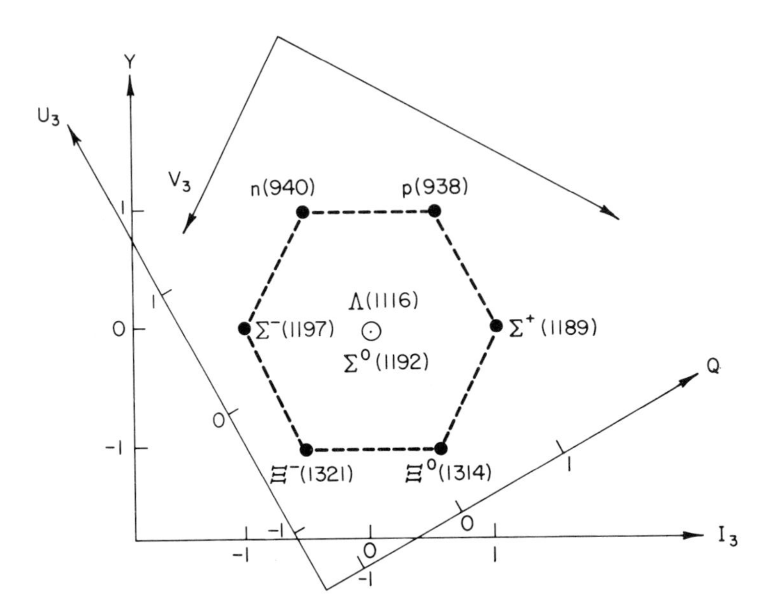

Fig. 9.4. Baryon octet, showing the *I*-spin, *U*-spin and *V*-spin $SU(2)$ subgroups. The states Λ and Σ^0 are eigenstates of *I*-spin, but not of *U*-spin or *V*-spin.

From the weight diagram of Fig. 9.4 we see that all members of a single *U*-spin multiplet have the same electric charge. Furthermore, the axis perpendicular to the U_3 axis is the charge Q. Thus, a *U*-spin eigenstate has a weight (U_3, Q), just as an *I*-spin eigenstate has a weight (I_3, Y). There is no such simple physical interpretation of the *V*-spin subgroup, and so we shall not discuss *V*-spin at any length.

Assuming that the Λ and Σ^0 are *I*-spin eigenstates, we now wish to obtain the linear combinations of Λ and Σ^0 which are eigenstates of *U*-spin. To do this, we shall introduce operators which act in the $SU(2)$ subspaces of $SU(3)$. We have defined the $SU(2)$ generators

$$h_1 = \begin{pmatrix} 1 & 0 \\ 0 & -1 \end{pmatrix}, \qquad e_{21} = \begin{pmatrix} 0 & 0 \\ 1 & 0 \end{pmatrix}, \qquad \text{and} \qquad e_{12} = \begin{pmatrix} 0 & 1 \\ 0 & 0 \end{pmatrix}. \tag{9.4}$$

We shall introduce the notation that these three operators will be called τ_3, τ_-, and τ_+, respectively when referred to *I*-spin, and σ_3, σ_-, and σ_+, respectively when referred to *U*-spin. We shall also define operators which act on states of more than one particle

$$\begin{aligned} I_3 &= \tfrac{1}{2}\sum_i \tau_3(i), \qquad & I_- &= \sum_i \tau_-(i), \qquad & I_+ &= \sum_i \tau_+(i), \\ U_3 &= \tfrac{1}{2}\sum_i \sigma_3(i), \qquad & U_- &= \sum_i \sigma_-(i), \qquad & U_+ &= \sum_i \sigma_+(i). \end{aligned} \tag{9.5}$$

Let us denote the two states of a fundamental double of I-spin by the notation α and β. We can construct the I-spin triplet Σ^+, Σ^0, Σ^- and the I-spin singlet Λ from two of these fundamental I-spin doublets. Then

$$\Sigma^+ = \alpha\alpha, \qquad \Sigma^0 = (\alpha\beta + \beta\alpha)/\sqrt{2}, \qquad \Sigma^- = \beta\beta, \qquad \Lambda = (\alpha\beta - \beta\alpha)/\sqrt{2}. \tag{9.6}$$

Let the triplet eigenstates of U-spin be n, χ, Ξ^0, and the singlet be φ, where χ and φ are orthogonal linear combinations of the I-spin eigenstates Λ and Σ^0

$$\chi = a\Sigma^0 + b\Lambda, \qquad \varphi = b\Sigma^0 - a\Lambda. \tag{9.7}$$

We want to find the coefficients a and b.

From the definitions of I_-, Σ^+, and Σ^0, we can directly verify that

$$I_-\Sigma^+ = \Sigma^0\sqrt{2}. \tag{9.8}$$

But U_- acting on the U-spin triplet eigenstate n must act analogously to I_- acting on the I-spin eigenstate Σ^+. Therefore, we have

$$U_-n = \chi\sqrt{2}. \tag{9.9}$$

Now we operate with the commutator $[U_-, I_+]$ on the state vector n of the neutron. Noting that

$$I_+n = p, \tag{9.10}$$

and using Eq. (9.9) we obtain

$$[U_-, I_+]n = U_-p - I_+\chi\sqrt{2}. \tag{9.11}$$

Then from Eq. (9.7) we obtain

$$[U_-, I_+]n = U_-p - I_+(a\Sigma^0 + b\Lambda)\sqrt{2}. \tag{9.12}$$

Now since

$$U_-p = \Sigma^+, \qquad I_+\Sigma^0 = \Sigma^+\sqrt{2}, \qquad \text{and} \qquad I_+\Lambda = 0, \tag{9.13}$$

Eq. (9.12) becomes

$$[U_-, I_+]n = (1 - 2a)\Sigma^+. \tag{9.14}$$

We now point out that the commutator $[U_-, I_+]$ must vanish. This can be proved as follows: If the commutator did not vanish, it would have to be equal to a linear combination of the $SU(3)$ generators, since U_- and I_+ are members of the Lie algebra of $SU(3)$. But I_+ leads to the changes $\Delta Y = 0$, $\Delta I_3 = 1$, and U_- leads to $\Delta Y = -1$, $\Delta I_3 = \frac{1}{2}$. Therefore the product U_-I_+ or I_+U_- leads to a change $\Delta I_3 = \frac{3}{2}$. But no generator of $SU(3)$ changes I_3 by $\frac{3}{2}$, so the commutator must be zero. Then from Eq. (9.14) we obtain

$a = \frac{1}{2}$. Because χ is a normalized function, we must have $b = \pm\frac{1}{2}\sqrt{3}$. Then, choosing the convention that the sign of b is positive, we obtain

$$\chi = \tfrac{1}{2}\,\Sigma^0 + \tfrac{1}{2}\sqrt{3}\Lambda, \qquad \varphi = \tfrac{1}{2}\sqrt{3}\,\Sigma^0 - \tfrac{1}{2}\Lambda. \tag{9.15}$$

In a similar manner we can obtain the linear combinations of Λ and Σ° which are eigenstates of V-spin. However, we shall not do so since we shall not make use of V-spin eigenstates.

9.5 Tests of *U*-Spin Invariance

Although the electromagnetic interaction does not conserve isospin, it ought to conserve U-spin, since all members of a given U-spin multiplet have the same electric charge. This is, at best, true only to a first approximation. It holds to the extent that U-spin is conserved in strong interactions. If U-spin is not conserved in strong interactions, different members of a U-spin multiplet might not have the same spatial wave functions. Then they would have different electric and magnetic form factors, and would interact differently with the electromagnetic field. Also, when members of a given U-spin multiplet have different masses, this will also break the conservation of U-spin. We cannot get rid of the effects of the violation of U-spin by strong interactions to see whether U-spin would then be strictly conserved in electromagnetic interactions.

Nevertheless, let us regard conservation of U-spin as holding for electromagnetic interactions. Then we can make predictions about the electromagnetic properties of the baryons. For example, since p and Σ^+ belong to the same U-spin multiplet, they should have the same magnetic moment μ. Thus, we have the prediction

$$\mu(\Sigma^+) = \mu(p). \tag{9.16}$$

Similarly, since Σ^- and Ξ^- belong to a U-spin doublet, we predict that

$$\mu(\Sigma^-) = \mu(\Xi^-). \tag{9.17}$$

Likewise, n, χ, and Ξ^0 belong to a U-spin triplet. The χ is a linear combination of Λ and Σ, and is therefore not a physical state. However, we do predict that

$$\mu(n) = \mu(\Xi^\circ). \tag{9.18}$$

Thus far, the magnetic moments of Σ^-, Ξ^0, and Ξ^- have not been measured, but experiments to do so seem feasible. The Σ^0 magnetic moment seems remote from experiment. The measured magnetic moments of p, n, and Σ^+ are

$$\mu(p) = 2.79, \qquad \mu(n) = -1.91, \qquad \mu(\Sigma^+) = 2.5 \pm 0.5, \tag{9.19}$$

in units of the proton magneton $e/(2M_p)$. From Eq. (9.19), we see that Eq. (9.16) holds within experimental error.

We can also obtain relations among the magnetic moments of the members of the decuplet, but since these states all decay strongly (except the Ω^-), their magnetic moments cannot be measured. We can also use U-spin invariance to obtain relations among the electromagnetic mass splittings of baryons. Let us assume that the mass of a baryon arises in part from a strong interaction which conserves isospin and in part from an electromagnetic interaction which conserves U-spin. Then since p, Σ^+ belong to a U-spin multiplet, the electromagnetic contributions to the masses of the proton and Σ^+ are given by

$$\delta M_p = \delta M_{\Sigma^+}. \tag{9.20}$$

Similarly we have

$$\delta M_n = \delta M_{\Xi^0}, \qquad \delta M_{\Sigma^-} = \delta M_{\Xi^-}. \tag{9.21}$$

From Eqs. (9.20) and (9.21) we obtain the relation

$$\delta M_n - \delta M_p + \delta M_{\Xi^-} - \delta M_{\Xi^0} = \delta M_{\Sigma^-} - \delta M_{\Sigma^+}. \tag{9.22}$$

But if the electromagnetic interaction is the only interaction which breaks the degeneracy of an isospin multiplet, we must have

$$\begin{aligned} \delta M_n - \delta M_p &= M_n - M_p, \\ \delta M_{\Xi^-} - \delta M_{\Xi^0} &= M_{\Xi^-} - M_{\Xi^0}, \\ \delta M_{\Sigma^-} - \delta M_{\Sigma^+} &= M_{\Sigma^-} - M_{\Sigma^+}. \end{aligned} \tag{9.23}$$

Substituting (9.23) into (9.22) we obtain

$$M_n - M_p + M_{\Xi^-} + M_{\Xi^0} = M_{\Sigma^-} - M_{\Sigma^+}. \tag{9.24}$$

This is known as the Coleman–Glashow (1961) relation. It was first obtained with somewhat different assumptions. The experimental mass splittings are (in MeV)

$$M_n - M_p = 1.3, \qquad M_{\Xi^-} - M_{\Xi^0} = 6.6 \pm 0.7, \qquad M_{\Sigma^-} - M_{\Sigma^+} = 7.9. \tag{9.25}$$

From these numbers, we see that Eq. (9.24) is well satisfied.

An equation analogous to (9.25) holds for the pseudoscalar meson octet. It is

$$M_{K^0} - M_{K^+} + M_{K^-} - M_{\bar{K}^0} = M_{\pi^-} - M_{\pi^+}.$$

However, since particle and antiparticle have the same mass, this equation says simply that $0 = 0$, and is not a useful check of U-spin conservation.

From U-spin invariance we also obtain relations along the electromagnetic mass differences of the members of the decuplet. These relations are

$$\begin{aligned} M_{\Delta^-} - M_{\Delta^0} &= M_{\Sigma^{*-}} - M_{\Sigma^{*0}} = M_{\Xi^{*-}} - M_{\Xi^{*0}}, \\ M_{\Delta^0} - M_{\Delta^+} &= M_{\Sigma^{*0}} - M_{\Sigma^{*+}}. \end{aligned} \tag{9.26}$$

The present experimental evidence is consistent with these relations, but is not accurate enough to provide a severe test of them.

Let us now see how to make use of the isospin and U-spin subgroups of $SU(3)$ to obtain relations among reaction cross sections. We consider the reactions

$$M + N \rightarrow M + D, \tag{9.27}$$

where M is a meson from the pseudoscalar octet, N is a nucleon (since no other members of the baryon octet are suitable as targets), and D is baryon from the decuplet.

First let us make use of the familiar isospin conservation. Consider the reactions

$$\pi^+ + p \rightarrow \Delta^{++} + \pi^0, \tag{9.28}$$

$$\pi^+ + p \rightarrow \Delta^+ + \pi^+. \tag{9.29}$$

The initial state has $I = \frac{3}{2}$. Therefore, if isospin is conserved, the final state must also be $I = \frac{3}{2}$. We need to use only the Clebsch–Gordan coefficients of $SU(2)$ to construct a state $\psi_{3/2}^{(3/2)}$ $(I = 3/2, I_3 = 3/2)$ from states of $I = 1$ and $I = 3/2$. Letting the symbol for a particle denote its isospin state function, we obtain

$$\psi_{3/2}^{(3/2)} = \sqrt{\tfrac{3}{5}}\,\Delta^{++}\pi^0 - \sqrt{\tfrac{2}{5}}\,\Delta^+\pi^+. \tag{9.30}$$

The ratio of the amplitudes for reactions (9.28) and (9.29) is just given by the ratio of the Clebsch–Gordan coefficients in Eq. (9.30), and the ratio of the cross sections is given by the ratio of the squares of the Clebsch–Gordan coefficients. Denoting the cross sections by $\sigma(MN \rightarrow DM)$, we thus obtain the prediction

$$\sigma(\pi^+ p \rightarrow \Delta^{++}\pi^0)/\sigma(\pi^+ p \rightarrow \Delta^+\pi^+) = \tfrac{3}{2}. \tag{9.31}$$

Next, we consider the following set of reactions

$$\begin{aligned} \pi^- + p &\rightarrow \Sigma^{*-} + K^+, \\ \pi^- + p &\rightarrow \Delta^- + \pi^+, \\ K^- - p &\rightarrow \Xi^{*-} + K^+, \\ K^- + p &\rightarrow \Sigma^{*-} + \pi^+. \end{aligned} \tag{9.32}$$

We cannot obtain any information about the ratios of these cross sections from isospin conservation, since π, K, Δ, Ξ^*, and Σ^* all belong to different isospin multiplets. However, if U-spin is conserved, we can make predictions. The states

$$(\pi^-, K^-), \qquad (p, \Sigma^+), \qquad (K^+, \pi^+),$$

are all U-spin doublets, while

$$(\Delta^-, \Sigma^{*-}, \Xi^{*-}, \Omega^-),$$

is a U-spin quartet. The initial state $\pi^- p$ is pure $U = 1$, while the initial state $K^- p$ is a linear combination of $U = 1$ and $U = 0$:

$$\begin{aligned} \pi^- p &= \psi_1^{(1)}, \\ K^- p &= (\psi_0^{(0)} - \psi_0^{(1)})/\sqrt{2}. \end{aligned} \tag{9.33}$$

The superscript on ψ gives the value of U and the subscript the value of U_3. The final state is composed of a quartet and doublet, and so can have $U = 2$ or 1. However, if U-spin is conserved, the final state must be $U = 1$, since there is no $U = 2$ in the initial state. The relevant U-spin eigenstates are

$$\begin{aligned} \psi_1^{(1)} &= \sqrt{\tfrac{3}{4}}\,\Delta^-\pi^+ - \sqrt{\tfrac{1}{4}}\,\Sigma^{*-}K^+, \\ \psi_0^{(1)} &= \sqrt{\tfrac{1}{2}}\,\Sigma^{*-}\pi^+ - \sqrt{\tfrac{1}{2}}\,\Xi^{*-}K^+. \end{aligned} \tag{9.34}$$

From Eqs. (9.33) and (9.34), we see that the ratios of the cross sections are predicted to be

$$\begin{aligned} &\sigma(\pi^- p \to \Sigma^{*-}K^+)/\sigma(\pi^- p \to \Delta^-\pi^+)/\sigma(K^- p \to \Xi^{*-}K^+)/\sigma(K^- p \to \Sigma^{*-}\pi^+) \\ &\quad = 1/3/1/1. \end{aligned} \tag{9.35}$$

If U-spin is conserved, these ratios should hold at all energies and at all angles. However, because the masses of the different members of a given U-spin multiplet may be very different, there will surely be a substantial correction near threshold. One might even do the experiment at an energy such that one process is allowed, whereas another is strictly forbidden by conservation of energy. Thus, because of the large amount of U-spin symmetry breaking, we should not expect Eq. (9.35) to hold very well. We can make a kinematic correction for the different masses involved in the reactions (9.32) by modifying Eq. (9.35) for the available phase space. However, there should also be significant dynamical corrections. We should expect (9.35) to hold best when the energy and momentum transfer are large compared to the mass splitting.

We can also assume V-spin conservation to obtain relations among cross sections. However, we do not obtain any useful relations which are independent of those obtained from isospin and U-spin conservation.

Conservation of U-spin also enables us to discuss photo-induced reactions. We consider the following two reactions:

$$\begin{aligned} \gamma + p &\to \Delta^0 + \pi^+, \\ \gamma + p &\to \Sigma^{*0} + K^+. \end{aligned} \tag{9.36}$$

If the electromagnetic interaction conserves U-spin, then a photon should have a definite U-spin multiplicity. Since there is only one photon, the photon must be a U-spin singlet.[4] We also know that the proton is one member of a U-spin doublet. Therefore, conservation of U-spin implies that the initial state of Eq. (9.36) has $U = \frac{1}{2}$. Therefore, assuming U-spin is conserved in strong interactions, the final states $\Delta^0\pi^+$ and $\Sigma^{*0} K^+$ should also have $U = \frac{1}{2}$. We see that both reactions are governed by a single amplitude since Δ^0 and Σ^{*0} both belong to a U-spin multiplet as do π^+ and K^+. The relevant $SU(2)$ Clebsch–Gordan coefficients appear in the following equation for the final state function:

$$\psi_{1/2}^{(1/2)} = \sqrt{\tfrac{2}{3}}\,\Delta^0\pi^+ - \sqrt{\tfrac{1}{3}}\,\Sigma^{*0}K^+. \tag{9.37}$$

Thus, the ratio of the cross sections is predicted to be

$$\sigma(\gamma p \to \Delta^0\pi^+)/\sigma(\gamma p \to \Sigma^* K^{0+}) = 2. \tag{9.38}$$

Again because of symmetry-breaking, this relation should hold best at large energies and large momentum transfers.

Now let us obtain a relation among electromagnetic decays from U-spin invariance. Let us consider as an example the decays

$$\pi^0 \to 2\gamma, \qquad \eta \to 2\gamma. \tag{9.39}$$

Since photons are U-spin singlets, we must construct a linear combination of π^0 and η which is a singlet. This has already been done for Σ and Λ, and the result is given in Eq. (9.15). Analogously, the U-spin singlet combination of π^0 and η is

$$\varphi = \tfrac{1}{2}(\sqrt{3}\,\pi^0 - \eta)$$

We thus obtain for the ratio of decay rates

$$\Gamma(\pi^0 \to 2\gamma)/\Gamma(\eta \to 2\gamma) = 3. \tag{9.40}$$

Equation (9.40) must be wrong as it stands, since there is much more available phase space for the η than for the π to decay into two photons. A more meaningful test of the theory would be to multiply both decay rates by the relative phase space and then compare these adjusted decay rates. We leave this as an exercise for the interested reader.

[4] We are referring only to internal degrees of freedom. Of course a photon has two spin orientations.

9.6 Gell-Mann–Okubo Mass Formula

Since $SU(3)$ is only an approximate symmetry, the problem arises as to how to identify the physical states belonging together in a single $SU(3)$ multiplet. This problem also exists for $SU(2)$, since isospin is also only approximately conserved.

However, it appears that isospin symmetry is broken only by the electromagnetic and weak interactions, and that therefore $SU(2)$ symmetry-breaking is relatively minor. Typically, the different members of an isospin multiplet with baryon number $|B| \leq 1$ have masses which differ by less than 10 MeV, a number which is small compared to the mass difference between different isospin multiplets of the same spin and parity. Therefore, we usually have no difficulty in identifying the different members of an isospin multiplet.

Even for isospin, however, the problem of identification of states is not always trivial. In heavy nuclei, for example, the Coulomb interaction is sufficiently large that states which are members of the same isospin multiplet may differ in energy by an amount which is much larger than the average spacing of energy levels in a nucleus. The recognition of the states belonging to a given isospin multiplet (called analog states) can be a major problem under these circumstances.

Returning to $SU(3)$, we find that the identification of states can be quite difficult, since the mass splitting within a multiplet may be several hundred MeV. Typically, this is comparable to the mass differences between multiplets. However, it turns out that the symmetry *breaking* of the masses exhibits a pattern in some cases, and we can use this pattern to help identify the multiplets.

Let us assume that the symmetry-breaking interaction, which is responsible for the mass splitting, although a strong interaction, is significantly weaker than the $SU(3)$-invariant interaction so that we may use perturbation theory in computing its effects.

The simplest assumption that leads to a splitting of the masses is that the mass operator M contains a term which is invariant under $SU(3)$ plus a term which is linear in the $SU(3)$ generators. If we neglect electromagnetic mass splitting, all members of an isospin multiplet have the same mass. Of the $SU(3)$ generators, only F_8 or the hypercharge Y does not change the third component of the isospin. Therefore, we write for the mass operator,

$$M = a + bY, \tag{9.41}$$

where a and b are constants within a given $SU(3)$ multiplet.

According to Eq. (9.41), there is a constant mass splitting among members of the decuplet, that is

$$M_{\Omega} - M_{\Xi^*} = M_{\Xi^*} - M_{\Sigma^*} = M_{\Sigma^*} - M_{\Delta}. \tag{9.42}$$

There is some difficulty in comparing this prediction with experiment, because of the electromagnetic mass differences among the members of the decuplet. It is plausible to assume that we can minimize the effect of the electromagnetic interaction if we consider the members of the decuplet which belong to a given U-spin multiplet. We therefore compare the masses of the negatively charged members of the decuplet, which belong to a U-spin quartet. We then have, using the experimental values of the masses from Table 9.3,

$$\begin{aligned} M_{\Omega^-} - M_{\Xi^{*-}} &= 138 \pm 2 \quad \text{MeV}, \\ M_{\Xi^{*-}} - M_{\Sigma^{*-}} &= 148 \pm 4 \quad \text{MeV}, \\ M_{\Sigma^{*-}} - M_{\Delta^-} &= 142 \pm 9 \quad \text{MeV}. \end{aligned} \tag{9.43}$$

The agreement between theory and experiment is rather good.

However, Eq. (9.41) fails for the baryon octet, as it does not break the mass degeneracy between the Λ and Σ hyperons. Experimentally

$$M_{\Sigma^0} - M_\Lambda = 76.9 \pm 0.1 \quad \text{MeV}.$$

The next simplest assumption is to add quadratic functions of the generators to the mass operator. We can add a term proportional to the Casimir operator of the $SU(2)$ isospin group and a term proportional to Y^2 without breaking isospin conservation. We then obtain

$$M = a + bY + cI(I+1) + dY^2, \tag{9.44}$$

where c and d are constant for an $SU(3)$ multiplet. For arbitrary c and d, Eq. (9.44) will not give a constant mass splitting for members of the decuplet and so will ruin the good agreement with experiment. However, we can obtain a relation between c and d by requiring that the constant mass splitting still hold. We require for members of the decuplet

$$cI(I+1) + dY^2 = x + yY, \tag{9.45}$$

where x and y are constants. Substituting the values of I and Y of the Δ, Σ^*, Ξ^*, and Ω in Eq. (9.45) and eliminating x and y, we obtain two equations relating c and d. They both give

$$d = -\tfrac{1}{4}c.$$

Then Eq. (9.44) becomes

$$M = a + bY + c[I(I+1) - \tfrac{1}{4}Y^2]. \tag{9.46}$$

This is the Gell-Man–Okubo mass formula, which was derived by Okubo (1962) by requiring that the mass operator transform like the eighth component of an octet vector. This formula leads to Eq. (9.42) for the decuplet masses. For the octet masses, Eq. (9.46) leads to the relation

$$\tfrac{1}{2}(M_N + M_\Xi) = \tfrac{3}{4}M_\Lambda + \tfrac{1}{4}M_\Sigma. \tag{9.47}$$

Let us compare Eq. (9.47) with experiment. In this case we cannot compare all members of the same U-spin multiplet, as the octet does not contain a U-spin quartet. We shall arbitrarily use the masses of the neutral members of the octet, hoping in this way to minimize the effect of the electromagnetic mass splitting. Using the experimental values of the masses from Table 9.2, we obtain

$$\begin{aligned} \tfrac{1}{2}M_n + \tfrac{1}{2}M_{\Xi^0} &= 1127.1 \pm 0.7 \quad \text{MeV} \\ \tfrac{3}{4}M_\Lambda + \tfrac{1}{4}M_{\Sigma^0} &= 1134.8 \pm 0.2 \quad \text{MeV}. \end{aligned} \tag{9.48}$$

The difference between the two values is 7.7 ± 0.7 MeV, and thus, there is a definite disagreement between theory and experiment. However, the discrepancy is small when compared to a baryon mass or even to a typical mass splitting, the latter being of order 100 MeV.

A more elegant way of deriving the Gell-Mann–Okubo mass formula is to assume, following Okubo (1962), that the mass operator M consists of two terms, one of which M_0 is a scalar under $SU(3)$ transformations, while the second M_8 transforms like the eighth component of an $SU(3)$ octet. The mass M_h of a hadron is then taken to be the expectation value of M. Let us evaluate this expectation value:

$$M_h = (\psi_m^{(j)}, M\psi_m^{(j)}) = (\psi_m^{(j)}, M_0\,\psi_m^{(j)}) + (\psi_m^{(j)}, M_8\,\psi_m^{(j)}).$$

Since M_8 is a vector, we can make use of Eq. (8.97) to write

$$(\psi_m^{(j)}, M_8\,\psi_m^{(j)}) + C_F(j)(\psi_m^{(j)}, F_8\,\psi_m^{(j)}) + C_D(j)(\psi_m^{(h)}, D_8\,\psi_m^{(j)}).$$

In order to simplify this expression further, we use the definition of D_8 given in Eq. (6.19):

$$D_8 = \tfrac{2}{3}\sum_{\alpha\beta} d_{8\alpha\beta}F_\alpha F_\beta.$$

Using the values of $d_{8\alpha\beta}$ from Table 6.2 and the definitions

$$I^2 = \sum_{i=1}^{3} F_i^2, \qquad Y = (2/\sqrt{3})F_8,$$

we obtain

$$D_8 = (I^2 - \tfrac{1}{3}F^2 - \tfrac{1}{4}Y^2)/\sqrt{3}.$$

Thus, we have written D_8 in terms of diagonal operators. Since M_0 and F_8 are already diagonal, we can write the expression for M_h in the form

$$M_h = a_j + b_j\,Y + c_j[I(I+1) - \tfrac{1}{4}Y^2]$$

where a_j, b_j, and c_j are constants for a given multiplet, here denoted by j. The values of these constants depend on the unknown matrix elements $(\psi_m^{(j)}, M_0\,\psi_m^{(j)})$ and the unknown constants $C_F(j)$ and $C_D(j)$, and so must be regarded as a set of parameters for each multiplet.

This derivation of the Gell-Mann–Okubo mass formula depends on the assumption that the symmetry-breaking term in the mass operator transforms like the eighth component of an octet. Since the formula is not exact, there must be another symmetry-breaking term in the mass operator (unless perturbation theory breaks down). But for some reason, the part that transforms like an octet is enhanced. The fact of *octet enhancement* appears to play a role in many other aspects of elementary particle physics, for example in the electromagnetic and weak interactions of hadrons.

If the assumptions underlying the Gell-Mann–Okubo mass formula are correct, then a formula analogous to Eq. (9.47) should also hold for meson octets. In applying this formula to mesons, it is customary to use the squares of the masses rather than the masses themselves. The custom is equivalent to the assumption that the M of Eq. (9.46) is the mass-squared operator when applied to mesons.

It is sometimes stated that mesons and baryons should be treated differently because a free meson may propagate according to the Kein–Gordon equation (which is quadratic in the mass), while a free baryon may propagate according to a Dirac equation (which is linear in the mass). Other reasons have also been given for using squares of masses for mesons. None of the reasons given seems compelling, but we shall adhere to the custom.

Let us neglect electromagnetic mass differences among the mesons in the following discussion. If we consider a meson octet, we note that it contains two isospin doublets, one with hypercharge $Y = 1$ and the other with $Y = -1$. Both these doublets have the same mass since they are antiparticles of each other.

If we let the mass of the isospin doublet, triplet, and singlet mesons be M_2, M_3, and M_1, the Gell-Mann–Okubo octet mass formula becomes

$$M_2{}^2 = \tfrac{1}{4}M_3{}^2 + \tfrac{3}{4}M_1{}^2. \tag{9.49}$$

We wish to compare this formula with the experimental masses of the pseudoscalar, vector, and tensor meson octets. There is a complication in doing so, however. As we have already discussed in Section 9.2, we find experimentally nine mesons of a given spin and parity, rather than eight. Two of the mesons of a given spin and parity have quantum numbers $I = Y = 0$. If $SU(3)$ is a good classification scheme, one of these should belong to an octet and the other to a singlet, but the question is which meson is which.

Since the Gell-Mann–Okubo mass formula works rather well for baryons, it is perhaps reasonable to use it to decide which of the two $I = Y = 0$ mesons belongs to the octet. The procedure we shall follow is to use the experimental values of the masses of the isospin doublet and triplet in Eq. (9.49) to predict the value of the mass of the isospin singlet. We shall then compare this prediction to the observed values of the masses of the two isospin singlets. We obtain the results shown in Table 9.10.

TABLE 9.10

COMPARISON OF GELL-MANN–OKUBO OCTET MASS FORMULA WITH EXPERIMENTAL MASSES OF MESONS

J^P	M_1 (MeV) predicted	Observed mesons and masses (MeV)		$SU(3)$ mixing angle
0^-	560	$\eta(549)$	$\eta'(958)$ or $E(1422)$	10° or 6°
1^-	930	$\varphi(1019)$	$\omega(783)$	40°
2^+	1460	$f'(1514)$	$f(1264)$	30°

We see that the values of the mass of the isospin singlet calculated from the Gell-Mann–Okubo mass formula do not agree with any of the experimental masses. If a linear mass formula is used, instead of a quadratic formula, the disagreement with experiment only increases.

However, it can be seen from Table 9.10 that in every case the predicted mass lies *between* the values of the masses of two mesons with the right quantum numbers. Because of this fact, we interpret the meson multiplets as nonets rather than as octets plus singlets. We assume that the $SU(3)$ symmetry-breaking interaction mixes the isosinglet member of the octet with the $SU(3)$ singlet. A mixing angle θ can be defined in the following way: Let M_1 be the mass of the isosinglet member of the octet as predicted by the Gell-Mann–Okubo mass formula, and m and M be the masses of the two physically observed isosinglet mesons. Then we define θ by the following formula

$$M_1{}^2 = m^2 \cos^2 \theta + M^2 \sin^2 \theta, \tag{9.50}$$

where m and M are identified so that $\theta \leq 45°$. The values of the mixing angle for the pseudoscalar, vector, and tensor mesons are given in Table 9.10. Because the mixing angle is so small in the pseudoscalar case, these mesons are often classified as an octet and a singlet. The vector and tensor mesons, however, are almost always classified as nonets.

It is clear that by introducing the angle θ we are able to obtain agreement with the experimental masses. However, at first sight it is not clear what is the use of such a fit, since we do not make any prediction about a relationship among the meson masses. Only if we can use the angle θ to make predictions about other processes is the idea of $SU(3)$ mixing useful. It turns out that we can indeed make a prediction about the decay modes of the φ and ω vector mesons, using $SU(3)$ mixing. We shall briefly discuss this question in Section 11.3 within the framework of the quark model.

Note that if a nonet is subsequently discovered with masses such that M_1 does not lie between m and M, Eq. (9.50) will still hold, but the angle θ will be imaginary. In such a case, the usefulness of the concept of $SU(3)$ mixing will be seriously diminished.

9.7 Meson–Baryon Coupling

Let us consider the coupling of a pseudoscalar meson to a baryon–antibaryon pair, assuming that the baryon, antibaryon, and meson all belong to octets. Then, to obtain an interaction which is invariant under $SU(3)$ transformations, we must couple $\mathbf{8} \otimes \mathbf{8}$ to give $\mathbf{8}$. As we have previously noted, there are two ways to do this, using F and D coupling.

We can use the Clebsch–Gordan coefficients of $SU(3)$ to obtain the relative coupling strengths in a straightforward way. However, we shall adopt a different procedure, in which we write the baryon and meson octets as 3×3 matrices instead of as eight-dimensional vectors. These matrices are

$$B = \begin{pmatrix} \Sigma^0/\sqrt{2} + \Lambda/\sqrt{6} & \Sigma^+ & p \\ \Sigma^- & -\Sigma^0/\sqrt{2} + \Lambda/\sqrt{6} & n \\ \Xi^- & \Xi^0 & -2\Lambda/\sqrt{6} \end{pmatrix}, \tag{9.51}$$

$$\bar{B} = \begin{pmatrix} \bar{\Sigma}^0/\sqrt{2} + \bar{\Lambda}/\sqrt{6} & \bar{\Sigma}^- & \bar{\Xi}^- \\ \bar{\Sigma}^+ & -\bar{\Sigma}^0/\sqrt{2} + \bar{\Lambda}/\sqrt{6} & \bar{\Xi}^0 \\ \bar{p} & \bar{n} & -2\bar{\Lambda}/\sqrt{6} \end{pmatrix}, \tag{9.52}$$

$$M = \begin{pmatrix} \pi^0/\sqrt{2} + \eta/\sqrt{6} & \pi^+ & K^+ \\ \pi^- & -\pi^0/\sqrt{2} + \eta/\sqrt{6} & K^0 \\ K^- & \bar{K}^0 & -2\eta/\sqrt{6} \end{pmatrix}. \tag{9.53}$$

The above matrices were originally written down by Ne'eman (1961). Our normalization differs from that of Ne'eman by a factor $\sqrt{2}$. In terms of these matrices, the F-type coupling C_F is given by a scalar formed with the antisymmetric combination of $\bar{B}$ and B, namely

$$C_F = -\sqrt{2}\,\mathrm{Tr}([\bar{B}, B]M), \tag{9.54}$$

where the minus sign and the $\sqrt{2}$ are inserted to conform to the convention of Gell-Mann (1961). On the other hand, the D-type coupling C_D is given by a scalar formed with the symmetric combination

$$C_D = \sqrt{2}\,\mathrm{Tr}(\{\bar{B}, B\}M). \tag{9.55}$$

Performing the matrix multiplication and carrying out the traces, we obtain

$$\begin{aligned}
C_F = {} & (\bar{p}p - \bar{n}n + 2\overline{\Sigma}{}^+\Sigma^+ - 2\overline{\Sigma}{}^-\Sigma^- + \overline{\Xi}{}^0\Xi^0 - \overline{\Xi}{}^-\Xi^-)\pi^0 \\
& + (\sqrt{2}\,\bar{p}n - \sqrt{2}\,\overline{\Xi}{}^0\Xi^- + 2\overline{\Sigma}{}^0\Sigma^- - 2\overline{\Sigma}{}^+\Sigma^0)\pi^+ \\
& + (\sqrt{2}\,\bar{n}p - \sqrt{2}\,\overline{\Xi}{}^-\Xi^0 + 2\overline{\Sigma}{}^-\Sigma^0 - 2\overline{\Sigma}{}^0\Sigma^+)\pi^- \\
& + \sqrt{3}(\bar{p}p + \bar{n}n + \overline{\Xi}{}^0\Xi^0 - \overline{\Xi}{}^-\Xi^-)\eta \\
& + (-\sqrt{3}\,\bar{p}\Lambda + \sqrt{3}\,\overline{\Lambda}\Xi^- - \bar{p}\Sigma^0 - \sqrt{2}\,\bar{n}\Sigma^- + \overline{\Sigma}{}^0\Xi^- + \sqrt{2}\,\overline{\Sigma}{}^+\Xi^0)K^+ \\
& + (-\sqrt{3}\,\overline{\Lambda}p + \sqrt{3}\,\overline{\Xi}{}^-\Lambda - \overline{\Sigma}{}^0 p - \sqrt{2}\,\overline{\Sigma}{}^- n + \overline{\Xi}{}^-\Sigma^0 + \sqrt{2}\,\overline{\Xi}{}^0\Sigma^+)K^- \\
& + (-\sqrt{3}\,\bar{n}\Lambda + \sqrt{3}\,\overline{\Lambda}\Xi^0 + \bar{n}\Sigma^0 - \sqrt{2}\,\bar{p}\Sigma^+ - \overline{\Sigma}{}^0\Xi^0 + \sqrt{2}\,\overline{\Sigma}{}^-\Xi^-)K^0 \\
& + (-\sqrt{3}\,\overline{\Lambda}n + \sqrt{3}\,\overline{\Xi}{}^0\Lambda + \overline{\Sigma}{}^0 n - \sqrt{2}\,\overline{\Sigma}{}^+ p - \overline{\Xi}{}^0\Sigma^0 + \sqrt{2}\,\overline{\Xi}{}^-\Sigma^-)\overline{K}{}^0
\end{aligned} \tag{9.56}$$

$$\begin{aligned}
C_D = {} & \{\bar{p}p - \bar{n}n + 2(\overline{\Sigma}{}^0\Lambda + \overline{\Lambda}\Sigma^0)/\sqrt{3} - \overline{\Xi}{}^0\Xi^0 + \overline{\Xi}{}^-\Xi^-\}\pi^0 \\
& + \{\sqrt{2}\,\bar{p}n + 2(\overline{\Sigma}{}^+\Lambda + \overline{\Lambda}\Sigma^-)/\sqrt{3} + \sqrt{2}\,\overline{\Xi}{}^0\Xi^-\}\pi^+ \\
& + \{\sqrt{2}\,\bar{n}p + 2(\overline{\Lambda}\Sigma^+ + \overline{\Sigma}{}^-\Lambda)/\sqrt{3} + \sqrt{2}\,\overline{\Xi}{}^0\Xi^0\}\pi^- \\
& + \{-\bar{p}p - \bar{n}n + 2(\overline{\Sigma}{}^+\Sigma^+ + \overline{\Sigma}{}^0\Sigma^0 + \overline{\Sigma}{}^-\Sigma^- - \overline{\Lambda}\Lambda) - \overline{\Xi}{}^0\Xi^0 - \overline{\Xi}{}^-\Xi^-\}\eta/\sqrt{3} \\
& + \{-\bar{p}\Lambda/\sqrt{3} + \bar{p}\Sigma^0 + \sqrt{2}\,\bar{n}\Sigma^- - \overline{\Lambda}\Xi^-/\sqrt{3} + \overline{\Sigma}{}^0\Xi^- + \sqrt{2}\,\overline{\Sigma}{}^+\Xi^0\}K^+ \\
& + \{-\overline{\Lambda}p/\sqrt{3} + \overline{\Sigma}{}^0 p + \sqrt{2}\,\overline{\Sigma}{}^- n - \overline{\Xi}{}^-\Lambda/\sqrt{3} + \overline{\Xi}{}^-\Sigma^0 + \sqrt{2}\,\overline{\Xi}{}^0\Sigma^+\}K^- \\
& + \{-\bar{n}\Lambda/\sqrt{3} - \bar{n}\Sigma^0 + \sqrt{2}\,\bar{p}\Sigma^+ - \overline{\Lambda}\Xi^0/\sqrt{3} - \overline{\Sigma}{}^0\Xi^0 + \sqrt{2}\,\overline{\Sigma}{}^-\Xi^-\}K^0 \\
& + \{-\overline{\Lambda}n/\sqrt{3} - \overline{\Sigma}{}^0 n + \sqrt{2}\,\overline{\Sigma}{}^+ p - \Xi^0\Lambda/\sqrt{3} - \Xi^0\Sigma^0 + \sqrt{2}\,\overline{\Xi}{}^-\Sigma^-\}\overline{K}{}^0.
\end{aligned} \tag{9.57}$$

In writing down the antibaryon–baryon–meson ($\overline{B}BM$) couplings C_D and C_F, we have included only the dependence on $SU(3)$ variables. In general, there will also be a dependence on spatial and spin variables as well.

In terms of C_D and C_F, a general $SU(3)$-invariant coupling C is given by

$$C = g[\alpha C_D + (1 - \alpha)C_F], \tag{9.58}$$

where g and α are constants. This expression is subject to the restriction that $\bar{p}p\pi^0$ coupling strength is g.

The quantity $\alpha/(1 - \alpha)$ is called the D/F ratio and cannot be determined from $SU(3)$-invariance alone. This ratio is important, not only in $\overline{B}BM$ coupling, but in any situation in which three octets are combined to form a scalar (or alternatively, two octets are coupled to form a third).

Because of the existence of two types of $SU(3)$-invariant couplings, the strength of the coupling of the hyperons to mesons, relative to that of nucleons to mesons, is not fixed. For example, in C_F there is no $\overline{\Lambda}\Sigma\pi$ coupling, while in C_D there is no $\overline{\Sigma}\Sigma\pi$ coupling.

We can use the matrix technique to obtain trilinear meson couplings as well. Let us use the symbol V for a vector meson octet and M for a pseudoscalar meson octet. There is no strong MMM coupling because such a coupling does not conserve parity. The strong MMV coupling must be pure F-type and the VVM coupling pure D-type because of the requirement of charge-conjugation invariance in strong interactions.

9.8 Hadron Decays

All the unstable members of the baryon octet decay into other baryons of the octet and ultimately into the proton plus leptons or photons. If $SU(3)$ were exact, all members of the octet would be mass degenerate and could not decay into other members of the same multiplet. Thus, since the baryon octet is the baryon multiplet of smallest average mass, all the octet decays are forbidden by $SU(3)$ invariance. However, this is not considered a difficulty for $SU(3)$, since no member of the baryon octet decays via strong interactions. In fact, all baryon octet decays proceed via the weak interaction except for the Σ^0 decay which goes via the electromagnetic interaction.

Although the members of the baryon octet cannot decay without $SU(3)$ violation, the members of the decuplet are allowed, in principle, to decay strongly in an $SU(3)$-conserving interaction. For example, $SU(3)$–invariance does not forbid the decay

$$D \to B + M,$$

where D is a member of the baryon decuplet, B a member of the baryon octet, and M a member of the meson pseudoscalar octet. This decay is allowed because **10** is contained in the Kronecker product of $\mathbf{8} \otimes \mathbf{8}$. Assuming $SU(3)$ invariance holds, then all decays $D \to B + M$ are governed by one amplitude. This is because $\mathbf{8} \otimes \mathbf{8}$ contains **10** only once. Thus, to obtain predictions concerning the ratio of decay rates of different members of the decuplet which are based on $SU(3)$ invariance, we need merely make use of the relevant Clebsch–Gordan coefficients of $SU(3)$.

In fact, the Δ, Σ^*, and Ξ^* of the decuplet are observed to decay strongly into a baryon and meson. Nevertheless, the decays violate $SU(3)$ invariance, at least in part because of mass differences among different members of a given $SU(3)$ multiplet. For example, the decay

$$\Sigma^* \to \Lambda + \pi$$

is allowed, but the decay

$$\Sigma^* \to N + \overline{K}$$

is forbidden by energy conservation.

As a first approximation, we may assume that the decay matrix elements are $SU(3)$ invariant, but that the kinematic or phase space factors break the symmetry. Such an assumption is in qualitative agreement with experiment in many cases.

On the other hand, the decay ratios $B^* \to B + M$, where B^* is an excited baryon octet, cannot be obtained from Clebsch–Gordan coefficients alone. This is because **8** appears twice in $\mathbf{8} \otimes \mathbf{8}$, once with F-type coupling and once with D-type coupling. One must know the D/F ratio to be able to calculate the relative decay rates. However, one can use the measured rates for two decays to determine the D/F ratio experimentally, and then predict other decay rates. This has been done, for example, by Glashow and Rosenfeld (1963).

An analogous situation occurs for the mesons. Because the pseudoscalar octet is the $SU(3)$ multiplet of lowest mass, the members of this octet are fordibben to decay; and in fact, all the decays of the octet go via electromagnetic or weak interactions. The vector meson octet, on the other hand, is allowed to decay into two pseudoscalar mesons without violating $SU(3)$ invariance, and the coupling is pure F-type. Such decays are observed to occur via the strong interactions, but again the mass splittings break the symmetry.

9.9 Weak Hadron Decays

The weak interactions are certainly not invariant under $SU(3)$ transformations; in fact they are not even invariant under the isospin $SU(2)$ subgroup. However, as has been pointed out by Gell-Mann (1962, 1964) and by Cabibbo (1963), there appear to be certain regularities in the way that $SU(3)$ is violated in the decays of hadrons.

We shall not discuss the weak interactions in any detail, but shall make a few remarks concerning the form of the interaction and its $SU(3)$-symmetry properties. The CP-conserving part of the weak interaction H_w appears to be fairly well described phenomenologically by a current–current interaction of the form

$$H_w = G \int J_\alpha^\dagger J_\alpha \, d^3x/\sqrt{2}, \tag{9.59}$$

where J_α is a "weak current," G is a coupling constant, and the $\sqrt{2}$ is inserted to conform to the conventional normalization. The current J_α is assumed to be composed of two parts, a leptonic part $J_\alpha^{(l)}$ and a hadronic part $J_\alpha^{(h)}$. We are here interested in the $SU(3)$-transformation properties of the hadronic current $J_\alpha^{(h)}$.

Before discussing the form of $J_\alpha^{(h)}$, we consider an analogous problem for the electromagnetic current j_α. The time component j_t of the four-vector

j_α is the charge density, and its integral over three-space is the total charge Q:

$$Q = \int j_t \, d^3x. \tag{9.60}$$

As we have stated earlier, the electric charge of a hadron, in units of the proton charge, is related to the hypercharge Y and the third component of the isospin I_3 by the Gell-Mann–Nishijima formula

$$Q = I_3 + \tfrac{1}{2}Y.$$

But I_3 and Y are given in terms of the $SU(3)$ generators F_3 and F_8 by

$$I_3 = F_3, \qquad Y = 2F_8/\sqrt{3}.$$

Therefore, the charge transforms as a component of an octet.

$$Q = F_3 + F_8/\sqrt{3}. \tag{9.61}$$

We now return to the weak hadronic current $J_\alpha^{(h)}$. This must contain an isospin-raising operator, since in the decay $n \to p + e^- + \bar{\nu}$, a neutron is converted into a proton. Likewise, $J_h^{(\alpha)}$ must contain a hypercharge (or strangeness) changing operator, since in the decay $\Lambda \to p + e^- + \bar{\nu}$, a Λ is converted into a proton. The isospin-raising operator can be written in terms of the $SU(3)$ generators as $F_1 + iF_2$, and the hypercharge-changing operator as $F_4 + iF_5$. The coupling strengths of these two parts are not equal, and the relative coupling is giving in terms of the Cabibbo angle θ. Putting all these ideas together we write for $J_\alpha^{(h)}$:

$$J_\alpha^{(h)} = \cos\theta(F_1 + iF_2) + \sin\theta(F_4 + iF_5). \tag{9.62}$$

Thus $J_\alpha^{(h)}$ can be considered as a component of an $SU(3)$ vector $J_{\mu\alpha}^{(h)}$, where the subscript μ refers to the $SU(3)$ index.

Now, just as we have written the electric charge as a spatial integral of j_t, we write the weak hadronic charge as a spatial integral of $J_{\mu t}^{(h)}$

$$\int J_{\mu t}^{(h)} \, d^3x = V_\mu(t) + A_\mu(t). \tag{9.63}$$

We have divided the expression for the weak charge into two terms, one of which V_μ is the time component of a vector, and the other A_μ, which is the time component of an axial vector. Both these terms can exist because parity is not conserved in weak interactions.

Gell-Mann has assumed the equal-time commutators of the V_μ and A_μ are

$$\begin{aligned} [V_\mu, V_\nu] &= if_{\mu\nu\kappa} V_\kappa, \\ [V_\mu, A_\nu] &= if_{\mu\nu\kappa} A_\kappa, \\ [A_\mu, A_\nu] &= if_{\mu\nu\kappa} V_\kappa. \end{aligned} \tag{9.64}$$

These equal-time commutation relations are the principal hypothesis of current algebra. They can be correct even if $SU(3)$ is badly broken. We can introduce two linear combinations of the V_μ and A_μ

$$2F_\mu^{\pm} = V_\mu \pm iA_\mu . \tag{9.65}$$

Then it can be directly verified that F_μ^+ and F_μ^- transform like two commuting F-spins, so that the current algebra is the algebra of the group $SU(3) \times SU(3)$. Gell-Mann (1964) has also pointed out a connection between current algebra and the group $SU(6)$.

A further discussion of weak interactions is outside the scope of this book. For further information about the CP-conserving interaction, see, for example, Adler and Dashen (1968) and Bernstein (1968). Little is known about the part of the weak interaction which does not conserve CP. For a phenomenological description, see for example, Wu and Yang (1964) and Lee and Wu (1966).

CHAPTER 10

APPROXIMATE $SU(6)$

10.1 Dynamical Symmetry

A number of years ago Wigner (1937) considered the group $SU(4)$ as an approximate symmetry of nuclei. This symmetry can come about as follows: The nucleon is a doublet of the $SU(2)$ isospin group and also a doublet of the $SU(2)$ ordinary spin group. Thus, the nucleon has four degrees of freedom, and can be considered to belong to a quartet representation of the direct-product group $SU(2) \times SU(2)$. This direct product is contained in $SU(4)$. If the forces between nucleons are approximately independent of spin and isospin, then $SU(4)$ is an approximate symmetry of nuclei. Under such circumstances, the nucleon can be considered to belong to the first fundamental (four-dimensional) representation of $SU(4)$.

However, unlike the isospin $SU(2)$ group, which may be an approximate symmetry no matter how high the energy, $SU(4)$ is at best a dynamical symmetry which cannot be valid at high energy. We see this as follows: If $SU(4)$ is a valid symmetry, spin and isospin indices are indistinguishable and can be transformed into each other. Now we perform a Lorentz boost on a system with $SU(4)$ symmetry. Such a boost mixes spin and orbital angular momentum. However, the $SU(2)$ isospin group is an internal symmetry group and cannot be mixed with orbital angular momentum by a boost. Therefore a boost distinguishes between spin and isospin and destroys $SU(4)$ invariance.

If we have a free massive single particle, we can always consider it in its rest system, in which case $SU(4)$ might be a good symmetry. Likewise, if we have two particles moving very slowly with respect to each other, $SU(4)$ might hold to a good approximation. But if we have two particles moving rapidly with respect to each other. There is no frame of reference in which $SU(4)$ can hold. Thus, $SU(4)$ can at best be a dynamical symmetry which holds in the nonrelativistic limit.

These same ideas have been generalized to the case of elementary particles by Gürsey and Radicati (1964) and by Sakita (1964). The symmetry as applied to elementary particles differs from the case of nuclei in two respects. In the first place the internal symmetry subgroup of elementary particles is $SU(3)$ rather than $SU(2)$ so that the direct-product group of the internal symmetry group and the spin group is $SU(3) \times SU(2)$. Then, if the interaction is approximately independent of spin and unitary spin, $SU(6)$ is an approximate dynamical symmetry which holds best at low energies. As we know, $SU(3)$ is a simple rank 2 group with I_3 and Y as the additive quantum numbers. Also $SU(2)$ is a simple rank 1 group with the third component of the angular momentum J_3 being the additive quantum number. The direct-product group $SU(2) \times SU(3)$ is therefore a semisimple rank 3 group. It should be obvious that the ranks of direct-product groups add. However, this direct product is only a subgroup of $SU(6)$, since $SU(6)$ is rank 5 and therefore has five additive quantum numbers. However, at the present time, we do not know of any useful physical interpretation of the two additive quantum numbers of $SU(6)$ other than those contained in $SU(3) \times SU(2)$.

A second difference between the $SU(4)$ of nuclear physics and the $SU(6)$ of elementary particle physics is that in the latter case the usual baryons and mesons do not belong to the fundamental six-dimensional representation of $SU(6)$. Of course this statement is just the $SU(6)$ analog of the statement that the usual hadrons do not belong to the fundamental three-dimensional representation of $SU(3)$. Since ordinary baryons and mesons belong only to zero-triality representations of $SU(3)$, we shall focus most of our attention in this chapter on representations of $SU(6)$ which contain zero-triality representations of $SU(3)$ as subgroups. However we shall also consider the fundamental six-dimensional representation, as this is the building block out of which all representations can be constructed.

Since spin is contained as a subgroup of $SU(6)$, we expect to find supermultiplets of $SU(6)$ containing states of different spin. In general, this is true, but the $SU(6)$ singlet contains only spin zero, and the $SU(6)$ sextet contains only spin $\frac{1}{2}$.

10.2 Classification of Hadrons

We can use Young tableaux to decide which multiplets of $SU(6)$ need to be considered so as to include only $SU(3)$ multiplets of zero triality. Recall that in Chapter 7 we defined an integer k which gives the class of a representation of $SU(n)$ specified by a particular Young tableau. The integer k is just the number of boxes of the tableau, modulo n. Let us confine our attention to representations of class $k = 0$, in which the number of boxes is an integral multiple of n. This class includes the so-called integral representations of $SU(2)$ and the zero-triality representations of $SU(3)$.

Now consider the $SU(n_1) \times SU(n_2)$ subgroup of $SU(n)$, where $n = n_1 n_2$. Suppose we wish to consider representations of $SU(n)$ which contain only $k = 0$ multiplets of $SU(n_1)$. Then we must consider only representations which have Young diagrams with ν boxes, where ν is an integral multiple of n_1. The proof follows from the fact that any Young diagram with ν boxes is contained in the Kronecker product of a single box with itself ν times. Now a single box of $SU(n)$ corresponds to a multiplet of n states. It also corresponds to the lowest dimensional multiplet of $SU(n_1) \times SU(n_2)$, which also has n states since $n = n_1 n_2$. Thus, a single box of $SU(n)$ contains the lowest-dimensional multiplet of $SU(n_1)$ only once. (If $n_1 = n_2$, we regard the two $SU(n_1)$ groups as distinct, and concentrate on the first of them.) Since this is so, a single box tableau of $SU(n)$ serves also as a single box of $SU(n_1)$. Then the Kronecker product of ν of them contains only multiplets of class $k = 0$, since by assumption ν is a multiple of n_1. This completes the proof.

Applying the theorem to $SU(6)$, we see that we must consider diagrams with 0, 3, 6, 9, ... boxes to insure that we obtain only $SU(3)$ multiplets of zero triality. We can also apply the theorem to the $SU(2)$ subgroup of $SU(6)$. Then we see that the $SU(2)$ multiplets with $k = 0$ (integral spin) must correspond to an even number of boxes. Since mesons have integral spin and zero triality, they must belong to multiplets with 0, 6, 12, ... boxes. Furthermore, since meson multiplets contain their own antiparticles, mesons must belong to self-conjugate diagrams of $SU(6)$. Baryons, on the other hand, with half-integral spin, cannot belong to multiplets with an even number of boxes and therefore must belong to multiplets with 3, 9, 15, ... boxes.

Let us first consider the baryons. It is simplest to consider $SU(6)$ multiplets corresponding to Young tableaux with only three boxes rather than to tableaux with 9, 15, or more boxes. There are only the following three possible arrangements of three boxes

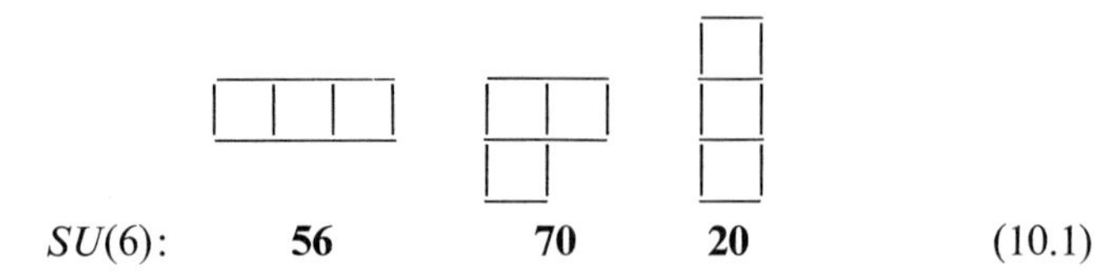

(10.1)

where we have also given the dimensionality of the corresponding representations of $SU(6)$. We see that the 56-dimensional representation is completely symmetric, the 20-dimensional representation is antisymmetric, and the 70-dimensional representation has mixed symmetry.

In order to classify the baryons according to $SU(6)$, we need to know which submultiplets of $SU(3) \times SU(2)$ are contained in a given multiplet of $SU(6)$. We call this the $SU(3) \times SU(2)$ *content* of the multiplet. To obtain the content of the three-box tableaux of $SU(6)$ we look at the three-box tableaux of the $SU(3)$ and $SU(2)$ subgroups. The possible $SU(3)$ tableaux are

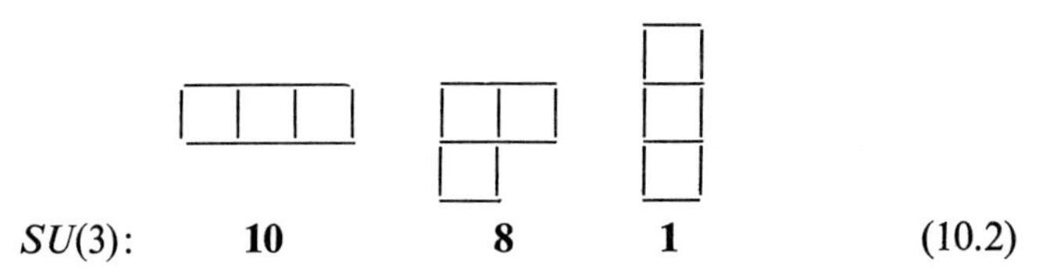

(10.2)

and the possible $SU(2)$ tableaux are

$SU(2)$: **4** **2** (10.3)

There is no antisymmetric $SU(2)$ tableau with three boxes, since $SU(2)$ has only two degrees of freedom.

Let us denote a multiplet of $SU(3) \times SU(2)$ by the symbol ${}^{2J+1}N$, where $2J + 1$ is the $SU(2)$ multiplicity and N is the $SU(3)$ multiplicity. We see from (10.2) and (10.3) that the only possible multiplets of $SU(3) \times SU(2)$ corresponding to three-box tableaux are the following

$$ {}^{4}10, \quad {}^{2}10, \quad {}^{4}8, \quad {}^{2}8, \quad {}^{4}1, \quad {}^{2}1. \tag{10.4} $$

The multiplet ${}^{4}10$ is completely symmetric in both $SU(3)$ and $SU(2)$ indices, and therefore is totally symmetric (see Table 10.1). Thus, the ${}^{4}10$ multiplet

TABLE 10.1

SYMMETRY PROPERTIES OF PRODUCTS (OR SUMS OF PRODUCTS) OF TWO FACTORS WHICH ARE BASIS FUNCTIONS OF THE SYMMETRIC GROUP S_3

Factors	Product
SS or AA[a]	S
SA	A
SM or AM	M
MM	S, A, or M

[a] The symbols S, A, and M stand for symmetric, antisymmetric, and mixed symmetry, respectively.

must belong to the symmetric **56** of $SU(6)$. The $^{4}1$ is antisymmetric in $SU(3)$ indices and symmetric in $SU(2)$ indices, so that it is overall antisymmetric. Therefore it must belong to the **20** of $SU(6)$. Likewise, we see that the $^{2}10$, $^{2}1$, and $^{4}8$, each being the product of a symmetric or antisymmetric tensor with a mixed tensor, must have mixed overall symmetry and belong in the **70**. This leaves only the $^{2}8$, which is the product of two mixed tensors, unclassified. But the product of two mixed tensors can be overall symmetric, antisymmetric, or mixed. Therefore, a $^{2}8$ can go into a **56**, a **70**, or a **20** of $SU(6)$. We thus have

$$\mathbf{56} \supset {}^{4}10 + {}^{2}8, \tag{10.5}$$

$$\mathbf{70} \supset {}^{2}10 + {}^{2}1 + {}^{4}8 + {}^{2}8, \tag{10.6}$$

$$\mathbf{20} \supset {}^{4}1 + {}^{2}8. \tag{10.7}$$

We can check these assignments by counting states. For example, the $^{4}10$ contains 40 states and the $^{2}8$ contains 16 for a total of 56.

Since, experimentally, the baryon octet is an $SU(2)$ doublet and the baryon decuplet is an $SU(3)$ quartet, both these multiplets can be combined into the **56** supermultiplet of $SU(6)$. The justification for combining the octet and decuplet into the **56** is not compelling, since both the **70** and the **20** of $SU(6)$ contain an octet of spin $\frac{1}{2}$. However, if $SU(6)$ is an approximate symmetry, the decuplet must be accompanied by an octet, and the observed baryon octet is a good candidate.

Since parity is a quantum number outside $SU(6)$, it is essential that all members of a supermultiplet have the same parity. Since the baryon octet and decuplet are both positive-parity multiplets, their assignment to the **56** is allowed.

We next consider classification of the mesons. For simplicity we restrict ourselves to multiplets with tableaux of zero and six boxes, rather than considering multiplets with 12 or more boxes. The tableau of zero boxes is an $SU(6)$ singlet which contains only an $SU(3)$ singlet of spin zero. The $\eta'(960)$ and $\eta'(1420)$ (or E meson) are likely candidates for this singlet. Of more interest are the tableaux with six boxes. The only self-conjugate ones are

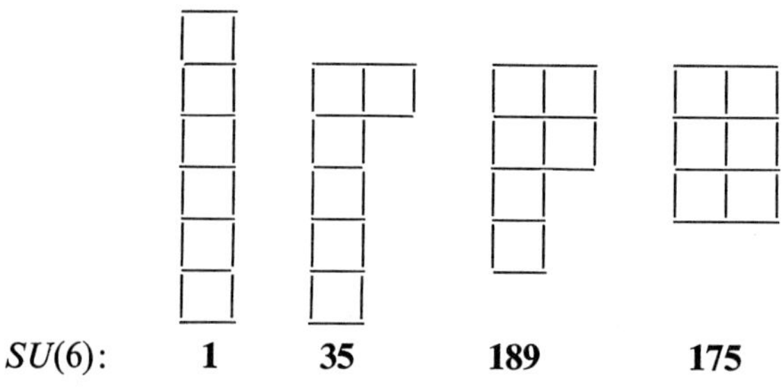

The singlet we have already considered. The tableau of next lowest multiplicity is the **35**. Let us obtain the $SU(3) \times SU(2)$ multiplets contained in this $SU(6)$ supermultiplet. This is most easily done by considering the Kronecker product of the first fundamental representation representation of $SU(6)$ and its conjugate. Using Eq. (7.38), we have

$$\mathbf{6} \otimes \bar{\mathbf{6}} = \mathbf{35} \oplus \mathbf{1}. \tag{10.8}$$

Making use of the fact that

$$\mathbf{6} \supset {}^2 3,\ \bar{6} \supset {}^2\bar{3}, \tag{10.9}$$

we obtain

$${}^2 3 \times {}^2\bar{3} = {}^{(2\times 2)}(3 \times \bar{3}) = {}^{(3+1)}(8 + 1) = {}^3 8 + {}^1 8 + {}^3 1 + {}^1 1. \tag{10.10}$$

Obviously the singlet of $SU(3) \times SU(2)$ must also go in the singlet of $SU(6)$, since the other representations of $SU(3) \times SU(2)$ have higher dimensions. Thus the 24-dimensional, eight-dimensional, and three-dimensional representations of $SU(3) \times SU(2)$ must all belong in the 35 representation of $SU(6)$:

$$35 \supset {}^3 8 + {}^1 8 + {}^3 1. \tag{10.11}$$

The arithmetic works out right: $24 + 8 + 3 = 35$.

If we look at the meson states of lowest mass, these belong to an octet and singlet with $J^P = 0^-$ and a mixed octet and singlet (or nonet) with $J^P = 1^-$. The pseudoscalar octet and the vector meson octet and singlet fill up the 35 supermultiplet nicely.

Note that these meson states all have negative parity. If the vector meson octet, the pseudoscalar octet, or the vector meson singlet had positive parity, $SU(6)$ would be violated but not $SU(2) \times SU(3)$. Thus, there is more predictive power in $SU(6)$ than in $SU(2) \times SU(3)$.

If $SU(6)$ is an approximate dynamical symmetry, we perhaps have a clue toward understanding why the vector meson octet and singlet are more mixed than the pseudoscalar octet and singlet. The vector octet and singlet are both in the same supermultiplet of $SU(6)$ while the pseudoscalar octet and singlet are in different supermultiplets. Therefore, the vector mesons are more nearly degenerate in energy than the pseudoscalar mesons, and the $SU(3)$ mixing interaction has a chance to be more effective in the vector meson case.

Thus, $SU(6)$ provides a good framework for classifying the lowest mass baryons and mesons into supermultiplets. In classifying the states of higher mass, it is useful to consider the multiplets of a still larger group: $SU(6) \times O(3)$. However, it is most convenient to consider this group within the framework of the quark model, and so we postpone these considerations to the next chapter.

A number of other predictions may be made using $SU(6)$. One of the most successful is the prediction that the ratio of the magnetic moment of proton to neutron is given by (Bég *et al.*, 1964)

$$\mu(p)/\mu(n) = -3/2. \tag{10.12}$$

This compares to the experimental value

$$\mu(p)/\mu(n) = -1.46. \tag{10.13}$$

The prediction follows from $SU(6)$ and the assumption that the electric charge transforms like a component of an $SU(3)$ octet operator. We shall derive Eq. (10.12) in Chapter 11, using the quark model.

10.3 Matrix Generators of $SU(6)$

In Section 8.3, we obtained a scheme for obtaining the $n^2 - 1$ generators of $SU(n)$ as $n \times n$ matrices. In fact we were able to write down the matrix elements of all the $n^2 - 1$ generators of $SU(n)$. However, this is only one representation for these generators, and it is not the most convenient one for our purposes. The representation of Section 8.3 is convenient for obtaining all the $SU(n-1)$ multiplets contained in a multiplet of $SU(n)$. It is a scheme which is good for the following hierarchy of symmetries:

$$SU(n) \supset SU(n-1) \supset SU(n-2) \supset \cdots \supset SU(2). \tag{10.14}$$

However, in considering $SU(6)$, we wish to emphasize the subgroup $SU(3) \times SU(2)$. Thus we want to construct the generators of $SU(6)$ from the generators of $SU(3) \times SU(2)$.

Let us begin with $SU(4)$ as an example. We shall construct the generators of $SU(4)$ from the generators of the $SU(2)$ isospin group, the generators of the $SU(2)$ spin group, and the unit matrices in these spaces. One representation for the isospin generators is by the 2×2 matrices τ_1, τ_2, τ_3. We add to these the 2×2 unit matrix τ_0. The same representation for spin is by the σ matrices, to which we add σ_0. Now consider the product generators $\varphi_{\mu\nu}$ given by

$$\varphi_{\mu\nu} = \sigma_\mu \tau_\nu, \qquad \mu, \nu = 1, \ldots, 4. \tag{10.15}$$

The 16 matrices $\varphi_{\mu\nu}$ are the generators of $U(4)$. If we eliminate the unit matrix $\sigma_0 \tau_0$, the remaining 15 are the generators of $SU(4)$. We can, of course, obtain another representation by using $\tau_+, \tau_-, \sigma_+,$ and σ_- instead of $\tau_1, \tau_2, \sigma_1,$ and σ_2.

In a similar fashion we can construct the generators of $SU(6)$, but instead of using the τ_μ we use the λ_μ of Gell-Mann. We define $\beta_{\mu\nu}$ to

$$\beta_{\mu\nu} = \sigma_\mu \lambda_\nu, \qquad \mu = 0, 1, \ldots, 4, \quad \nu = 0, 1, \ldots, 9 \tag{10.16}$$

Then the 36 $\beta_{\mu\nu}$ are the generators of $U(6)$. Only the unit generator β_{00} has a trace different from zero. This element generates $U(1)$. The remaining 35 $\beta_{\mu\nu}$ are the generators of $SU(6)$. These generators are written in a product notation. However, we can obtain a 6×6 representation of these matrices by taking the outer product of the 3×3 with the 2×2 matrices.

The matrix elements of the *outer product* of two matrices A and B are defined in the following way:

$$(A \times B)_{ij,kl} = A_{ik} B_{jl}. \tag{10.17}$$

We can order these matrix elements in a square array as follows:

$$A \times B = \begin{pmatrix} A_{11}B_{11} & A_{11}B_{12} & \cdots & A_{11}B_{1m} & A_{12}B_{11} & \cdots & A_{1n}B_{1m} \\ A_{11}B_{21} & A_{11}B_{22} & \cdots & A_{11}B_{2m} & & & \vdots \\ \vdots & & & & & & \\ A_{11}B_{m1} & & & & & & \\ A_{21}B_{11} & & & & & & \\ \vdots & & & & & & \vdots \\ A_{n1}B_{m1} & \cdots & & & & \cdots & A_{nn}B_{mm} \end{pmatrix}. \tag{10.18}$$

In other words, each matrix element of the matrix A is multiplied on the right by the entire matrix B:

$$A \times B = \begin{pmatrix} A_{11}B & A_{12}B & \cdots & A_{1n}B \\ \vdots & & & \vdots \\ A_{n1}B & \cdots & \cdots & A_{nn}B \end{pmatrix},$$

where

$$A_{ij}B = A_{ij}\begin{pmatrix} B_{11} & B_{12} & \cdots & B_{1m} \\ \vdots & & & \vdots \\ B_{m1} & & \cdots & B_{mm} \end{pmatrix}.$$

An alternative definition which leads to a different representation is that each matrix element of B is multiplied on the left by the entire matrix A. We shall stick with the first definition. For example, the matrix $\beta_{18} = \sigma_1 \lambda_8$ written as an outer product is given by

$$\beta_{18} = \sqrt{\frac{1}{3}}\left(\begin{array}{ccc|ccc} & & & 1 & 0 & 0 \\ & 0 & & 0 & 1 & 0 \\ & & & 0 & 0 & -2 \\ \hline 1 & 0 & 0 & & & \\ 0 & 1 & 0 & & 0 & \\ 0 & 0 & -2 & & & \end{array}\right).$$

There are many other useful representations for the generators of $SU(6)$. For example, instead of using the nondiagonal λ_μ and σ_ν, we can use the raising and lowering operators of $SU(3)$ and $SU(2)$. Then the 30 nondiagonal $SU(6)$ generators are just the e_{ij} of $SU(6)$ with matrix elements

$$(e_{ij})_{\alpha\beta} = \delta_{i\alpha}\,\delta_{j\beta}, \tag{10.19}$$

which satisfy

$$e_{ij}^\dagger = e_{ji}. \tag{10.20}$$

The five diagonal generators in this representation are the same as in the representation by the $\beta_{\mu\nu}$ matrices. This representation of the generators is most useful for obtaining the Clebsch–Gordan coefficients of $SU(6)$. The six states of $SU(6)$ on which the generators act have the following interpretation: The states $u_1\ u_2\ u_3$ of the first fundamental representation are the states with the same weights as the three states of $SU(3)$ and also with $J_3 = \frac{1}{2}$, while the states $u_4\ u_5\ u_6$ correspond to the $SU(3)$ states with $J_3 = -\frac{1}{2}$.

At first glance it is surprising that we have all the generators of $SU(6)$ from just the generators of $SU(3) \times SU(2)$. After all, the group $SU(3) \times SU(2)$ has 11 generators, whereas the group $SU(6)$ has 35 generators. Yet there appears to be no difference between $SU(3) \times SU(2)$ and $SU(6)$ when acting on a six-dimensional state vector. We conclude that for a single *free* particle with six degrees of freedom there is no difference between $SU(3) \times SU(2)$ and $SU(6)$.

Generalizing, the state vector of a free particle with n degrees of freedom can always be considered as a basis function of the fundamental representation of $SU(n)$. It is only when we consider the interactions between two or more such particles that there is a difference between, for example, $SU(n)$ and $SU(n_1) \times SU(n_2)$, where $n = n_1 n_2$.

10.4 Troubles with $SU(6)$

Using $SU(6)$, we can make a number of predictions about hadron–hadron scattering. For example, consider meson–baryon scattering

$$M + B \to M + B, \tag{10.21}$$

where M is a meson belonging to the **35** and B is a baryon belonging to the **56**. To see how many independent amplitudes are necessary to describe this process, we decompose the Kronecker product by the method of Young tableaux. We obtain

$$\mathbf{56} \otimes \mathbf{35} = \mathbf{56} \oplus \mathbf{70} \oplus \mathbf{700} \oplus \mathbf{1134}. \tag{10.22}$$

We see that all scatterings of the form (10.21) (and there are many such processes accessible to experiment) are predicted to be given in terms of only four amplitudes. The relations between the various amplitudes can be obtained by using the $SU(6)$ Clebsch–Gordan coefficients.

Unfortunately, however, there is no reason to expect these $SU(6)$ predictions to agree with experiment. We can see this as follows: Since $SU(3)$ is a subgroup of $SU(6)$, whenever $SU(3)$ is badly broken, $SU(6)$ will also fail to give good predictions. Now at low energy, the mass splitting of the $SU(3)$ multiplets will lead to large violations of $SU(3)$ symmetry. An extreme example is the case in which the energy is sufficiently low so that some reactions are forbidden by energy conservation, while other reactions related by $SU(3)$ symmetry are allowed. So $SU(6)$ is expected to give poor results at low energy. But in Section 10.1 we argued that $SU(6)$ cannot be a good symmetry at high energy. Therefore, we cannot expect the predictions of $SU(6)$ for hadron–hadron scattering to be good at any energy.

The situation with respect to the strong decays of unstable hadrons is even worse. Since the $SU(2)$ subgroup of $SU(6)$ is the spin group, spin and orbital angular momentum are separately conserved in any interaction which is invariant under $SU(6)$.

Suppose we consider the decay of the ρ meson, which has spin 1. Experimentally, the ρ is observed to decay strongly into two pions with a decay width Γ of more than 100 MeV. However, pions have spin 0. Thus, in the decay, the spin of the ρ is converted into orbital angular momentum of the pions. But nonrelativistic $SU(6)$ has no mechanism to convert spin into orbital angular momentum, and the strong $\rho \to 2\pi$ decay is forbidden by the symmetry. Similarly the Δ baryon has spin $\frac{3}{2}$ and is observed to decay strongly into a nucleon and a pion with total spin $\frac{1}{2}$. Again the extra unit of spin is converted into orbital angular momentum, with the pion and nucleon emerging in a state of relative orbital angular momentum $L = 1$. But the nonrelativistic $SU(6)$ again forbids this decay. We conclude that although $SU(6)$ may be useful in the classification of states, it certainly does not lead to the correct selection rules for decays. One should regard $SU(6)$ as holding at best in the rest frame of a particle. A particle in its rest frame has no orbital angular momentum, and so for such states $SU(6)$ has a chance to be good.

Because of this difficulty, and others related to the nonrelativistic character of $SU(6)$, numerous attempts have been made to find a relativistic generalization. Most of these attempts have involved consideration of noncompact groups. Some of the defects of $SU(6)$ have in fact been overcome in this way, but only at the expense of other difficulties of principle. One possible exception is the group $SU(6)_W$, introduced by Lipkin and Meshkov (1965). This group allows the decays $\rho \to 2\pi$ and $\Delta \to N\pi$ to occur. Furthermore, it gives a number of moderately successful predictions concerning relative hadron decay rates and hadron–hadron scattering cross sections in the forward and backward directions. Pais (1966), in his excellent review, discusses the many attempts to make relativistic generalizations of $SU(6)$, and gives many references.

In Section 9.1, we assumed that the Hamiltonian H describing elementary particles consists of two parts H_0 and H', where H_0 is invariant under the direct product group $SU(3) \times P$ (where P is the Poincaré group) and H' breaks the symmetry. Unfortunately, we cannot extend this idea to $SU(6)$, because we do not have any physical interpretation of the direct-product group $SU(6) \times P$. If a group is to be a direct product, all the generators of one of the groups of the product must commute with all the generators of the other. But if the $SU(2)$ contained in $SU(6)$ is interpreted as the spin group, its generators do not commute with the angular momentum generators of P. This is just a reflection of the fact that the total angular momentum is conserved, but not spin and orbital angular momentum separately.

Since $SU(6) \times P$ does not make sense physically, a number of people have considered the possibility of embedding $SU(6)$ and P in a group G which is not a direct product. But as we have already remarked in Section 9.1, O'Raifeartaigh (1968) and others have shown that such a scheme does not work if G is a finite Lie group.

An approach which we favor is to recognize that $SU(6)$ is an approximate dynamical symmetry, and therefore cannot be a symmetry of the strong interaction Hamiltonian. Rather, we assume that the Hamiltonian is of such a form that when certain static properties of hadrons are calculated with it, the results are the same as those given by approximate $SU(6)$ invariance.

In the next chapter, we shall discuss a model which does not incorporate $SU(6)$ invariance, but nevertheless can be used to obtain some of the good predictions of $SU(6)$ without giving its bad predictions. This the quark model.

CHAPTER 11

THE QUARK MODEL

11.1 Sakata Triplets

Some time ago Fermi and Yang (1949) proposed a model in which a pion is a bound state of a nucleon and antinucleon. This model can be generalized to include a large number of other mesons with zero hypercharge. In this scheme, the nucleon isospin doublet and the antinucleon doublet are the fundamental building blocks out of which meson states of zero hypercharge are constructed.

We can fit this model into the framework of group theory. The state vectors of the nucleon

$$u_1 = p, \qquad u_2 = n, \tag{11.1}$$

can be considered as a fundamental doublet of $U(2)$, where the $SU(2)$ subgroup is the isospin and the $U(1)$ subgroup is the baryon number. (In this case $I = \frac{1}{2}$, $B = 1$.) Likewise the antinucleon state vectors

$$v_1 = \bar{n}, \qquad v_2 = -\bar{p} \tag{11.2}$$

can be considered to be a second fundamental doublet of $U(2)$, again with $I = \frac{1}{2}$, but with baryon number $B = -1$. The meson isospin multiplicities in this model are obtained from the Clebsch–Gordan series of two doublets:

$$\mathbf{2} \otimes \overline{\mathbf{2}} = \mathbf{3} \oplus \mathbf{1} \tag{11.3}$$

We distinguish between **2** and $\bar{\mathbf{2}}$ because the group is $U(2)$, not $SU(2)$. However, the multiplicities of the Clebsch–Gordan series are the same for **2** and $\bar{\mathbf{2}}$.

We see from Eq. (11.3) that the model leads to a prediction that all mesons with $Y=0$ should be either isospin triplets or singlets. This prediction is in agreement with the known facts up to the present except for the possible occurrence of relatively weakly coupled exotic mesons.

The model can be generalized to include excited baryon states with hypercharge $Y=1$. Such excited baryons are considered as bound states of two nucleons and an antinucleon. The Clebsch–Gordan series is

$$\mathbf{2} \otimes \mathbf{2} \otimes \bar{\mathbf{2}} = \mathbf{2} \oplus \mathbf{2} \oplus \mathbf{4}. \tag{11.4}$$

Thus, the model predicts that all baryons of hypercharge $Y=1$ should be isospin doublets and quartets. This prediction is also in accord with the presently known facts except possibly for weakly coupled exotic baryons.

Recall that the strangeness S of a hadron is defined by $S = Y - B$. Use of the quantum number S rather than Y lets us state the predictions of the model succinctly: The nucleon and antinucleon doublets are the building blocks of all hadrons of zero strangeness.

This scheme, while attractive in many ways, is insufficient to account for particles with strangeness different from zero. Sakata (1956) suggested the simplest extension of the Fermi–Yang model to include strangeness different from zero. Sakata's idea was to add one additional state to the fundamental building blocks: namely the Λ hyperon, which has the quantum numbers $I=0$, $B=1$, $S=-1$. Sakata postulated that all hadrons can be constructed from the triplet p, n, Λ, and their antiparticles.

Ikeda *et al.* (1959) took the next step by regarding the proton, neutron, and Λ to be a fundamental triplet of $U(3)$. As with the Fermi–Yang model, a meson is given by a bound state of a baryon and antibaryon. The relevant Clebsch–Gordan series is

$$\mathbf{3} \otimes \bar{\mathbf{3}} = \mathbf{8} \oplus \mathbf{1}. \tag{11.5}$$

Thus, the Sakata model leads to the successful prediction that mesons should occur in octets and singlets. Furthermore, if the interaction between baryon and antibaryon depends only weakly on F-spin (unitary spin), the octet and singlet will not differ appreciably in mass. With the assumption that this is the case, the model predicts that mesons should occur in multiplets of nine. Thus, within the framework of the model, we can obtain a dynamical explanation for the existence of nonets.

Although the Sakata model provides a good description of meson multiplicities, it leads to difficulties with the baryons. The reason is as follows: Since according to the model, the p, n, and Λ fill the fundamental triplet, the Σ and Ξ must belong to another multiplet. We can form other multiplets with

$B = 1$ from two Sakata triplets and an antitriplet. We can readily obtain the Clebsch–Gordan series with Young tableaux. We have

$$
\begin{array}{|c|}\hline \; \\ \hline \end{array} \times \begin{array}{|c|}\hline 1 \\ \hline \end{array} \times \begin{array}{|c|}\hline 1 \\ \hline 2 \\ \hline \end{array} = \left(\begin{array}{|c|c|}\hline \; & 1 \\ \hline \end{array} + \begin{array}{|c|}\hline \; \\ \hline 1 \\ \hline \end{array} \right) \times \begin{array}{|c|}\hline 1 \\ \hline 2 \\ \hline \end{array}
$$

$$
= \left(\begin{array}{|c|c|c|}\hline \; & \; & 1 \\ \hline 2 \\ \cline{1-1} \end{array} + \begin{array}{|c|c|}\hline \; & \; \\ \hline 1 \\ \cline{1-1} 2 \\ \cline{1-1} \end{array} \right) + \begin{array}{|c|c|}\hline \; & 1 \\ \hline \; \\ \cline{1-1} 2 \\ \cline{1-1} \end{array} + \begin{array}{|c|c|}\hline \; & 1 \\ \hline \; & 2 \\ \hline \end{array}
$$

or, removing columns of three, we obtain

$$
\begin{array}{|c|}\hline \; \\ \hline \end{array} \times \begin{array}{|c|}\hline \; \\ \hline \end{array} \times \begin{array}{|c|}\hline \; \\ \hline \; \\ \hline \end{array} = \begin{array}{|c|c|c|}\hline \; & \; & \; \\ \hline \; \\ \cline{1-1} \end{array} + \begin{array}{|c|}\hline \; \\ \hline \end{array} + \begin{array}{|c|}\hline \; \\ \hline \end{array} + \begin{array}{|c|c|}\hline \; & \; \\ \hline \; & \; \\ \hline \end{array}
$$

Then we have

$$
\mathbf{3} \otimes \mathbf{3} \otimes \bar{\mathbf{3}} = \mathbf{15} \oplus \mathbf{3} \oplus \mathbf{3} \oplus \bar{\mathbf{6}}. \tag{11.6}
$$

In the Sakata model, we must decompose the Kronecker product of $\mathbf{3} \otimes \mathbf{3} \otimes \bar{\mathbf{3}}$ because we want states with baryon number $B = 1$. In the model, each $\mathbf{3}$ has $B = 1$ and $\bar{\mathbf{3}}$ has $B = -1$. Then, since baryon number is an additive quantum number, from $\mathbf{3} \otimes \mathbf{3} \otimes \bar{\mathbf{3}}$ we get states with $B = 1$. From Eq. (11.6), we should expect to find baryon multiplets belonging to the $\mathbf{15}$, $\bar{\mathbf{6}}$, or $\mathbf{3}$ multiplets of $U(3)$. The Σ and Ξ must belong to a 15 with $J^P = \frac{1}{2}^+$ in this scheme, as the $\bar{\mathbf{6}}$ and $\mathbf{3}$ do not have the right quantum numbers. But the other members of this multiplet seem to be missing.

Similarly the Δ must belong to a 15 with $J^P = \frac{3}{2}^+$, since there is no $I = \frac{3}{2}$ state in the $\bar{\mathbf{6}}$ or $\mathbf{3}$ representations. But the $\mathbf{15}$ contains, in addition to these multiplets an $I = 1$ multiplet of strangeness $S = 1$ which has not been observed. Another difficulty is that there is no place for the Ω in the $\mathbf{15}$, $\bar{\mathbf{6}}$, or $\mathbf{3}$ representations. We shall not discuss the other difficulties with this baryon classification scheme except to note that there seems no way to obtain good agreement with experiment.

11.2 Properties of Quarks

It was because of the failure of the Sakata model to give the correct baryon multiplicities that Gell-Mann and Ne'eman put the nucleon and Λ into an octet together with the Σ and Ξ. But if the p, n, and Λ do not belong to a fundamental triplet of $U(3)$, perhaps other, still undiscovered particles do. Gell-Mann (1964a) and Zweig (1964) independently proposed a model in which baryons and mesons are composites of a fundamental triplet of $U(3)$.[1] Gell-Mann called these particles "quarks"; Zweig called them "aces."

Many physicists have searched for quarks. Jones (1969) has given a description, with references, of many of the attempts to find quarks at accelerators, in seawater, in rocks, and in cosmic rays. McCusker and Cairns (1969) claim to have observed quarks in very energetic cosmic-ray showers. As of the time of this writing, many physicists feel that additional experimental work is needed to check the findings of McCusker and Cairns. However, independently of whether quarks exist, the quark model is useful as a tool enabling us to obtain many results in a simple way. We shall consider the model in this spirit.

It is convenient to consider the group $SU(3)$ rather than $U(3)$, and to treat the baryon number separately. Consider the decomposition of the Kronecker product $\mathbf{3} \otimes \mathbf{3} \otimes \mathbf{3}$. Using Young tableaux, we obtain

$$\square \times \square \times \square = \left(\boxed{}\boxed{1} + \begin{array}{c} \boxed{} \\ \boxed{1} \end{array} \right) \times \boxed{1}$$

or

$$\square \times \square \times \square = \boxed{}\boxed{}\boxed{} + \begin{array}{l} \boxed{}\boxed{} \\ \boxed{} \end{array} + \begin{array}{l} \boxed{}\boxed{} \\ \boxed{} \end{array} + \begin{array}{c} \boxed{} \\ \boxed{} \\ \boxed{} \end{array} \tag{11.7}$$

This reduction holds for any $SU(n)$ with $n > 2$. In particular, the $SU(3)$ multiplicities are

$$\mathbf{3} \otimes \mathbf{3} \otimes \mathbf{3} = \mathbf{10} \oplus \mathbf{8} \oplus \mathbf{8} \oplus \mathbf{1}. \tag{11.8}$$

We see from Eq. (11.8) that if we regard a baryon as made up of three triplets, rather than two triplets and an antitriplet, we obtain a baryon decuplet, two octets, and a singlet. Baryon multiplicities of 10, 8, and 1 have been observed experimentally. Because of the inequivalence of the $\mathbf{3}$ and $\bar{\mathbf{3}}$ representations of $SU(3)$, the decomposition of $\mathbf{3} \otimes \mathbf{3} \otimes \mathbf{3}$ is quite different from the decomposition of $\mathbf{3} \otimes \mathbf{3} \otimes \bar{\mathbf{3}}$. [Compare Eqs. (11.6) and (11.8)].

In the model a meson is made of a quark and antiquark. Thus, the Clebsch–Gordan series for mesons is given in the quark model by Eq. (11.5), just as in the Sakata model. As a consequence, mesons are predicted to occur in octets and singlets, or nonets. These are just the meson multiplets which have been prominently observed thus far.

[1] Actually, Goldberg and Ne'eman (1963) were the first to construct baryons out of a fundamental triplet with baryon number $B = \frac{1}{3}$. Gell-Mann and Zweig first raised the possibility that this triplet might consist of actual particles.

Let us denote the state vectors of the three quarks by the symbols q_1, q_2, and q_3. (Quarks are often denoted by the symbols p, n, and λ, but we are using p and n for the states of the nucleon.)

Let us consider the baryon decuplet, which has the Young tableau $\boxed{\ \ |\ \ |\ \ }$, and is therefore a symmetric state of three quarks. We can make the following identification between the baryons of the decuplet and the standard arrangements of this tableau:

$$\boxed{1|1|1} = \Delta^{++}, \quad \boxed{1|1|2} = \Delta^{+}, \quad \boxed{1|2|2} = \Delta^{\circ}, \quad \boxed{2|2|2} = \Delta^{-}$$
$$\boxed{1|1|3} = \Sigma^{**}, \quad \boxed{1|2|3} = \Sigma^{*0}, \quad \boxed{2|2|3} = \Sigma^{*-}$$
$$\boxed{1|3|3} = \Xi^{*0}, \quad \boxed{2|3|3} = \Xi^{*-} \tag{11.9}$$
$$\boxed{3|3|3} = \Omega^{-}.$$

These basis functions are fully written out in Eq. (8.87) with $u_i \rightarrow q_i$. In particular, we see from Eq. (11.9) that the Δ^{++} is made from three identical quarks of type q_1, the Δ^{-} from three identical quarks of type q_2, and the Ω^{-} from three identical quarks of type q_3. Then, from the values of the additive quantum numbers of Δ^{++}, Δ^{-}, and Ω, we can deduce the corresponding quantum numbers of the quarks, which must have $\frac{1}{3}$ these values. Since the Δ^{++} has $I_3 = \frac{3}{2}$, $Y = 1$, $Q = 2$, the first quark q_1 has $I_3 = \frac{1}{2}$, $Y = \frac{1}{3}$, $Q = \frac{2}{3}$. Similarly, since Δ^{-} has $(I_3\, Y Q) = (-\frac{3}{2}, 1, -1)$ the second quark q_2 has $(I_3\, YQ) = (-\frac{1}{2}, \frac{1}{3}, -\frac{1}{3})$. Likewise, since Ω has $(I_3\, YQ) = (0, -2, -1)$ the third quark q_3 has $(I_3\, YQ) = (0, -\frac{2}{3}, -\frac{1}{3})$. Since all baryons have baryon number $B = 1$, the quarks must have baryon number $B = \frac{1}{3}$.

The spin of a quark must be half integral, as we cannot obtain particles of half-integral spin (baryons) from composites of particles of integral spin. In this chapter we shall make the customary assumption that a quark has spin $\frac{1}{2}$. We shall consider other possibilities in Chapter 12. All quarks must have the same parity, as this quantum number is outside the group. We assume it is positive.

The next property of quarks we wish to discuss is their statistics. As we have remarked in Section 1.5, so far as is known from experiment, the wave functions of identical particles of half-integral spin are antisymmetric under the interchange of any two of them. It is natural to assume that this property is true for quarks as well. However, as we shall see later, baryon wave functions are in some ways simpler if quarks are not fermions.

We next wish to make some assumptions about the masses and magnetic moments of the quarks. We assume that q_1 has mass m, q_2 has mass $m + \varepsilon$, where ε is an electromagnetic mass splitting parameter, and q_3 has mass

$m + \delta$, where δ is a parameter arising from a medium-strong, isospin-conserving interaction which breaks $SU(3)$. We assume the magnetic moments of the quarks are proportional to their charges, with a proportionality constant μ_0. It is also possible to assume that the quark q_3 in view of its heavier mass, has its magnetic moment reduced by the factor $m/(m + \delta)$. With this assumption we can obtain a somewhat better prediction for the magnetic moment of the Λ than with our assumption. However, for simplicity, we shall calculate the baryon magnetic moments assuming the quark moments are strictly proportional to their charges.

We list in Table 11.1 the masses, magnetic moments, and quantum numbers of the three quarks.

TABLE 11.1

QUANTUM NUMBERS, MASSES, AND MAGNETIC MOMENTS OF THE QUARKS[a]

Symbol	Isospin		Hypercharge	Charge	Mass	Magnetic moment
	I	I_3	Y	Q		
q_1	$\frac{1}{2}$	$\frac{1}{2}$	$\frac{1}{3}$	$\frac{2}{3}$	m	$\frac{2}{3}\mu_0$
q_2	$\frac{1}{2}$	$-\frac{1}{2}$	$\frac{1}{3}$	$-\frac{1}{3}$	$m + \varepsilon$	$-\frac{1}{3}\mu_0$
q_3	0	0	$-\frac{2}{3}$	$-\frac{1}{3}$	$m + \delta$	$-\frac{1}{3}\mu_0$

[a] All quarks have baryon number $B = \frac{1}{3}$ and spin and parity $J^P = \frac{1}{2}^+$.

11.3 Baryon and Meson Wave Functions

We can make a number of predictions about the properties of baryons and mesons by making simple assumptions about the properties and interactions of quarks. For example, let us assume that the mass-splitting terms δ and ε, as well as the $SU(3)$ symmetry-breaking interactions, are sufficiently small that perturbation theory can be used in calculating their effects. In order to proceed further, we need expressions for the unperturbed $SU(3)$-invariant functions (or wave functions, as we shall often call them) of baryons and mesons in terms of the quark basis vectors.

From the quark basis vectors, we can construct wave functions of the baryon and meson multiplets, using the methods of Chapter 8. In Table 11.2 we give the decuplet wave functions. These are easily obtained from Eqs. (11.9) and (8.87). Our notation is that the symbol for a particle denotes its wave function.

In Table 11.2 are given only the $SU(3)$ part of the decuplet wave functions. We must multiply these by functions of the spin and spatial variables in order to obtain the complete wave functions.

TABLE 11.2

BARYON DECUPLET WAVE FUNCTIONS CONSTRUCTED FROM THE BASIS VECTORS OF THREE QUARKS

Baryon	Wave function
Δ^{++}	$q_1q_1q_1$
Δ^{+}	$(q_1q_1q_2+q_1q_2q_1+q_2q_1q_1)/\sqrt{3}$
Δ^{0}	$(q_1q_2q_2+q_2q_1q_2+q_2q_2q_1)/\sqrt{3}$
Δ^{-}	$q_2q_2q_2$
Σ^{*+}	$(q_1q_1q_3+q_1q_3q_1+q_3q_1q_1)/\sqrt{3}$
Σ^{*0}	$(q_1q_2q_3+q_1q_3q_2+q_2q_1q_3+q_2q_3q_1+q_3q_1q_2+q_3q_2q_1)/\sqrt{6}$
Σ^{*-}	$(q_2q_2q_3+q_2q_3q_2+q_3q_2q_2)/\sqrt{3}$
Ξ^{*0}	$(q_1q_3q_3+q_3q_1q_3+q_3q_3q_1)/\sqrt{3}$
Ξ^{*-}	$(q_2q_3q_3+q_3q_2q_3+q_3q_3q_2)/\sqrt{3}$
Ω^{-}	$q_3q_3q_3$

We obtain a simple spin wave function if we assume that the orbital angular momentum of the three-quark system is zero. Then, if quarks have spin $\frac{1}{2}$, the spin wave functions of the decuplet $\chi_m^{(3/2)}$ are (for $m=\frac{3}{2}$ and $\frac{1}{2}$)

$$\chi_{3/2}^{(3/2)} = \alpha\alpha\alpha,$$
$$\chi_{1/2}^{(3/2)} = (\alpha\alpha\beta + \alpha\beta\alpha + \beta\alpha\alpha)/\sqrt{3}, \tag{11.10}$$

where α and β denote quark spin functions with third component $+\frac{1}{2}$ and $-\frac{1}{2}$, respectively.

The properties of functions of three variables can be classified according to their behavior under transformations of the symmetric group S_3. We have seen in Chapter 4 that such functions may be symmetric S, antisymmetric A, or have mixed symmetry M. The functions of mixed symmetry are basis functions of two-dimensional irreducible representations of S_3. The decuplet $SU(3)$ functions of Table 11.2 and the spin wave functions of Eq. (11.10) are symmetric functions.

We next consider the spatial wave functions of the baryon decuplet. The symmetry properties of these spatial wave functions will depend on the symmetry properties of the $SU(3)$ and spin wave functions. In Table 10.1 we list the overall symmetry properties of functions which are products (or sums of products) of two functions of definite symmetry properties.

Since quarks have half-integral spin, it is natural to assume that they are fermions, i.e., that their wave functions are antisymmetric under the interchange of all the coordinates of any two of them. But we have constructed decuplet $SU(3)$ and spin state functions which are totally symmetric. Therefore, if quarks are to be fermions, the spatial wave function of the decuplet must be antisymmetric under the interchange of any two quarks.

It is certainly possible to construct an antisymmetric spatial wave function with zero orbital angular momentum. One such choice is

$$\psi(\mathbf{r}_1\mathbf{r}_2\mathbf{r}_3) = (\mathbf{r}_1 \cdot \mathbf{r}_2 \times \mathbf{r}_3 + \mathbf{r}_2 \cdot \mathbf{r}_3 \times \mathbf{r}_1 + \mathbf{r}_3 \cdot \mathbf{r}_1 \times \mathbf{r}_2)\phi(r_1 r_2 r_3) \qquad (11.11)$$

where $\mathbf{r}_1$, $\mathbf{r}_2$, and $\mathbf{r}_3$ are the spatial coordinates of the quarks and ϕ is totally symmetric in r_1, r_2, and r_3. We assume ψ is normalized to unity, i.e., $(\psi, \psi) = 1$. The wave function of Eq. (11.11) is more complicated than a symmetric spatial wave function. Although the total orbital angular momentum of $\psi(\mathbf{r}_1\mathbf{r}_2\mathbf{r}_3)$ is zero, the relative angular momentum of any two quarks is greater than zero. The parity of this wave function is positive.

We next construct the baryon octet and singlet $SU(3)$ wave functions. We see from Eq. (11.8) that two octets can be constructed from the basis vectors of three quarks. Because of this, the Clebsch–Gordan coefficients are not determined from the group properties alone. We shall choose the first octet to be symmetric in the indices of the first two quarks and the second octet to be antisymmetric in these indices. With this choice, the basis functions of the two baryon octets are given in Table 11.3. We can obtain the unitary

TABLE 11.3

BARYON OCTET STATE FUNCTIONS CONSTRUCTED FROM THE STATE VECTORS OF THREE QUARKS[a]

Baryon	State function (octet 1)
p_1	$(2q_1q_1q_2 - q_1q_2q_1 - q_2q_1q_1)/\sqrt{6}$
n_1	$(q_1q_2q_2 + q_2q_1q_2 - 2q_2q_2q_1)/\sqrt{6}$
Σ_1^+	$(q_1q_3q_1 + q_3q_1q_1 - 2q_1q_1q_3)/\sqrt{6}$
Σ_1^0	$(q_1q_3q_2 + q_3q_1q_2 + q_2q_3q_1 + q_3q_2q_1 - 2q_1q_2q_3 - 2q_2q_1q_3)/\sqrt{12}$
Λ_1	$\frac{1}{2}(q_1q_3q_2 + q_3q_1q_2 - q_2q_3q_1 - q_3q_2q_1)$
Σ_1^-	$(q_2q_3q_2 + q_3q_2q_2 - 2q_2q_2q_3)/\sqrt{6}$
Ξ_1^0	$(2q_3q_3q_1 - q_1q_3q_3 - q_3q_1q_3)/\sqrt{6}$
Ξ_1^-	$(2q_3q_3q_2 - q_2q_3q_3 - q_3q_2q_3)/\sqrt{6}$
	State function (octet 2)
p_2	$(q_1q_2q_1 - q_2q_1q_1)/\sqrt{2}$
n_2	$(q_1q_2q_2 - q_2q_1q_2)/\sqrt{2}$
Σ_2^+	$(q_3q_1q_1 - q_1q_3q_1)/\sqrt{2}$
Σ_2^0	$\frac{1}{2}(q_3q_1q_2 - q_1q_3q_2 + q_3q_2q_1 - q_2q_3q_1)$
Λ_2	$(2q_1q_2q_3 - 2q_2q_1q_3 + q_3q_2q_1 - q_2q_3q_1 + q_1q_3q_2 - q_3q_1q_2)/\sqrt{12}$
Σ_2^-	$(q_3q_2q_2 - q_2q_3q_2)/\sqrt{2}$
Ξ_2^0	$(q_3q_1q_3 - q_1q_3q_3)/\sqrt{2}$
Ξ_2^-	$(q_3q_2q_3 - q_2q_3q_3)/\sqrt{2}$

[a] There are two different linearly independent octets, the one denoted by the subscript 1 being symmetric in the indices of the first two quarks and the one denoted by the subscript 2 being antisymmetric in these indices.

singlet baryon wave function by constructing a function orthogonal to Σ^{*0}, $\Sigma_1{}^0$, Λ_1, $\Sigma_2{}^0$, and Λ_2 (see Tables 11.2 and 11.3). This function, which can be identified with the wave function of the $\Lambda(1405)$, is given by

$$\Lambda^* = (q_1q_2q_3 - q_2q_1q_3 + q_2q_3q_1 - q_1q_3q_2 + q_3q_1q_2 - q_3q_2q_1)/\sqrt{6}. \quad (11.12)$$

The members of the baryon octet have spin $\frac{1}{2}$. Just as there are two ways to construct an octet from three $SU(3)$ triplets, there are two ways to construct a spin-$\frac{1}{2}$ wave function from three doublets. The Clebsch–Gordan series is

$$\mathbf{2} \otimes \mathbf{2} \otimes \mathbf{2} = \mathbf{4} \oplus \mathbf{2} \oplus \mathbf{2}.$$

Because of the equivalence of $\mathbf{2}$ and $\bar{\mathbf{2}}$ of $SU(2)$, we obtain the same spin multiplicities as in the Fermi–Yang model [see Eq. (11.4)]. We choose as the two possibilities wave functions χ_m and χ_m' which are symmetric and antisymmetric respectively under the interchange of the spin coordinates of the first two quarks. These functions are given by (for $m = \frac{1}{2}$)

$$\begin{aligned} \chi_{1/2} &= (2\alpha\alpha\beta - \alpha\beta\alpha - \beta\alpha\alpha)/\sqrt{6}, \\ \chi_{1/2}' &= (\alpha\beta\alpha - \beta\alpha\alpha)/\sqrt{2}. \end{aligned} \quad (11.13)$$

We now consider the question of the symmetry properties of the spatial wave functions of the baryon octet. To do this we must first consider the symmetry properties of the $SU(3)$ and spin parts of the baryon wave functions. The $SU(3)$ octet functions of Table 11.3 are neither symmetric nor antisymmetric, but have mixed symmetry. Likewise, the spin $\frac{1}{2}$ wave functions of Eq. (11.13) have mixed symmetry.

From Table 10.1, we see that the products (or sums of products) of two functions of mixed symmetry may be symmetric, antisymmetric, or mixed. We therefore have the problem of deciding in which of these categories to classify the combined baryon $SU(3)$-spin wave functions. We can resolve this question if we accept the $SU(6)$ classification scheme of Chapter 10 in which the baryon octet is most naturally classified in the **56**. If the combined $SU(3)$-spin wave function of a baryon is symmetric and quarks are fermions, then the spatial wave function must be antisymmetric, for example, of the form of Eq. (11.11). In the following sections, and especially in Section 11.6, we shall further discuss the $SU(6)$ classification of baryons within the framework of the quark model.

We construct the meson wave functions from q and $\bar{q}$ states. The octet functions are obtained from Eqs. (8.82), (8.83), and (8.84), using Eq. (8.79). These functions are given in Table 11.4. We have identified the mesons with the pseudoscalar octet, since the vector and tensor mesons are more conveniently classified as nonets because of $SU(3)$ mixing. The unitary singlet

TABLE 11.4

STATE FUNCTIONS OF THE PSEUDOSCALAR MESON OCTET CONSTRUCTED FROM THE STATE VECTORS OF QUARK AND ANTIQUARK

Meson	State function
K^+	$\bar{q}_3 q_1$
K^0	$\bar{q}_3 q_2$
π^+	$\bar{q}_2 q_1$
π^0	$(-\bar{q}_1 q_1 + \bar{q}_2 q_2)/\sqrt{2}$
η	$(-2\bar{q}_3 q_3 + \bar{q}_1 q_1 + \bar{q}_2 q_2)/\sqrt{6}$
π^-	$-\bar{q}_1 q_2$
$\bar{K}^0$	$\bar{q}_2 q_3$
K^-	$-\bar{q}_1 q_3$

function η' is orthogonal to π^0 and η of Table 11.4. It is given by

$$\eta' = (\bar{q}_1 q_1 + \bar{q}_2 q_2 + \bar{q}_3 q_3)/\sqrt{3}. \tag{11.14}$$

Except for the isosinglet state, the $SU(3)$ octet functions of Table 11.4 hold for the vector and tensor mesons as well as for the pseudoscalars. However, the two isospin zero members of the vector or tensor nonets are mixed. Let us consider the vector case. The φ and ω can be written as

$$\begin{aligned} \varphi &= \cos\theta\,(2\bar{q}_3 q_3 - \bar{q}_1 q_1 - \bar{q}_2 q_2)/\sqrt{6} + \sin\theta\,(\bar{q}_1 q_1 + \bar{q}_2 q_2 + \bar{q}_3 q_3)/\sqrt{3}, \\ \omega &= -\sin\theta\,(2\bar{q}_3 q_3 - \bar{q}_1 q_1 - \bar{q}_2 q_2)/\sqrt{6} + \cos\theta\,(\bar{q}_1 q_1 + \bar{q}_2 q_2 + \bar{q}_3 q_3)/\sqrt{3}. \end{aligned} \tag{11.15}$$

If $\sin\theta = 1/\sqrt{3}$, or $\theta \approx 35°$, Eq. (3.15) becomes particularly simple. We obtain

$$\begin{aligned} \varphi &= \bar{q}_3 q_3, \\ \omega &= (\bar{q}_1 q_1 + \bar{q}_2 q_2)/\sqrt{2}. \end{aligned} \tag{11.16}$$

Similarly, with a mixing angle of 35°, the tensor mesons would be given by

$$\begin{aligned} f' &= \bar{q}_3 q_3, \\ f &= (\bar{q}_1 q_1 + \bar{q}_2 q_2)/\sqrt{2}. \end{aligned} \tag{11.17}$$

As we have noted, experimentally $\theta = 40°$ for the vector mesons and $\theta = 30°$ for the tensor mesons. Thus, to a fairly good approximation, the physical states can be represented by the wave functions of Eqs. (11.16) and (11.17).

To the extent that meson decays proceed through the interaction of a single quark, the transition $\bar{q}_3 q_3 \leftrightarrow \bar{q}_1 q_1$ or $\bar{q}_3 q_3 \leftrightarrow \bar{q}_2 q_2$ are forbidden. Then, if the approximation of single-quark interaction is a good one, we can use Eqs. (11.16) and (11.17) to make predictions about the decays of the isosinglet mesons. In particular the φ and f' should decay chiefly into strange mesons, while ω and f should decay principally into nonstrange ones. From Tables 9.7 and 9.8, we can see that this qualitative feature holds. Our conclusion is that the concept of $SU(3)$ mixing is useful and that the mixing angle can be used to make predictions. A further discussion of this topic can be found in Dalitz (1968).

The spins and parities of the meson nonets are accounted for quite naturally by the model. It is convenient to use spectroscopic notation to describe the quantum numbers of a bound state of a quark and antiquark. In this notation the orbital angular momentum is denoted by $S, P, D, F, \ldots$ for $l = 0, 1, 2, 3, \ldots$. The spin multiplicity (either singlet or triplet) is denoted by a left superscript and the total angular momentum is denoted by a right subscript. With this notation the pseudoscalar nonet is most simply classified as a ${}^1S_0$ state of the quark–antiquark system, the vector nonet as a ${}^3S_1$ state, and the tensor nonet as a ${}^3P_2$ state. The parities of these nonets are then correctly given, since the parity P of a fermion–antifermion pair with orbital angular momentum l is given by

$$P = (-1)^{l+1}. \tag{11.18}$$

We next consider the charge-conjugation properties of the $I_3 = Y = 0$ members of the meson nonets within the model. A particle–antiparticle pair with $I_3 = Y = 0$ has charge-conjugation parity C given by

$$C = (-1)^{l+s}, \tag{11.19}$$

where s is the spin of the pair. The proof of Eq. (11.19), which we shall not give, makes use of the usual relation between spin and statistics.

In Section 9.1, we defined a hadron state with $|B| \leq 1$ as *exotic* if it does not belong to a **1**, **8**, or **10** multiplet of $SU(3)$. We now give a definition of an exotic hadron within the framework of the quark model. A baryon is exotic if its quantum numbers are such that it cannot be a bound state of three quarks, and a meson is exotic if its quantum numbers are such that it cannot be a bound state of a quark and antiquark. With this definition all nonexotic baryons belong to **1**, **8**, and **10**, multiplets, and all nonexotic mesons belong to **1** and **8**.

Furthermore, nonexotic mesons with $Y = I = 0$ must also satisfy Eqs. (11.18) and (11.19). Thus, any meson with odd spin and parity and $C = +$ would be considered exotic even if it belonged to a singlet or octet. This is seen as follows: Since P is odd, from Eq. (11.18), l is even. Now, since a quark has spin $\frac{1}{2}$, the only two possibilities for s are $s = 0$ or 1. But, by conservation of angular momentum, if the meson spin is odd and l is even, s must be

1. Therefore, by (11.19) C must be negative, in contradiction to our assumption. Therefore, the meson cannot be made from a quark and antiquark, and so must be exotic. Exotic states can be accomodated within the framework of the quark model if amplitudes including additional quark–antiquark pairs are added to the hadron wave functions. Of course, considerable simplicity is lost in this way. As we have stated in Section 9.1, exotic states have not been produced prominently in experiments performed up to this time.

In the model, the pseudoscalar mesons are assigned $(ls) = (00)$, the vectors (01) and the tensors (11). Thus, the model predicts that the π^0, η, η', $A_2{}^0$, f, and f' have $C = +$, while the ρ, ω, and φ have $C = -$. These predictions are in agreement with experiment.

11.4 Baryon Magnetic Moments

We can use the meson and baryon wave functions obtained in the previous section to make predictions about the properties and interactions of hadrons. In order to obtain some of these predictions we need to make further assumptions about the interactions of quarks. Many authors have applied the quark model to obtain predictions about the hadrons. Many of the references to the original papers are given in reviews by Pais (1966), Dalitz (1967, 1968, 1969), and Morpurgo (1968).

In this section we shall consider the baryon magnetic moments. Some of the results we shall obtain will depend on the overall symmetry properties of the baryon wave functions, while others will depend only on the fact that the baryons belong to an octet. We have assumed in Section 11.2 that the magnetic moment operator $\boldsymbol{\mu}_q$ of a quark is proportional to its charge. Then $\boldsymbol{\mu}_q$ is given by

$$\boldsymbol{\mu}_q = \mu_0 \boldsymbol{\sigma} Q, \tag{11.20}$$

where $\boldsymbol{\sigma}$ and Q are the Pauli spin operator and the charge operator, respectively, of a quark. It is simplest to assume that the magnetic moment operator $\boldsymbol{\mu}$ of a baryon is the sum of the magnetic moment operators of the constituent quarks. Then we have

$$\boldsymbol{\mu} = \mu_0 \sum_i \boldsymbol{\sigma}(i) Q(i), \tag{11.21}$$

where the sum is over the three quarks in a baryon. In writing down this expression, we are assuming in effect that the orbital angular momentum of the quarks is zero and are neglecting any contribution to $\boldsymbol{\mu}$ from relativistic effects such as exchange currents. The value of the magnetic moment μ_B of any baryon is the expectation value of μ_3 (the third or z component of $\boldsymbol{\mu}$)

with respect to a baryon wave function maximally polarized along the z axis, that is

$$\mu_B = \mu_0\Big(B, \sum_i \sigma_3(i)Q(i)B\Big). \tag{11.22}$$

where B denotes a baryon wave function which is maximally polarized in the z direction. It is possible to evaluate μ_B for any baryon once its spin and unitary spin wave functions are specified. As an example, let us suppose that the proton wave function p is given by

$$p = p_2\, \chi'_{1/2}\psi, \tag{11.23}$$

where p_2 is given in Table 11.3, $\chi'_{1/2}$ is given in Eq. (11.13), and the normalized function ψ depends on spatial variables. We can ignore ψ when computing the proton magnetic moment μ_p, since the baryon magnetic moment operator μ_3 acts only on charge and spin variables. We shall explicitly evaluate μ_p using the proton function $p_2\, \chi'_{1/2}$ to illustrate how such calculations are done. Using the definition of μ_3, we have

$$\mu_p = \mu_0(p_2\, \chi'_{1/2}, [\sigma_3(1)Q(1) + \sigma_3(2)Q(2) + \sigma_3(3)Q(3)]p_2\, \chi'_{1/2}) \tag{11.24}$$

Let us first evaluate the term $(p_2\, \chi'_{1/2}, \sigma_3(1)Q(1)p_2\, \chi'_{1/2})$. We have

$$\begin{aligned}\sigma_3(1)Q(1)p_2\, \chi'_{1/2} = \tfrac{1}{2}\sigma_3(1)Q(1)(&q_1q_2\, q_1\alpha\beta\alpha - q_2\, q_1q_1\alpha\beta\alpha \\ &- q_1q_2\, q_1\beta\alpha\alpha + q_2\, q_1q_1\beta\alpha\alpha).\end{aligned} \tag{11.25}$$

Recall that the ordering of quark basis vectors in a product is significant; for example, $q_1q_2\, q_1\ \alpha\beta\alpha$ is shorthand for

$$q_1(1)q_2(2)q_1(3)\alpha(1)\beta(2)\alpha(3).$$

Then since

$$\begin{aligned} Q(i)q_1(i) &= \tfrac{2}{3}q_1(i), &\quad Q(i)q_2(i) &= -\tfrac{1}{3}q_2(i), \\ \sigma_3(i)\alpha(i) &= \alpha(i), &\quad \sigma_3(i)\beta(i) &= -\beta(i), \end{aligned} \tag{11.26}$$

Eq. (11.25) becomes

$$\begin{aligned}\sigma_3(1)Q(1)p_2\chi'_{1/2} = \tfrac{1}{2}(&\tfrac{2}{3}q_1q_2\, q_1\alpha\beta\alpha + \tfrac{1}{3}q_2\, q_1q_1\alpha\beta\alpha \\ &+ \tfrac{2}{3}q_1\, q_2\, q_1\beta\alpha\alpha + \tfrac{1}{3}q_2\, q_1q_1\beta\alpha\alpha).\end{aligned} \tag{11.27}$$

But Eq. (11.27) is orthogonal to $p_2\, \chi'_{1/2}$, and thus

$$(p_2\, \chi'_{1/2}, \sigma_3(1)Q(1)p_2\, \chi'_{1/2}) = 0. \tag{11.28}$$

By symmetry of the wave function under the interchange of the first and second quarks, we also have

$$(p_2\, \chi'_{1/2}, \sigma_3(2)Q(2)p_2\, \chi'_{1/2}) = 0. \tag{11.29}$$

Next we evaluate $(p_2 \chi'_{1/2}, \sigma_3(3)Q(3)p_2 \chi'_{1/2})$. Using Eq. (11.26), we obtain

$$\sigma_3(3)Q(3)p_2 \chi'_{1/2} = \tfrac{2}{3} p_2 \chi'_{1/2},$$

so that

$$(p_2 \chi'_{1/2}, \sigma(3)Q(3)p_2 \chi'_{1/2}) = \tfrac{2}{3}. \qquad (11.30)$$

Substituting Eqs. (11.28)–(11.30) in Eq. (11.24), we obtain

$$\mu_p = \tfrac{2}{3}\mu_0 . \qquad (11.31)$$

This result is uninteresting by itself, since the proton magnetic moment is given in terms of a free parameter μ_0. However, we can evaluate all other baryon magnetic moments in terms of the same parameter to obtain relations among the baryon magnetic moments.

In evaluating the magnetic moments of the octet, we shall use baryon wave functions which are more general than the proton function used in the example. Specifically, we assume that the baryon state function B is given by (De Souza and Lichtenberg, 1967)

$$B = f_1 B_1 \chi_{1/2} + f_2 B_2 \chi_{1/2} + f_3 B_1 \chi'_{1/2} + f_4 B_2 \chi'_{1/2}, \qquad (11.32)$$

where the B_1 and B_2 are the octet functions of Table 11.3 and the f_i $(i = 1, \ldots, 4)$ are spatial wave functions which satisfy the normalization condition

$$\sum_i (f_i, f_i) = 1. \qquad (11.33)$$

Using the methods illustrated in our example, we obtain the following expressions for the magnetic moment of the proton and neutron:

$$\mu_p = (2\mu_0/3)\{(f_1, f_1) + (f_4, f_4) + \operatorname{Re}[(f_1, f_4) + (f_2, f_3)]\} \qquad (11.34)$$

$$\mu_n = -\mu_p + \tfrac{1}{3}\mu_0 . \qquad (11.35)$$

The magnetic moments of the other baryons are given in terms of these by the following relations (or sum rules as they are often called)

$$\mu_p = \mu_{\Sigma^+}, \qquad \mu_n = \mu_{\Xi^0}, \qquad \mu_{\Sigma^-} = \mu_{\Xi^-}, \qquad (11.36)$$

$$\mu_\Lambda = -\mu_{\Sigma^0} = \tfrac{1}{2}\mu_n, \qquad \mu_{\Sigma^-} = -\mu_p - \mu_n . \qquad (11.37)$$

These sum rules are independent of the f_i and are therefore independent of the overall symmetry properties of the baryon wave functions. They do depend, however, on the assumption that baryons belong to an octet of $SU(3)$ and on the properties of quarks given in Section 11.2. If we compare with Eqs. (9.16)–(9.18) we see that the relations of Eq. (11.36) follow from U-spin invariance alone. However, the sum rules of Eq. (11.37) are not so general. These sum rules can be derived without the use of the quark model. See, for example, Coleman and Glashow (1961).

The magnetic moments of the Λ and Σ^+ hyperons have been measured to be

$$\mu_\Lambda = (0.38 \pm 0.08)\mu_n$$
$$\mu_{\Sigma^+} = (0.90 \pm 0.18)\mu_p .$$

The value of μ_{Σ^+} agrees with the relation of Eq. (11.36), while the value of μ_Λ is 1.5 standard deviations away from agreeing with the relation of Eq. (11.37).

We see from Eq. (11.35) that the relation between the magnetic moments of the proton and neutron involves the unknown parameter μ_0. Only if we make some assumption about the overall symmetry properties of the nucleon wave function can we obtain a prediction for the ratio. If the part of the nucleon wave function which depends on spin and unitary spin is totally symmetric, we have

$$f_1 = f_4 = (1/\sqrt{2})\psi, \qquad f_2 = f_3 = 0, \tag{11.38}$$

where ψ is, for example, given by Eq. (11.11). Using Eq. (11.38) in Eq. (11.34), we obtain

$$\mu_p = \mu_0 .$$

Then Eq. (11.35) becomes

$$\mu_n = -2\mu_p/3, \tag{11.39}$$

a result which follows from assigning the baryon octet to the **56** of $SU(6)$, as we have noted in Chapter 10. In fact, if we substitute Eq. (11.38) into Eq. (11.32), we obtain the $SU(6)$ baryon wave functions

$$B = \psi(B_1\chi_{1/2} + B_2\chi'_{1/2})/\sqrt{2}, \tag{11.40}$$

as we shall verify in Section 11.6

We have obtained the relations of Eqs. (11.36), (11.37), and (11.39) in a somewhat tedious, but in a straightforward way in order to illustrate the use of operators on the state functions of Table 11.3. We can obtain these same results in a simpler fashion which we shall now illustrate. Instead of using $SU(3)$ octet functions of Table 11.3 we shall assume that quarks of type 1, 2, and 3 are distinguishable. See, for example, Franklin (1968). In those cases in which a baryon contains two identical quarks, we let them be the first two. Of the members of the octet, the p, n, Σ^+, Σ^-, Ξ^0, Ξ^- contain two identical quarks. We denote these baryon wave functions by B_3 (the subscript 3 is to distinguish them from the $SU(3)$ functions of Table 11.3). The B_3 functions are

$$\begin{aligned} p_3 &= q_1 q_1 q_2, & n_3 &= q_2 q_2 q_1, \\ \Sigma_3{}^+ &= q_1 q_1 q_3, & \Sigma_3{}^- &= q_2 q_2 q_3, \\ \Xi_3{}^0 &= q_3 q_3 q_1, & \Xi_3{}^- &= q_3 q_3 q_2 . \end{aligned} \tag{11.41}$$

The Σ^0 and Λ contain one quark of each type and therefore the first two quarks are not automatically symmetric under the interchange of their unitary spin indices. Let the first two quarks be the two with isospin $\frac{1}{2}$. Then, since the Σ^0 and Λ have isospin 1 and 0, respectively, the Σ^0 function is symmetric and the Λ is antisymmetric under the interchange of the $SU(3)$ indices of the first two quarks. We shall omit this symmetrization or antisymmetrization between the first two quarks and write

$$\Lambda_3 = \Sigma_3{}^0 = q_1 q_2 q_3 . \tag{11.42}$$

Then all the information about the symmetry properties must be carried by the spin and spatial functions.

We then let the baryon wave function B be given by

$$B = B_3(g_1 \chi_{1/2} + g_2 \chi'_{1/2}), \tag{11.43}$$

where g_1 and g_2 are spatial functions normalized such that

$$(g_1, g_1) + (g_2, g_2) = 1. \tag{11.44}$$

Equation (11.43) is assumed to hold for all baryons of the octet except the Λ. The Λ is singled out, because with our choice of ordering of the quarks, the Λ wave function cannot be made symmetric under interchange of the $SU(3)$ coordinates of the first two quarks. We assume that the Λ wave function is given by

$$\Lambda = \Lambda_3(g_2 \chi_{1/2} - g_1 \chi'_{1/2}). \tag{11.45}$$

Because g_1 and g_2 are bound-state wave functions, they must be real (neglecting electromagnetic and weak interactions). This ensures that the Λ and Σ° wave functions are orthogonal. If we use the functions of Eq. (11.43) in Eq. (11.22), we obtain the following expression for the proton moment:

$$\mu_p = \mu_0[1 - \tfrac{4}{3}(g_2, g_2)]. \tag{11.46}$$

The other magnetic moments are given in terms of μ_p and μ_0 exactly as previously by Eqs. (11.35)–(11.37). From Eqs. (11.35) and (11.46) we obtain the following expression for the ratio $R \equiv -\mu_p/\mu_n$:

$$R = \frac{3 - 4(g_2, g_2)}{2 - 4(g_2, g_2)}. \tag{11.47}$$

The experimental value of R is given by R (experiment) = 1.46. It is interesting that the form of Eq. (11.47) is such that we cannot obtain exact agreement with experiment for any value of (g_2, g_2) consistent with the normalization condition

$$0 \leq (g_2, g_2) \leq 1.$$

If $(g_2, g_2) = 0$, i.e., $g_2 = 0$ we obtain $R = 1.5$, only 3% higher than the experimental value.

The 3% discrepancy between the prediction and experiment can arise either from relativistic effects (such as exchange currents) or from an admixture of nonzero orbital angular momentum in the spatial wave function. If $g_2 \neq 0$ the disagreement with experiment is increased. If $g_2 = 0$, we have a function which is symmetric under the interchange of spin and unitary spin indices. This is a function belonging to the **56** of *SU*(6). This same result that a symmetric function gives best agreement with experiment can be obtained in a somewhat more complicated fashion using Eqs. (11.34) and (11.35) instead of Eq. (11.47).

Another interesting feature of the expressions for the proton and neutron magnetic moments of Eqs. (11.46) and (11.35) is that they do not contain any interference terms between the spatial wave functions g_1 and g_2. This means that if the nucleon wave functions contain a small admixture of terms which are not symmetric, the change in their magnetic moments will be second order in these small terms. If $g_2 = 0$, the spin and unitary spin functions of Eq. (11.43) and (11.45) are given by

$$\begin{gathered} p = q_1 q_1 q_2 \chi_{1/2}, \qquad n = q_2 q_2 q_1 \chi_{1/2}, \\ \Sigma^+ = q_1 q_1 q_3 \chi_{1/2}, \qquad \Sigma^0 = q_1 q_2 q_3 \chi_{1/2}, \qquad \Sigma^- = q_2 q_2 q_3 \chi_{1/2}, \\ \Lambda = q_1 q_2 q_3 \chi'_{1/2}, \\ \Xi^0 = q_3 q_3 q_1 \chi_{1/2}, \qquad \Xi^- = q_3 q_3 q_3 \chi_{1/2}. \end{gathered} \tag{11.48}$$

We should point out that even if *SU*(6) is badly broken, the proton and neutron wave functions could be symmetric under the interchange of the spin functions of two identical quarks. Thus, our assumptions are weaker than those of *SU*(6). See, for example, Greenberg (1964) and Franklin (1968) for further discussions of the symmetry properties of quarks.

We can also calculate the magnetic moments of the members of the decuplet. However, there seems to be no way to measure these moments, except for that of the Ω, because of the instability of the other members of the decuplet against strong decay. The magnetic moment of the Ω is given by

$$\mu_\Omega = -\mu_p.$$

Also of some interest are the so-called transition moments, defined by $(B', \mu_3 B)$, where B and B' are different baryons. We obtain the following predictions for two transition moments which can be measured indirectly:

$$(\Delta^+, \mu_3 p) = -2\sqrt{\tfrac{2}{3}}(\Sigma^0, \mu_3 \Lambda) = \tfrac{2}{3}\sqrt{2}\,\mu_p.$$

11.5 Hadron Mass Splittings

We shall assume that the mass of a hadron is given by the expectation value of a mass operator M with respect to an $SU(3)$-invariant hadron wave function. In order to compute the masses of the hadrons, we must select a form for the mass operator M. We take M to be a sum of three terms:

$$M = T + M_s + V \tag{11.49}$$

where T is the total kinetic energy of the constituent quarks, M_s is the sum of the quark masses and V is the sum of the quark interactions.

If the kinetic energy term T is small, as we shall assume, then it is reasonable to neglect $SU(3)$ breaking terms in T. This means that the value of T will be a constant for all members of a multiplet. Since we shall be interested only in hadron mass differences, we can omit this constant. In the next section we shall further discuss the assumption that the bound quarks have a small kinetic energy.

The mass term M_s is simply given by a sum of quark masses:

$$M_s = \sum_{i=1}^{3} m(i). \tag{11.50}$$

The interaction term V we take (for simplicity) to be a sum of two-body quark interactions. In general the interaction between two quarks will contain an $SU(3)$-invariant part and an $SU(3)$-breaking part. The $SU(3)$-invariant part is a constant for a given multiplet and can be omitted, since we are interested in mass differences. The symmetry-breaking interaction will, in general, contain an isospin-conserving part which can depend on the total isospin and the total hypercharge Y of the two-quark system and an electromagnetic part which breaks isospin symmetry. We then write for V

$$V = \sum_{i<j} [V_{IY}(ij) + v(ij)] \tag{11.51}$$

where $V_{IY}(ij)$ is an isospin-conserving interaction and $v(ij)$ the electromagnetic interaction between the ith and jth quarks.

Let us first consider the isospin-conserving mass splitting. Then we can neglect ε in the quark masses of Table 11.1 and also the $v(ij)$. It should be clear that, with an arbitrary symmetry-breaking interaction V_{IY}, we will not obtain the Gell-Mann–Okubo mass formula. Our procedure will be to compute the decuplet and octet mass splittings to see what restrictions must be placed on V_{IY} so that the Gell-Mann–Okubo mass formula holds.

The mass of a baryon is given by

$$M_B = (B, MB) = C + (B, M_s B) + (B, VB) \tag{11.52}$$

where C is a constant for an $SU(3)$ multiplet and includes the contributions from the kinetic energy term and the $SU(3)$-invariant interaction. We can evaluate the remaining terms $(B, M_s B)$ and (B, VB) using the baryon wave functions of Tables 11.2 and 11.3. We introduce the following notation for the expectation values of $V_{IY}(ij)$:

$$\begin{aligned} V_{IY} &= V_1 \quad \text{if} \quad I = 1, \quad Y = \tfrac{2}{3}, \\ V_{IY} &= V_2 \quad \text{if} \quad I = 0, \quad Y = \tfrac{2}{3}, \\ V_{IY} &= V_3 \quad \text{if} \quad I = \tfrac{1}{2}, \quad Y = -\tfrac{1}{3}, \\ V_{IY} &= V_4 \quad \text{if} \quad I = 0, \quad Y = -\tfrac{4}{3}. \end{aligned} \tag{11.53}$$

Also we see from Table 11.1 that the expectation values $(q_j, m(i)q_j)$ of $m(i)$ are (neglecting ε)

$$\begin{aligned} &m \quad && \text{if} \quad j = 1, 2, \\ &m + \delta \quad && \text{if} \quad j = 3. \end{aligned} \tag{11.54}$$

Then, the masses of the members of the baryon decuplet are

$$\begin{aligned} M_\Delta &= C + 3m + 3V_1, \\ M_{\Sigma^*} &= C + 3m + \delta + V_1 + 2V_3, \\ M_{\Xi^*} &= C + 3m + 2\delta + 2V_3 + V_4, \\ M_\Omega &= C + 3m + 3\delta + 3V_4. \end{aligned} \tag{11.55}$$

We can directly verify that the masses of Eq. (11.55) satisfy the sum rule

$$M_\Omega - M_{\Xi^*} + M_{\Sigma^*} - M_\Delta = 2(M_{\Xi^*} - M_{\Sigma^*}). \tag{11.56}$$

independently of any assumption of how the two-body force depends on I and Y. We also see from Eq. (11.55) that in order for the decuplet masses to satisfy the equal spacing rule, the expectation values V_i must satisfy the relations

$$V_1 + V_4 = 2V_3. \tag{11.57}$$

The expressions we get for the baryon octet masses do not depend on whether we use the functions B_1 or B_2 from Table 11.3 or any linear combination of them. We obtain

$$\begin{aligned} M_N &= C' + 3m + 3(V_1 + V_2)/2, \\ M_\Lambda &= C' + 3m + \delta + V_2 + 2V_3, \\ M_\Sigma &= C' + 3m + \delta + V_1 + 2V_3, \\ M_\Xi &= C' + 3m + 2\delta + 2V_3 + V_4, \end{aligned} \tag{11.58}$$

where we have put a prime on the C to emphasize that it is different from the C of Eq. (11.55). For arbitrary V_i, we obtain no sum rule. However, if

$V_1 + V_4 = 2V_3$, we obtain the Gell-Mann–Okubo octet formula

$$2(M_N + M_\Xi) = 3M_\Lambda + M_\Sigma .$$

Moreover, we also obtain the relation

$$M_\Xi - M_\Sigma = M_{\Sigma*} - M_\Delta. \tag{11.59}$$

We emphasize that this derivation of the Gell-Mann–Okubo mass formula depends on two approximations: first, that three-body quark forces are negligible; and second that it is adequate to calculate to lowest order perturbation theory in the symmetry-breaking terms.

Equation (11.59) is in disagreement with experiment by 22 MeV, which is about 18% of the $\Xi - \Sigma$ mass difference. If the V_{IY} are spin dependent, then Eq. (11.59) does not hold, in general, as the spin wave functions of the decuplet and octet are different. However, in the special case that the octet wave functions are symmetric in the simultaneous interchange of $SU(3)$ and spin indices, Eq. (11.59) still holds. Since it is usually assumed that the octet wave functions are indeed symmetric under this exchange, the prediction that Eq. (11.59) holds is a defect of the model. We could cure this defect by relaxing the assumption about the symmetry of the wave functions, but then we would lose other good predictions, such as that the ratio of the proton to neutron magnetic moment is -1.5. Alternatively, we could include three-body forces among the quarks, but we would lose simplicity in this way.

Note that in obtaining Eqs. (11.55) and (11.58), we used the baryon wave functions of Table 11.3 rather than the simplified wave functions of Eq. (11.48). In order to use the latter wave functions, we must rewrite the expression for the interaction V in terms of spin variables instead of isospin variables. In that way, we can relax the assumption of isospin invariance of the interaction. We shall not carry this out, but leave it as an exercise.

If we consider meson octet mass splittings, we find a situation analogous to that for the baryon octet. We do not obtain the Gell-Mann–Okubo mass formula unless there is a special condition among the symmetry-breaking interactions of the quark and antiquark. We shall not reproduce the results of the meson calculation here. Instead, we shall consider the simpler problem of obtaining the mass formula which follows from the assumption that the quark–antiquark interaction is invariant under $SU(3)$ so that just the quark masses break the symmetry. Again neglecting electromagnetic mass differences, we obtain the following expressions for the masses of the pseudoscalar octet:

$$\begin{aligned} M_K &= 2m + \delta + W_0 , \\ M_\pi &= 2m + W_0 , \\ M_\eta &= 2m + \tfrac{4}{3}\delta + W_0 , \end{aligned} \tag{11.60}$$

where W_0 is the quark–antiquark octet interaction energy in the $^1S_0$ state. From Eq. (11.60) we see that we obtain the Gell-Mann–Okubo mass formula for mesons

$$4M_K = M_\pi + 3M_\eta . \tag{11.61}$$

Furthermore, if δ is positive, we obtain the following inequality

$$M_\eta > M_K > M_\pi . \tag{11.62}$$

in agreement with experiment. The mass of the η' is unrelated to the octet masses because this particle is an $SU(3)$ singlet.

Now let us obtain the mass relations among the members of the vector meson nonet. The unitary spin wave functions of the K^* and ρ are just the same as those for the K and π, respectively, and are given in Table 11.4. However, the wave functions of the ω and φ are not pure $SU(3)$ basis functions, but are given by the mixed wave functions of Eq. (11.16). The masses of the vector mesons are

$$\begin{aligned} M_{K*} &= 2m + \delta + W_1, \\ M_\rho &= 2m + W_1, \\ M_\omega &= 2m + W_1, \\ M_\varphi &= 2m + 2\delta + W_1, \end{aligned} \tag{11.63}$$

where W_1 is the quark–antiquark nonet interaction energy in the $^3S_1$ state. From Eq. (11.63), we do not obtain the Gell-Mann–Okubo octet formula, but instead get the relations

$$\begin{aligned} M_\rho &= M_\omega , \\ 2M_{K*} &= M_\varphi + M_\rho . \end{aligned} \tag{11.64}$$

We also obtain the inequalities

$$M_\varphi > M_{K*} > M_\rho , \tag{11.65}$$

in agreement with experiment.

The relations of Eqs. (11.61) and (11.64) are in fair agreement with experiment, the disagreement being only a few percent. The agreement is somewhat improved in the case of the pseudoscalar mesons if a mass-squared relation is used.

We next consider the electromagnetic mass splitting of the baryons within a given isospin multiplet. To calculate this splitting, we need a form for the electromagnetic interaction $v(ij)$ between two quarks. It is plausible that $v(ij)$ should consist of the usual Coulomb interaction plus a spin-dependent magnetic interaction. First we shall obtain a set of sum rules which result if we neglect the magnetic interaction. We write the Coulomb potential between two quarks as

$$v(ij) = \frac{Q(i)Q(j)}{r_{ij}}, \tag{11.66}$$

where r_{ij} is the distance between the ith and jth quark. With an interaction of the form of Eq. (11.66), the electromagnetic contribution to the baryon mass splitting is independent of the symmetry of the spin and unitary spin wave functions, and depends only on the charges of the constituent quarks. Therefore, we can use the simple baryon wave functions of Eqs. (11.41) and (11.42) to evaluate the electromagnetic contributions to the masses of the baryons. If we assume that the expectation value of $1/r_{ij}$ is a constant K for the multiplet, we obtain

$$\begin{aligned} &dM_p = \varepsilon, && dM_n = 2\varepsilon - \tfrac{1}{3}K, && \\ &dM_{\Sigma^+} = 0, && dM_{\Sigma^0} = \varepsilon - \tfrac{1}{3}K, && dM_{\Sigma^-} = 2\varepsilon + \tfrac{1}{3}K, \\ &dM_{\Xi^0} = -\tfrac{1}{3}K, && dM_{\Xi^-} = \varepsilon + \tfrac{1}{3}K, && \end{aligned} \tag{11.67}$$

where dM_B denotes the electromagnetic contribution to the mass of the baryon B. According to the form of the interaction we have chosen, the electromagnetic contribution to the mass of a baryon is the only quantity which breaks the isospin degeneracy. Therefore we obtain from Eq. (11.67)

$$\begin{aligned} dM_n - dM_p = M_n - M_p &= \varepsilon - \tfrac{1}{3}K, \\ M_{\Sigma^0} - M_{\Sigma^+} &= \varepsilon - \tfrac{1}{3}K, \\ M_{\Sigma^-} - M_{\Sigma^0} &= \varepsilon + \tfrac{2}{3}K, \\ M_{\Xi^-} - M_{\Xi^0} &= \varepsilon + \tfrac{2}{3}K. \end{aligned} \tag{11.68}$$

We can eliminate the parameters ε and K from Eq. (11.68) to obtain the sum rules

$$\begin{aligned} M_{\Sigma^-} - M_{\Sigma^+} &= M_n - M_p + M_{\Xi^-} - M_{\Xi^0}, \\ M_{\Sigma^0} - M_{\Sigma^+} &= M_n - M_p, \\ M_{\Xi^-} - M_{\Xi^0} &= M_{\Sigma^-} - M_{\Sigma^0}. \end{aligned} \tag{11.69}$$

The first of these sum rules is the Coleman–Glashow relation, which agrees quite well with experiment. The second and third relations, however, definitely disagree with experiment. Our conclusion is that the interaction of Eq. (11.66) is too simple.

We shall obtain improved agreement with experiment if we include a magnetic interaction. Since in general, a magnetic interaction depends both on charge and spin, the results will depend on the symmetry of the wave functions. We shall consider the case in which the octet wave functions are symmetric $SU(6)$ functions of Eq. (11.40). Furthermore, for ease in calculation, we shall assume that these functions are multiplied by a symmetric spatial wave function. Then the baryon wave function is totally symmetric under the interchange of all the coordinates of any two quarks. Thus, if the assumption is correct, quarks cannot be fermions. Neither, however, can quarks be bosons, since otherwise a baryon would be a boson. We shall postpone a discussion of this difficulty until Section 11.9. In the meantime, we shall often assume that the spatial wave function of quarks bound in a baryon is symmetric, as in this way we obtain in a simple fashion results which agree quite well with experiment.

If the spatial wave function is symmetric, the quarks in a baryon can be in states of relative orbital angular momentum zero. Then the magnetic interaction between two quarks can be taken to be a spin-dependent contact interaction of the type responsible for atomic hyperfine structure. (See, for example, Abragam, 1961, p. 170.) The magnetic contribution to $v(ij)$ is then given by

$$v_{\mathrm{m}}(ij) = -(8\pi/3)\boldsymbol{\mu}_q(i) \cdot \boldsymbol{\mu}_q(j)\delta(r_{ij}), \qquad (11.70)$$

where $\delta(r_{ij})$ is the Dirac delta-function. If we take the expectation value of $v_{\mathrm{m}}(ij)$ with respect to a symmetric S-wave spatial function g, we obtain a quantity which we call $w(ij)$ given by

$$w(ij) = (g, v_{\mathrm{m}}(ij)g) = -b\boldsymbol{\mu}_q(i) \cdot \boldsymbol{\mu}_q(j),$$

where the positive parameter b is given by

$$b = (8\pi/3)(g, \delta(r_{ij})g). \qquad (11.71)$$

The quantity $(g, \delta(r_{ij})g)$ does not vanish in general if the relative orbital angular momentum is zero between quarks i and j. Furthermore, because g is symmetric, $(g, \delta(r_{ij})g)$ is independent of i and j.

We shall assume that the decuplet spatial wave functions are symmetric just like those of the octet, and that the expectation values of the Coulomb and magnetic potentials are given in terms of the parameters K and b. With these assumptions we can proceed to calculate the electromagnetic mass splittings of the baryon octet and decuplet. We shall not give the details of the calculation, but only the sum rules obtained by eliminating the parameters ε, K, and b. These sum rules are

$$\begin{aligned} M_n - M_p + M_{\Xi^-} - M_{\Xi^0} &= M_{\Sigma^-} - M_{\Sigma^+}, \\ M_{\Delta^0} - M_{\Delta^+} &= M_{\Sigma^{*0}} - M_{\Sigma^{*+}}, \\ M_{\Delta^-} - M_{\Delta^0} &= M_{\Sigma^{*-}} - M_{\Sigma^{*0}}, \\ M_{\Delta^-} - M_{\Delta^0} &= M_{\Xi^{*-}} - M_{\Xi^{*0}}. \end{aligned} \qquad (11.72)$$

$$\begin{aligned} M_{\Delta^-} - M_{\Delta^{++}} &= 3(M_{\Delta^0} - M_{\Delta^+}), \\ 3(M_{\Sigma^-} - M_{\Sigma^+}) - 6(M_{\Sigma^0} &- M_{\Sigma^+}) + 9(M_n - M_p) \\ &= 5(M_{\Xi^{*-}} - M_{\Xi^{*0}}) + 2(M_{\Delta^0} - M_{\Delta^{++}}). \end{aligned} \qquad (11.73)$$

$$M_n - M_p + M_{\Xi^{*-}} - M_{\Xi^{*0}} = M_{\Sigma^{*-}} - M_{\Sigma^{*+}}. \qquad (11.74)$$

The first four of these sum rules [Eqs. (11.72)] follow from U-spin invariance alone, the next two [Eqs. (11.73)] follow from any $SU(3)$-invariant baryon wave functions and the last [Eq. (11.74)] follows from $SU(6)$-invariance. In addition, because K and b are positive parameters, we obtain the following inequalities

$$M_{\Sigma^-} - M_{\Sigma^0} > M_{\Sigma^0} - M_{\Sigma^+} > M_n - M_p. \tag{11.75}$$

All these sum rules and inequalities are in good agreement with the data.

For further information about hadron mass splittings plus a discussion of how some of the assumptions can be relaxed, see Rubinstein *et al.* (1967).

11.6 Quark Model and $SU(6)$

If a quark has spin $\frac{1}{2}$, it is a doublet of the spin group $SU(2)$ as well as a triplet of $SU(3)$ and thus is a fundamental sextet of $SU(3) \times SU(2)$. But the quark basis functions can equally well be regarded as a fundamental sextet of $SU(6)$. See, for example, the discussion in Section 10.3. We choose as basis functions of $SU(6)$, the following six quark states u_i:

$$\begin{array}{lll} u_1 = q_1\alpha, & u_2 = q_2\alpha, & u_3 = q_3\alpha, \\ u_4 = q_1\beta, & u_5 = q_2\beta, & u_6 = q_3\beta. \end{array} \tag{11.76}$$

where as before the q_i are the quark basis functions of $SU(3)$ and α and β are states with spin "up" and "down," respectively.

When we consider the interactions of two or more quarks, we find there is a difference between $SU(6)$ and $SU(3) \times SU(2)$, since one multiplet of $SU(6)$ may contain several multiplets of $SU(3) \times SU(2)$. The quark model provides a natural way for us to incorporate $SU(6)$ as an approximate dynamical symmetry. If the interactions of quarks are taken to be Lorentz invariant (as we believe they should be), then $SU(6)$ will automatically be broken in high energy processes. This is because a Lorentz transformation acts on the states of the $SU(2)$ spin subgroup of $SU(6)$, but does not affect the internal symmetry subgroup $SU(3)$. Then, any symmetry transformation such as $SU(6)$, which allows transformations which mix spin and internal symmetry coordinates, must be broken. However, the interaction could be such as to preserve $SU(6)$ invariance in a low-energy approximation in which the quark velocities are negligibly small.

In the model, a baryon is supposed to consist of a bound state of three quarks. If the bound quarks are moving at relativistic velocities inside the baryon, it is difficult to see how $SU(6)$ can be an approximate dynamical symmetry of baryons. Therefore relativistic effects which break $SU(6)$ symmetry must be neglected. Morpurgo (1965) has discussed under what conditions such an approximation might be valid. Briefly, Morpurgo argues that if quarks are bound by a potential of range a, then a typical momentum k of a bound quark is $k \approx 1/a$. Now if the range of the potential is much greater than the inverse of a quark mass m, i.e., $a \gg 1/m$, then $k \ll m$, and a nonrelativistic approximation might be a good one. This argument does not hold for a singular potential, for example, a Yukawa potential $\sim(a/r)e^{-r/a}$. It is not known whether the argument is good for any system with a binding energy comparable to the masses of the constituents, as has been conjectured to be the case for quarks. We shall simply assume that a nonrelativistic approximation is adequate to describe quarks inside hadrons.

We have remarked that the reduction of the Kronecker product of three quark state vectors given in Eq. (11.7) holds for any $SU(n)$ with $n > 2$. In particular for $SU(6)$, using the methods described in Chapter 7, we obtain the following multiplicities

$$\mathbf{6} \otimes \mathbf{6} \otimes \mathbf{6} = \mathbf{56} \oplus \mathbf{70} \oplus \mathbf{70} \oplus \mathbf{20}. \tag{11.77}$$

Regarding the quark also as a fundamental multiplet $^2 3$ of $SU(3) \times SU(2)$, Eq. (11.77) becomes

$$\begin{aligned} {}^2 3 \times {}^2 3 \times {}^2 3 &= {}^{(2\times 2\times 2)}(3 \times 3 \times 3) \\ &= {}^{(4+2+2)}(10 + 8 + 8 + 1) \\ &= {}^4 10 + {}^4 8 + {}^4 8 + {}^4 1 + {}^2 10 + {}^2 8 \\ &\quad + {}^2 8 + {}^2 1 + {}^2 10 + {}^2 8 + {}^2 8 + {}^2 1. \end{aligned} \tag{11.78}$$

We have already used symmetry arguments in Section 10.2 to show how these multiplets of $SU(3) \times SU(2)$ are classified into multiplets of $SU(6)$. See Eqs. (10.5), (10.6), and (10.7).

The $SU(6)$ states of the **56** are particularly easy to construct from the basis vectors of three quarks, since the **56** belongs to the totally symmetric tableau $\boxed{||}$. We shall not give all 56 of these baryon wave functions. However, it is instructive to write down several of these functions as examples. The function denoted by the Young tableau $\boxed{1 | 1 | 4}$ is given by

$$\boxed{1 | 1 | 4} = (u_1 u_1 u_4 + u_1 u_4 u_1 + u_4 u_1 u_1)/\sqrt{3}.$$

Using Eq. (11.76), this becomes

$$\boxed{1 | 1 | 4} = q_1 q_1 q_1 (\alpha\alpha\beta + \alpha\beta\alpha + \beta\alpha\alpha)/\sqrt{3}. \tag{11.79}$$

Thus, this function factors into an $SU(3)$ decuplet function Δ^{++} times the spin $\frac{3}{2}$ function of Eq. (11.10). Thus we have

$$\boxed{1 | 1 | 4} = \Delta^{++} \chi_{3/2}^{(3/2)}. \tag{11.80}$$

Two examples which are somewhat more interesting are given by the Young tableaux $\boxed{1\,|\,1\,|\,5}$ and $\boxed{1\,|\,2\,|\,4}$. These functions are

$$
\begin{aligned}
\boxed{1\,|\,1\,|\,5} &= (u_1 u_1 u_5 + u_1 u_5 u_1 + u_5 u_1 u_1)/\sqrt{3} \\
&= (q_1 q_1 q_2\, \alpha\alpha\beta + q_1 q_2 q_1 \alpha\beta\alpha + q_2 q_1 q_1 \beta\alpha\alpha)/\sqrt{3}, \\
\boxed{1\,|\,2\,|\,4} &= (u_1 u_2 u_4 + u_1 u_4 u_2 + u_2 u_1 u_4 + u_2 u_4 u_1 + u_4 u_1 u_2 \\
&\quad + u_4 u_2 u_1)/\sqrt{6} \\
&= (q_1 q_2 q_1 \alpha\alpha\beta + q_1 q_1 q_2\, \alpha\beta\alpha + q_2 q_1 q_1 \alpha\alpha\beta \\
&\quad + q_2 q_1 q_1 \alpha\beta\alpha + q_1 q_1 q_2\, \beta\alpha\alpha + q_1 q_2 q_1 \beta\alpha\alpha)/\sqrt{6}.
\end{aligned}
\tag{11.81}
$$

These $SU(6)$ functions can be written in terms of the $SU(3)$ and spin functions defined in Section 11.3. We obtain

$$
\begin{aligned}
\boxed{1\,|\,1\,|\,5} &= (\Delta^+ \chi_{1/2}^{(3/2)} + p_1 \chi_{1/2} + p_2 \chi'_{1/2})/\sqrt{3}, \\
\boxed{1\,|\,2\,|\,4} &= (2\Delta^+ \chi_{1/2}^{(3/2)} - p_1 \chi_{1/2} - p_2 \chi'_{1/2})/\sqrt{6}.
\end{aligned}
\tag{11.82}
$$

From Eq. (11.82) we see that the proton $SU(6)$ wave function is

$$
p = (p_1 \chi_{1/2} + p_2 \chi'_{1/2})/\sqrt{2}. \tag{11.83}
$$

This verifies (for the proton) the statement made in Section 11.4 that the $SU(6)$ baryon wave functions are given by Eq. (11.40). The same procedure works for all the baryons. Note from Eq. (11.82) that an alternative way of writing the proton $SU(6)$ wave function is

$$
p = \sqrt{\tfrac{2}{3}}\,\boxed{1\,|\,1\,|\,5} - \sqrt{\tfrac{1}{3}}\,\boxed{1\,|\,2\,|\,4} \tag{11.84}
$$

Since this proton function is a linear combination of the symmetric functions $\boxed{1\,|\,1\,|\,5}$ and $\boxed{1\,|\,2\,|\,4}$, it is obviously totally symmetric. However, we wish to reemphasize that we can assume that the proton wave function is symmetric without assuming all the consequences of $SU(6)$ invariance. To take an extreme example, the function of Eq. (11.83) might hold even if strange particles did not exist. More practically, Eq. (11.83) can hold even though the Λ, Σ, and Ξ hyperons are not degenerate in mass with the proton. This kind of $SU(6)$-breaking is a consequence of $SU(3)$-breaking rather than a consequence of the difference between internal and spin degrees of freedom. This means that the proton function of Eq. (11.83) follows if we make the lesser assumption that the proton is a basis function of a symmetric representation of $SU(4)$, where $SU(4)$ contains the spin and isospin $SU(2)$ groups. If we apply Eq. (11.7) to $SU(4)$, we obtain

$$
\mathbf{4} \otimes \mathbf{4} \otimes \mathbf{4} = \mathbf{20}_S \oplus \mathbf{20}_M \oplus \mathbf{20}_M \oplus \mathbf{4}. \tag{11.85}
$$

where we have distinguished the symmetric multiplet from the multiplets of mixed symmetry by the letters S and M, since they have the same multiplicities. Regarded as states of $SU(2) \times SU(2)$, Eq. (11.85) becomes

$$^{2}2 \times {}^{2}2 \times {}^{2}2 = {}^{(4+2+2)}(4 + 2 + 2). \tag{11.86}$$

Using arguments analogous to those used in Section 10.2, we find

$$\begin{aligned} \mathbf{20}_S &\supset {}^{4}4 + {}^{2}2, \\ \mathbf{20}_M &\supset {}^{4}2 + {}^{2}4 + {}^{2}2, \\ \mathbf{4} &\supset {}^{2}2. \end{aligned} \tag{11.87}$$

These multiplicities apply to the baryons of hypercharge $Y = 1$ (or non-strange baryons). The spin-$\frac{3}{2}$ baryon Δ and the spin-$\frac{1}{2}$ nucleon N fit nicely into the $\mathbf{20}_S$. Thus, we see that a symmetric nucleon wave function does not depend on $SU(6)$. However, the $SU(4)$ nucleon wave function is identical to the one predicted from $SU(6)$.

We now point out that we can have a symmetric nucleon wave function even if $SU(4)$ is broken. For example, it is very possible for an $SU(4)$ symmetry-breaking interaction to split the mass degeneracy of the Δ and N without disturbing the symmetry of either the Δ or N wave functions under interchange of quark indices. Therefore, the assumption that the nucleon wave function is symmetric is less restrictive than the assumption of either $SU(4)$ or $SU(6)$. It may be for this reason that the predicted value of the ratio of the proton to neutron magnetic moment is so close to the experimental value despite the fact that $SU(4)$ and $SU(6)$ are broken symmetries.

11.7 Orbital Excitations

We have dwelt in considerable detail on the octet and decuplet baryons and on the pseudoscalar and vector meson nonets. Within the framework of the quark model, these are all states with total orbital angular momentum zero. Many of the mesons and baryons of higher mass have angular momentum which is sufficiently high that the quarks must be in states with orbital angular momentum greater than zero. We list in Table 11.5 the possible values of the total orbital angular momentum of the quarks in mesons and baryons with a given spin and parity.

As we have already noted, the group $SU(6)$ includes spin but not orbital angular momentum. Therefore, to the extent that $SU(6)$ is a good symmetry, the interactions between quarks should not couple spin and orbital angular momentum. This suggests that as a first approximation we should consider states of the direct-product group $SU(6) \times O(3)$ where $O(3)$ is the orthogonal

TABLE 11.5

POSSIBLE VALUES OF THE TOTAL ORBITAL ANGULAR MOMENTUM L OF QUARKS BOUND IN BARYONS OR QUARK–ANTIQUARK PAIRS BOUND IN MESONS[a]

Mesons		Baryons		Mesons		Baryons	
J^P	L	J^P	L	J^P	L	J^P	L
0^-	0	$\frac{1}{2}^+$	0, 1, 2,	2^-	2	$\frac{5}{2}^+$	1, 2, 3, 4
0^+	1	$\frac{1}{2}^-$	1, 2,	2^+	1, 3	$\frac{5}{2}^-$	1, 2, 3, 4
1^-	0, 2	$\frac{3}{2}^+$	0, 1, 2, 3	3^-	2, 4	$\frac{7}{2}^+$	2, 3, 4, 5
1^+	1	$\frac{3}{2}^-$	1, 2, 3	3^+	3	$\frac{7}{2}^-$	2, 3, 4, 5

[a] The spin and parity of a hadron is denoted by J^P.

group in three dimensions and includes both proper rotations and reflections. We shall call the basis functions of an irreducible representation of $SU(6) \times O(3)$ a *supermultiplet* and denote such a supermultiplet by (N, L^P) where N is the $SU(6)$ multiplicity, L is the orbital angular momentum, and P is the parity. Many authors, among them Greenberg (1964) and Dalitz (1967, 1968, 1969), have considered hadrons within this framework.

Of course $SU(6) \times O(3)$ is a broken symmetry. If the symmetry-breaking interaction between quarks includes a spin–orbit term, then a supermultiplet will split into a number of multiplets of different mass, each with a different value of J. These multiplets in turn will be split by an $SU(3)$-breaking interaction.

For simplicity, we shall let that part of the quark–quark or quark–antiquark interaction which is invariant under $SU(6) \times O(3)$ be a harmonic oscillator potential. The energy levels of a harmonic oscillator are of course equally spaced and increase according to the number v of excitation quanta. In the meson case the relative motion of the quark–antiquark pair is described by a single three-dimensional harmonic oscillator. The ordering of the energy levels according to the number v of excitation quanta is

$$\begin{aligned} v &= 0, & L^P &= 0^-, \\ v &= 1, & L^P &= 1^+, \\ v &= 2, & L^P &= 2^-, \quad 0^-, \\ v &= 3, & L^P &= 3^+, \quad 1^+. \end{aligned} \tag{11.88}$$

Each of these levels should correspond to two supermultiplets of $SU(6) \times O(3)$, a $(35, L^P)$ and a $(1, L^P)$. We have already considered the ground state $L = 0^-$ supermultiplets $(35, 0^-)$ and $(1, 0^-)$. Since the quark–antiquark system can have spin 0 or 1, the spin–orbit interaction splits these levels into nonets with $J^{PC} = 0^{-+}, 1^{--}$. The charge-conjugation quantum number applies only to the $Y = I_3 = 0$ members of the nonets. Next consider the $v = 1$ supermultiplets with $L^P = 1^+$ which are split into the following nonets

$$J^{PC} = 0^{++}, \quad 1^{++}, \quad 2^{++}, \quad 1^{+-}. \tag{11.89}$$

The experimental evidence for the existence of these nonets is given in Tables 9.8 and 9.9. As a last example we consider the $v = 2$ supermultiplets with $L^P = 2^-, 0^-$. These split into the following nonets

$$J^{PC} = 1^{--}, 2^{--}, \quad 3^{--}, \quad 2^{-+}, \quad 0^{-+}, \quad 1^{--}. \tag{11.90}$$

There is only fragmentary experimental evidence concerning the existence of these nonets.

In the case of the baryons, we consider three quarks interacting by means of harmonic oscillator potentials. The level ordering appears in a number of places, for example, in Dalitz (1969). We shall make the usual assumption that the three-quark wave function is totally symmetric under the interchange of all the coordinates of any two quarks. We shall discuss this assumption in Section 11.9. With this restriction on the wave functions, the ordering of the supermultiplets is

$$\begin{aligned}
v = 0:& \quad (56, 0^+), \\
v = 1:& \quad (70, 1^-), \\
v = 2:& \quad (56, 2^+), \quad (56, 0^+), \quad (70, 2^+), \quad (70, 0^+), \quad (20, 1^+).
\end{aligned} \tag{11.91}$$

A spin–orbit interaction breaks these supermultiplets into singlets, octets and decuplets. In addition, $SU(3)$ mixing can occur. We shall not give any detailed comparison with experiment except to note that there is substantial evidence for the existence of a $(56, 0^+)$ supermultiplet with $v = 0$, a $(70, 1^-)$ with $v = 1$, and a $(56, 2^+)$ and $(56, 0^+)$ with $v = 2$. There is at present little evidence for the existence of the other supermultiplets of Eq. (11.91) with $v = 2$.

A number of authors have calculated decay rates of baryons within the framework of the harmonic-oscillator quark model. Agreement with experiment for decay via both pion and photon emission is rather good. See, for example, Faiman and Hendry (1969).

11.8 High Energy Scattering

Lipkin and Scheck (1966) have shown that some features of hadron–hadron collisions at high energy can be correctly predicted in terms of the quark model with one important assumption. Briefly, these authors assumed that the hadron–hadron scattering amplitude in the forward direction is the sum of all possible two-body quark–quark or quark–antiquark scattering amplitudes. There is a theorem in quantum mechanics, known as the optical theorem, which states that the imaginary part of the forward scattering amplitude is proportional to the total cross section. Therefore, the assumption of Lipkin and Scheck applies to the total cross sections.

Since there are only three quarks but many hadrons, it is possible to obtain sum rules relating various hadron–hadron cross sections. We let the total cross section for the collision of particles a and b be $\sigma(ab)$. If the quark–quark and quark–antiquark amplitudes are $SU(3)$-invariant, then, following Lipkin and Scheck, we obtain the following sum rules (among others)

$$\begin{aligned}
\sigma(pp) - \sigma(np) &= \sigma(K^+p) - \sigma(K^+n), \\
\sigma(pp) + \sigma(\bar{p}p) &= \tfrac{3}{2}[\sigma(\pi^+p) + \sigma(\pi^-p)] \\
&\qquad + \tfrac{1}{2}[\sigma(K^+p) + \sigma(K^-p) - \sigma(K^+n) - \sigma(K^-n)], \\
\sigma(pp) + \sigma(\bar{p}p) &= 2[\sigma(\pi^+p) + \sigma(\pi^-p)] - \tfrac{1}{2}[\sigma(K^+p) + \sigma(K^-p)], \\
\sigma(\bar{p}p) - \sigma(\bar{p}n) &= \sigma(K^-p) - \sigma(K^-n), \\
\sigma(\Lambda p) - \sigma(pp) &= \sigma(K^-n) - \sigma(\pi^+p),
\end{aligned} \tag{11.92}$$

What is remarkable about these rules is that they relate meson–baryon to baryon–baryon cross sections, and thus cannot be obtained from considerations of symmetry (unless we consider a group in which mesons and baryons belong in the same multiplet). Other sum rules involving only meson–baryon cross sections are derivable from the model, but they can be obtained from $SU(6)$-invariance without quarks. The agreement with experiment at high energy $E \gtrsim 10$ GeV is rather good. See Lipkin and Scheck (1966) for further details. If, in the limit of very high energy, all quark–quark and quark–antiquark cross sections are equal, then we obtain a very simple relation between the baryon–baryon and meson–baryon total cross sections:

$$\sigma(BB) = \tfrac{3}{2}\sigma(MB). \tag{11.93}$$

These ideas have been extended by many authors. For example, Franco (1967) considered the corrections that would arise from double scattering of quarks in hadrons, using the theory of Glauber (1959). Applying double-scattering corrections, De Souza *et al.* (1968) obtained sum rules which are quadratic and cubic rather than linear in the hadron–hadron cross sections. We do not have the space to discuss other ramifications of these ideas.

11.9 Troubles with the Quark Model

We are not concerned here with the question of the existence of real quarks, but with the difficulties of interpretation of the quark model.

First, we consider the mass of a quark as estimated from the properties of hadrons. The fact that the Gell-Mann–Okubo mass formula holds quite well for baryons seems to indicate that perturbation theory is a good approximation. But perturbation theory is expected to be good only if the $SU(3)$-invariant forces are much larger than the symmetry-breaking forces. Assuming that the quark mass arises chiefly from an $SU(3)$-invariant interaction, we are led to the conclusion that a quark is very massive, i.e., that

$$m \gg M_p, \tag{11.94}$$

where m is the mass of the quark and M_p is the mass of the proton. It has been conjectured, for example, that the mass of a quark is $\gtrsim 10$ GeV. With such a mass, the $SU(3)$ symmetry-breaking interaction, which is of the order of several hundred MeV, would be small compared to the $SU(3)$-invariant interaction.

But the assumption that quarks are very heavy conflicts with all estimates of the quark mass based on the known properties of hadrons. We shall discuss two of these estimates here.

We have seen that if $SU(6)$ invariance holds, then the constant μ_0 which measures the magnetic moment of a quark is just the magnetic moment of a proton. [See Eq. (11.46) with $g_2 = 0$.] If a quark has no anomalous magnetic moment but only a Dirac moment, then the constant μ_0 is a direct measure of the mass of a quark. The expression is $\mu_0 = e/(2m)$ (as usual, $\hbar = c = 1$), where e is the charge of the proton. But the proton magnetic moment is observed to be $\mu_p = 2.79e/(2M_p)$. Then, since $\mu_0 = \mu_p$, we have $e/(2m) = 2.79e/(2M_p)$, or

$$m = M_p/2.79. \tag{11.95}$$

Thus, according to this argument, a quark is a relatively light object, of mass about $\frac{1}{3}$ the mass of the proton.

A second estimate comes from high energy hadron–hadron collisions. The predictions of Lipkin and Scheck discussed in the previous section are based on the assumption that quarks inside hadrons scatter as free particles. If this is so, it is plausible that the binding energy of quarks in hadrons is small. Thus, the mass of a quark ought to be a little more than $\frac{1}{3}$ the mass of a baryon or $\frac{1}{2}$ the mass of a meson. With either estimate, a quark is light.

Thus, according to some estimates, quarks are very heavy, while according to other estimates, they are light. It is beyond the scope of this work to discuss in detail the many proposals for a way out of this dilemma. We shall merely remark that the proposals fall into two main categories: (1) Quarks are heavy ($m \gg M_p$), but when bound act like quasi particles with small effective masses. (2) Quarks are light ($m \approx \frac{1}{3}M_p$), but have special properties or interactions which make it very difficult (or impossible) to knock them out of baryons and mesons.

A second difficulty with quarks is the problem of saturation of forces. It is not clear why we observe bound states of three quarks (baryons), but not bound states of two, four, or five quarks. Likewise, it is mysterious why bound states of a quark and antiquark (mesons) are observed, but not bound states of two quarks and an antiquark. Again, we can only speculate about the answers. One possibility is that repulsive many-body quark interactions exist. This assumption is not satisfactory for two reasons: (1) It is of an *ad hoc* character. (2) Predictions which agree rather well with experiment follow from the assumption that quarks have principally two-body interactions.

Another difficulty with quarks is that although they have half-integral spin, their wave functions appear to be symmetric when bound in baryons. Again we speculate on possible solutions to this problem. One possibility is that quarks are fermions, but have such complicated interactions that the ground state of a three-quark system has an antisymmetric spatial wave function. If this is so, the entire picture of harmonic oscillator potentials is wrong. Indeed, any ordinary attractive potential between quarks leads to a prediction that the ground state is spatially symmetric. However, if the quark–quark potential is, for example, more attractive in $L = 1$ states than in $L = 0$ states, the situation could be reversed.

A more intriguing possibility has been suggested by Greenberg (1964). According to Greenberg, quarks are not fermions, but obey parastatistics; specifically they are parafermions of order three. The wave functions of identical particles which obey parastatistics are basis functions of multidimensional representations of the symmetric group S_n. In particular, identical parafermions of order three satisfy the rule that at most three of them can be put in a given state. Thus three paraquarks could be bound in a symmetric state, but not a fourth. This assumption does not by itself explain why antisymmetric combinations of three quarks (or combinations with mixed symmetry) are not also observed. It must be postulated that the forces are such that the symmetric states be lowest in energy. This *symmetric quark model* has been used in most of the nonrelativistic treatments of quarks bound in hadrons. See, for example, Greenberg (1964), Dalitz (1967), and Morpurgo (1968).

With this additional assumption that symmetric states lie lowest, one can account for the fact that bound states of four quarks are not observed as low-energy hadrons. However, the assumption does not prevent the deep binding of two quarks and an antiquark.

As a last difficulty with quarks, we mention the aesthetic objection that it is unattractive to introduce particles with fractional charge, hypercharge, and baryon number. However, we do not know of any difficulty in principle with fractional values for these quantum numbers.

CHAPTER 12

VARIANTS OF THE QUARK MODEL

12.1 Examples of Models

One of the pitfalls in describing physical phenomena using a number of different inequivalent models is that at most one of these models can be right. Thus, from the point of view of physics, a discussion of different models has an ephemeral quality about it. Nevertheless, we shall treat several models in this chapter in order to point out some of their common aspects and some of their differences. In Section 12.4, we shall amplify our reasons for discussing various models.

Once we allow the assumed fundamental constituents of hadrons to have extra degrees of freedom, or allow more than three types of constituents, many different models become possible. However, not all of these give as many predictions as the usual quark model. Discussions of variants of the quark model have been given by many authors, for example, by Gürsey *et al.* (1964) and by Han and Nambu (1965).

A model which is similar to Greenberg's model of parafermion quarks is one in which there exist nine fermion quarks, three of each type. We distinguish the three different quarks of a given type by the subscripts r, w, and b (for red, white, and blue). The quarks $q_{1\mathrm{r}}$, $q_{1\mathrm{w}}$, and $q_{1\mathrm{b}}$, for example, have $Q = \frac{2}{3}$, $Y = \frac{1}{3}$, $I = I_3 = \frac{1}{2}$, $B = \frac{1}{3}$. However, if these three quarks are distinguishable, the Δ^{++} baryon could contain one q_{1r}, one q_{1w}, and one q_{1b} all in symmetric states. A fourth quark would have to go into a state which is antisymmetric with respect one of the quarks. Then, if the interaction is attractive in symmetric states but repulsive in antisymmetric states, the fourth quark would not be bound. Thus, in this rwb model, we introduce a new discrete degree of freedom which takes on three values. If this degree of freedom has no consequences other than to distinguish the quarks in the Pauli sense, then this model is equivalent to the parastatistics model. However, there might exist a relatively weak interaction which depends on this degree of freedom to give a dynamical meaning to the distinguishability of the different quarks of the same type. At present, there seems no need to consider such distinctions.

Let us next consider a model with only one additional particle. This gives us enough freedom to enable us to give the constituents integral charge, hypercharge, and baryon number. To be specific, we consider a model with an $SU(3)$ triplet u_i $(i = 1, 2, 3)$ of spin $\frac{1}{2}$ and an $SU(3)$ singlet u_0 of spin 0. The quantum numbers of the triplet and singlet are given in Table 12.1.

TABLE 12.1

QUANTUM NUMBERS OF AN $SU(3)$ SINGLET u_0 AND TRIPLET u_i $(i=1, 2, 3)$

Symbol	$SU(3)$ multiplicity	Hypercharge Y	Baryon number B	Charge Q	Spin S	Isospin I	Isospin I_3	Mass	Magnetic moment
u_0	1	-2	1	-1	0	0	0	m_0	0
u_1	3	1	0	1	$\frac{1}{2}$	$\frac{1}{2}$	$\frac{1}{2}$	m	α_1
u_2	3	1	0	0	$\frac{1}{2}$	$\frac{1}{2}$	$-\frac{1}{2}$	$m+\varepsilon$	α_2
u_3	3	0	0	0	$\frac{1}{2}$	0	0	$m+\delta$	α_3

We want the additive quantum numbers of the triplet and singlet to fit on the usual $SU(3)$ weight diagrams. Therefore, we let the weight m_2 of any $SU(3)$ multiplet in this model be given by

$$m_2 = \tfrac{1}{2}\sqrt{3}\, Y + b, \tag{12.1}$$

where b is a new quantum number which depends on the multiplet. For the triplet, $b = -1/\sqrt{3}$, while for the singlet $b = \sqrt{3}$. Quarks have $b = 0$, so that in the usual quark model this quantum number plays no role.

In this triplet–singlet model, a baryon is composed of four particles, three triplets and a singlet. From the three triplets we construct baryon octet and decuplet wave functions just as in Tables 11.2 and 11.3. With such wave functions, the baryons have all the correct $SU(3)$ properties. The addition of the singlet u_0 gives the baryons the correct values of charge, hypercharge, and baryon number. Mesons are constructed from a triplet and an anti-triplet, just as in the quark model and the Sakata model.

With our construction, the baryons and mesons all have $b = 0$. We then add the dynamical principle that the interactions of triplets and singlets are such that the only states of low mass are those with $b = 0$. This principle is analogous to the usual quark-model hypothesis that only states with zero triality have low mass. Our dynamical principle, just as the quark model hypothesis, is made on an *ad hoc* basis to avoid the appearance of states which are not experimentally observed. The principal advantage of the triplet–singlet model is that it may appeal to those who, on aesthetic grounds, are opposed to particles with fractional charge, hypercharge, and baryon number. In nearly all other respects, the model either shares the defects of the usual quark model or has additional defects of its own. As an example of a shared defect, since the triplet has spin $\frac{1}{2}$, it is expected to be a fermion, but the simplest baryon states are still constructed from symmetric wave functions.

There are additional differences from the quark model. For example, let us calculate the baryon magnetic moments in the triplet–singlet model, using baryon wave functions analogous to those of Eqs. (11.48). The baryon wave functions will differ from those in Eqs. (11.48) in two respects: (1) Each q_i of Eq. (11.48) is replaced by a corresponding u_i of the triplet. (2) Each wave function contains a singlet wave function u_0 as a multiplicative factor. However, the singlet does not have a magnetic moment, since it has spin zero. Then, since all constituent particles are assumed to be in states with orbital angular momentum zero, the magnetic moment operator $\boldsymbol{\mu}$ of a baryon will be the sum of the moments of the constituent triplets. Thus we have

$$\mu_3 = \sum_{i=1}^{3} \alpha(i)\sigma_3(i), \tag{12.2}$$

where $\alpha(i)$ is the magnetic moment of the ith triplet. The eigenvalue of $\alpha(i)$ with respect to a triplet of type j is given by α_j, i.e.,

$$\alpha(i)q_j = \alpha_j q_j .$$

We obtain the magnetic moments of the eight baryons in terms of the magnetic moments α_1, α_2, and α_3 of the three triplets. The result is

$$\begin{aligned}
\mu_p &= \tfrac{1}{3}(4\alpha_1 - \alpha_2), & \mu_n &= \tfrac{1}{3}(4\alpha_2 - \alpha_1), \\
\mu_\Lambda &= \alpha_3 , & \mu_{\Sigma^0} &= \tfrac{1}{3}(2\alpha_1 + 2\alpha_2 - \alpha_3), \\
\mu_{\Sigma^+} &= \tfrac{1}{3}(4\alpha_1 - \alpha_3), & \mu_{\Sigma^-} &= \tfrac{1}{3}(4\alpha_2 - \alpha_3), \\
\mu_{\Xi^0} &= \tfrac{1}{3}(4\alpha_3 - \alpha_1), & \mu_{\Xi^-} &= \tfrac{1}{3}(4\alpha_3 - \alpha_2), \\
\mu_{\Omega^-} &= 3\alpha_3 .
\end{aligned} \tag{12.3}$$

We can use the experimental values of the proton, neutron, and Λ magnetic moments to solve for the values of α_1, α_2, and α_3. We obtain [in units of $e/(2M_p)$]

$$\alpha_1 = 1.85, \qquad \alpha_2 = -0.97, \qquad \alpha_3 = -0.73 \pm 0.16. \tag{12.4}$$

But the triplet charges are $Q_1 = 1$, $Q_2 = Q_3 = 0$. Therefore, to obtain agreement with experiment the triplets must have anomalous magnetic moments which are not proportional to their charges. This means that we need three independent moment parameters rather than one as in the quark model. In fact, the results of Eqs. (12.3) are identical to those of Franklin (1968), who considered a quark model with anomalous magnetic moments.

From Eqs. (12.3) we obtain the following sum rules:

$$\begin{aligned} 15\mu_{\Sigma^+} &= 16\mu_p + 4\mu_n - 5\mu_\Lambda, \\ \mu_{\Sigma^+} + \mu_{\Sigma^-} &= 2\mu_{\Sigma^0}, \\ 15\mu_{\Xi^-} &= 20\mu_\Lambda - \mu_p - 4\mu_n, \\ 3(\mu_{\Xi^0} + \mu_{\Xi^-}) &= 8\mu_\Lambda - \mu_p - \mu_n, \\ \mu_{\Omega^-} &= 3\mu_\Lambda. \end{aligned} \tag{12.5}$$

In the usual quark model with magnetic moments proportional to their charges, we obtain these sum rules plus two others which are

$$2\mu_p = -3\mu_n, \qquad \mu_n = 2\mu_\Lambda.$$

As we have remarked previously, the first of these is within 3% of experiment, while the second disagrees by 1.5 standard deviations.

We shall just mention two other differences between the triplet–singlet model and the quark model. The first is that in the triplet–singlet model there will, in general, be more baryon excited states than in the quark model because four-particle states have more energy levels than three-particle states unless special restrictions are placed on the forces. A second difference is that baryon–baryon high-energy cross sections cannot be related to meson-baryon cross sections because of the extra singlet in the baryon.

Another model which is of some interest is the *SUB* model of Cabibbo *et al.* (1967), which is based on a model of Han and Nambu (1965). In this model, there are three sets of fundamental triplets of fermions, denoted by S_i, U_i, and B_i, for a total of nine particles, just as in the rwb model. However, there are two related differences: First, in the *SUB* model all particles have integral, rather than fractional charge; second, the values of Q, I_3, and Y of the *SUB* particles are not related by the Gell-Mann–Nishijima formula

$$Q = I_3 + \tfrac{1}{2}Y.$$

Instead, these quantum numbers satisfy a formula

$$Q = I_3 + \tfrac{1}{2}Y + \tfrac{1}{3}C, \tag{12.6}$$

where the additional quantum number C is known as the charm. The necessity of a new quantum number is the price paid for the requirement that the fundamental states have integral charge, as we already saw in the triplet–singlet model. In the *SUB* model, baryons and mesons have zero charm. We list in Table 12.2 the quantum numbers of the *SUB* particles.

TABLE 12.2

QUANTUM NUMBERS OF THE NINE STATES OF THE *SUB* MODEL[a]

Symbol	Charge Q	Isospin I	Isospin I_3	Hyper-charge Y	Charm C	Magnetic moment
S_1, U_1	1	$\frac{1}{2}$	$\frac{1}{2}$	$\frac{1}{3}$	1	μ_0
S_2, U_2	0	$\frac{1}{2}$	$-\frac{1}{2}$	$\frac{1}{3}$	1	0
S_3, U_3	0	0	0	$-\frac{2}{3}$	1	0
B_1	0	$\frac{1}{2}$	$\frac{1}{2}$	$\frac{1}{3}$	-2	0
B_2	-1	$\frac{1}{2}$	$-\frac{1}{2}$	$\frac{1}{3}$	-2	$-\mu_0$
B_3	-1	0	0	$-\frac{2}{3}$	-2	$-\mu_0$

[a] All states are assumed to have spin $\frac{1}{2}$ and baryon number $\frac{1}{3}$.

We see from this table that the S and U particles have identical quantum numbers. The reason they are taken to be distinguishable particles is so that symmetric baryon wave functions are allowed. Note that the magnetic moments of the *SUB* particles are proportional to their charges, as with quarks.

If we compute the baryon magnetic moments, we get the same answers as in the quark model with analogous wave functions. We can understand this as follows: We see from Table 12.2 that the *average* charge of S_i, U_i, and B_i is the same as the charge of the quark q_i. Then, since the baryon magnetic moment operator is linear in the charges of the constituent particles, magnetic moment of a baryon will depend only on the average charge of the constituents.

In fact, many of the properties of baryons and mesons which we have discussed are the same in this model as in the quark model. One possible way to distinguish between the models is that in the *SUB* model we might expect at very high energy to see baryons with charm different from zero, arising, for example, from antisymmetric states of *SSS*. Another difference is that Cabibbo *et al.* (1967) have calculated the radiative corrections to β

decay and found them to be finite in the *SUB* model but infinite in the usual quark model. However, we do not regard this as compelling evidence against the quark model, since the results depend on the assumption that quarks are point particles.

In the previous chapter, we assumed, for simplicity, that quarks had spin $\frac{1}{2}$. However, we found that with this assumption we were led to consider quarks with additional degrees of freedom in order to have a model with fermion quarks in *S* states. As an alternative to assigning additional internal degrees of freedom to quarks, we can assume that they have higher spin. If the quarks have spin $\frac{3}{2}$, for example, we can construct baryon octet and decuplet wave functions which are antisymmetric in the spin and unitary spin degrees of freedom. Franklin (1969) has considered models in which quarks have spin $\frac{3}{2}$, $\frac{5}{2}$, ... and found as good agreement with experiment as with quarks of spin $\frac{1}{2}$. The quarks must have anomalous magnetic moments, however, in order for such models to give the correct values of the proton and neutron magnetic moments.

12.2 Two-Particle Model of Baryons

In the previous section we discussed a small sample of a large number of models which are quark-like in spirit but differ in details. Most of these models have one feature in common: A baryon is a composite of three (or more) fundamental particles. Thus, in order to obtain properties of baryons in these models, we must solve (at best) a three-body problem.

In order to simplify calculations of the properties of baryons, we now consider a model in which a baryon is a composite of two particles, a boson and a fermion, rather than of three fermions (Lichtenberg and Tassie, 1967). One possibility is to assign the fermion the quantum numbers of a quark and the boson the quantum numbers of a bound state of two quarks or diquark. The possibility that a diquark might be a stable particle was already considered by Gell-Mann (1964). In our model, however, we can regard the boson and fermion as different elementary particles, so that a baryon can be a bound state of the two particles with orbital angular momentum zero. A meson is considered to be a bound state of a quark and antiquark as in the usual quark model. Alternatively, we can consider a model with integral charge. For example, we can assign to the fermion the quantum numbers of the *B* particle of Table 12.2 and assign to the boson the quantum numbers of a bound state of an *S* and a *U* particle.

One use of the two-particle model is that dynamical calculations are simple enough to enable us to evaluate many properties of hadrons in approximations that go beyond the lowest-order perturbation theory. Thus, it is relatively easy, for example, to obtain corrections to the Gell-Mann–Okubo mass formula. We shall not describe any of these calculations, but instead shall discuss some of the essential features of the model.

To be specific, let us consider a version of a quark–diquark model which incorporates $SU(6)$ invariance (Lichtenberg *et al.*, 1968). We let the quark as usual belong to a six-dimensional representation of $SU(6)$ and the diquark belong to a 21-dimensional representation. The **21** contains the following $SU(3) \times SU(2)$ multiplets:

$$\mathbf{21} \supset {}^{3}6 + {}^{1}\bar{3}. \tag{12.7}$$

Thus, the diquark consists of an $SU(3)$ sextet s_i of spin one and an $SU(3)$ triplet t_i of spin zero. The quantum numbers of the diquark are given in Table 12.3.

TABLE 12.3

QUANTUM NUMBERS OF DIQUARK CONSISTING OF AN $SU(3)$ SEXTET s_i OF SPIN 1 AND AN $SU(3)$ TRIPLET t_i OF SPIN 0 [a]

Symbol	Charge Q	Hypercharge Y	Isospin I	I_3
s_1	$\frac{4}{3}$	$\frac{2}{3}$	1	1
s_2	$\frac{1}{3}$	$\frac{2}{3}$	1	0
s_3	$-\frac{2}{3}$	$\frac{2}{3}$	1	-1
s_4	$\frac{1}{3}$	$-\frac{1}{3}$	$\frac{1}{2}$	$\frac{1}{2}$
s_5	$-\frac{2}{3}$	$-\frac{1}{3}$	$\frac{1}{2}$	$-\frac{1}{2}$
s_6	$-\frac{2}{3}$	$-\frac{4}{3}$	0	0
t_1	$\frac{1}{3}$	$\frac{2}{3}$	0	0
t_2	$\frac{1}{3}$	$-\frac{1}{3}$	$\frac{1}{2}$	$\frac{1}{2}$
t_3	$-\frac{2}{3}$	$-\frac{1}{3}$	$\frac{1}{2}$	$-\frac{1}{2}$

[a] The baryon number is $\frac{2}{3}$.

We construct the baryon states from a **21** (diquark) and **6** (quark) as follows:

$$\mathbf{21} \otimes \mathbf{6} = \mathbf{56} \oplus \mathbf{70}. \tag{12.8}$$

We can assign the baryon octet and decuplet to the **56** as in the usual quark model. The $SU(6)$ baryon wave functions can be constructed according to the methods of Chapter 8. Then we can obtain sum rules for the masses and for the magnetic moments. See Carroll *et al.* (1968) for details.

The quark–diquark model and the three-quark model give quite different predictions for the baryon excited states classified according to the representations of $SU(6) \times O(3)$. For example, according to the quark–diquark model, baryons should belong either to the **56** or **70** multiplets of $SU(6)$, whereas in the quark model **20** multiplets should exist as well. [Compare Eqs. (11.77) and (12.8).] Also, in a quark–diquark model, the orbital angular momentum L and parity P are related by $P = (-1)^L$, whereas in a three-quark model $P = (-1)^{l+l'}$, where l and l' are two internal angular momenta. Thus, in the three-quark model one cannot predict the parity from the total orbital angular momentum. However, Mitra (1967) has proposed a three-quark model with special restrictions on the forces. It then turns out that spin and parity are related by $P = (-1)^L$, where $L = l + l'$, and also that baryons must belong to 56- or 70-dimensional multiplets.

In the three-quark model, the forces are assumed to have the property that only states which are symmetric under quark interchange have reasonably low mass. In the quark–diquark model, on the other hand, no such symmetry principle exists, and unwanted states must be eliminated by more detailed assumptions about the forces, for example, by the assumption that quark–diquark exchange forces exist. See Lichtenberg (1969) for details.

12.3 Dyon Model

Schwinger (1969) has proposed a model which has certain features of the quark model and other features of the three-triplet model of Han and Nambu (1965). In Schwinger's model, like that of Han and Nambu, there are three sets of triplets, for a total of nine particles. These particles, like quarks, are fractionally charged. Furthermore, not only do they have electric charge, but also magnetic charge (i.e., they are magnetic monopoles). Since the particles carry two types of charge, Schwinger has named them dyons.

If magnetic charge exists, then, as shown by Dirac (1931, 1948), one has a natural explanation for the quantization of electric charge. Schwinger requires the electric charge e and the magnetic charge g_0 to satisfy quantization condition

$$eg_0/\hbar c = 2n, \tag{12.9}$$

where n is an integer. (This quantization condition differs by a factor four from the one originally proposed by Dirac: $eg_0/\hbar c = \frac{1}{2}n$.) Since the magnitude of the unit of electric charge is given by

$$e^2/\hbar c \simeq 1/137, \tag{12.10}$$

it follows from Schwinger's quantization condition that

$$g_0{}^2/\hbar c \simeq 4(137). \tag{12.11}$$

Furthermore, since dyons are fractionally charged with charges

$$2e_0, \qquad -e_0, \qquad -e_0, \tag{12.12}$$

where $e_0 = \frac{1}{3}e$, it follows that the unit of magnetic charge g is given by

$$g^2/\hbar c \simeq 16(137), \tag{12.13}$$

or $g = 3g_0$. Schwinger further argues that dually charged particles may have fractional magnetic charge, and that dyons have magnetic charges

$$2g_0, \qquad -g_0, \qquad -g_0. \tag{12.14}$$

From Eq. (12.11) we see that the magnetic charge of a dyon is very large. Also, from Maxwell's equations we expect that at large distances the force between two dyons goes as $1/r^2$. Thus, dyons interact via long range, very strong forces. Ordinary baryons and mesons are considered as magnetically neutral bound states of dyons.

This model can be incorporated into a scheme of broken $SU(3) \times SU(3)$. The first $SU(3)$ is the usual one we have discussed in this book, with the weight (I_3, Y) related to the electric charge by the formula $Q = I_3 + \frac{1}{2}Y$. The second $SU(3)$ refers to the magnetic charge variables. For further discussion, see Han and Biedenharn (1970).

The dyon model has a number of attractive features:

(1) Since dyons have magnetic charge, they can account in a natural way for the quantization of electric charge.

(2) Although dyons are fermions, nevertheless a baryon can be composed of three dyons in a symmetric state with respect to the usual $SU(3)$, spin, and spatial variables. This is because the dyons can be in an antisymmetric state with respect to magnetic variables.

(3) Since dyons have fractional electric charge, the charm quantum number, which does not have any apparent physical significance, is not needed.

(4) Since the magnetic forces are superstrong, a natural mechanism is provided to bind dyons tightly in baryons and mesons.

(5) In a model with both electric and magnetic charges, CP symmetry can be broken. In fact, the problem is how to keep the CP violation small.

12.4 Usefulness of the Various Models

Although we can construct many different possible models in which hadrons are composed of fundamental constituents, many of the predictions we obtain are the same for all. Furthermore, many of the models are flexible enough so that predictions can be changed by changing the details of the largely unknown interactions. Due to this fact, the present experiments do not lead us uniquely to one model rather than another. Therefore, we should use any convenient model as a tool to obtain results in a simple way. For most purposes, the most convenient model is the usual quark model with symmetric baryon wave functions.

Why, then, have we introduced the other models? One reason is to try to overcome some of the conceptual difficulties of the quark model connected with statistics. Another reason is to illustrate that at the present time we cannot really distinguish between models. A third reason is to point out that there may exist new quantum numbers such as charm, or that magnetic monopoles might exist. A fourth reason is to help bring out what features of the models depend on their $SU(3)$ structure and what features depend on dynamics.

The two-particle model of baryons has an additional nice feature: in order to use it, we must construct boson–fermion wave functions. Thus, we have an opportunity to evaluate some new $SU(3)$ and $SU(6)$ Clebsch–Gordan coefficients. These are given in the paper of Lichtenberg *et al.* (1968). However, we hope the reader does not look up the coefficients, but is sufficiently interested in the methods described in this book to evaluate them himself.

REFERENCES

ABRAGAM, A. (1961). "Principles of Nuclear Magnetism." Oxford Univ. Press (Clarendon), London and New York.

ADLER, S. L., AND DASHEN, R. F. (1968). "Current Algebras." Benjamin, New York.

AHARONOV, Y., AND SUSSKIND, L. (1967). *Phys. Rev.* **155**, 1428.

BAIRD, G. E., AND BIEDENHARN, L. C. (1964). "Proc. 1st Coral Gables Conf. on Symmetry Principles at High Energy," p. 58. Freeman, San Francisco.

BARGMANN, V., (1964). *J. Math. Phys.* **5**, 682.

BÉG, M. A. B., LEE, B. W., AND PAIS, A. (1964). *Phys. Rev. Letters* **13**, 514.

BEHRENDS, R. E., DREITEIN, J., FRONSDAL, C., AND LEE, B. W. (1962). *Rev. Mod. Phys.* **34**, 1.

BERNSTEIN, J., FEINBERG, G., AND LEE, T. D. (1965). *Phys. Rev.* **139**, B1650.

BERNSTEIN, J. (1968). "Elementary Particles and Their Currents." Freeman, San Francisco.

CABIBBO, N. (1963). *Phys. Rev. Letters* **10**, 531 (reprinted in Gell-Mann and Ne'eman, 1964).

CABIBBO, N., MAIANI, L., AND PREPARATA, G. (1967). *Phys. Letters* **25B**, 132.

CARROLL, J., LICHTENBERG, D. B., AND FRANKLIN, J. (1968). *Phys. Rev.* **174**, 1681.

CARTAN, E. (1933). "Sur la Structure des Groupes de Transformations Finis et Continue." Thèse, Paris, 1894; 2nd ed., 1933. Vuibert, Paris.

CASELLA, R. C. (1969). *Phys. Rev. Letters* **22**, 554.

CHEW, G. F., GELL-MANN, M., AND ROSENFELD, A. H. (1964). *Sci. Am.* **210**, No. 2, 74.

CHRISTENSON, J., CRONIN, J., FITCH, V., AND TURLAY, R. (1964). *Phys. Rev. Letters* **13**, 138.

COLEMAN, S., AND GLASHOW, S. L. (1961). *Phys. Rev. Letters* **6**, 423 (reprinted in Gell-Mann and Ne'eman, 1964).

CONDON, E. U., AND SHORTLEY, G. H. (1935). "The Theory of Atomic Spectra." Cambridge Univ. Press, London and New York.

DALITZ, R. H. (1967). "Proc. 13th Intern. Conf. on High Energy Phys." Univ. of California Press, Berkeley.

DALITZ, R. H. (1968). *In* "Meson Spetroscopy" (C. Baltay and A. Rosenfeld, eds.), p. 497. Benjamin, New York.

DALITZ, R. H. (1969). *In* "Pion Nucleon Scattering" (G. Shaw and D. Wong, eds.). Wiley, New York.

DE SOUZA, P. D., AND LICHTENBERG, D. B. (1967). *Phys. Rev.* **161**, 1513.

DE SOUZA, P. D., HEINZ, R. M., AND LICHTENBERG, D. B. (1968). *Phys. Rev.* **169**, 1185.

DE SWART, J. J. (1963). *Rev. Mod. Phys.* **35**, 916 (reprinted in Gell-Mann and Ne'eman,1964).

DIRAC, P. A. M. (1931). *Proc. Roy. Soc. (London) Ser. A* **133**, 60.

DIRAC, P. A. M. (1948). *Phys. Rev.* **74**, 817.

EDMONDS, A. (1957. "Angular Momentum in Quantum Mechanics." Princeton Univ. Press, Princeton, New Jersey.

EDMONDS, A. (1962). *Proc. Roy. Soc. (London) Ser. A* **268**, 567.

Faiman, D. and Hendry, A. W. (1969). *Phys. Rev.* **180**, 1572, 1609.
Fermi, E., and Yang, C. N. (1949). *Phys. Rev.* **76**, 1739.
Franco, V. (1967). *Phys. Rev. Letters* **18**, 1159.
Franklin, J. (1968). *Phys. Rev.* **172**, 1807.
Franklin, J. (1969). *Phys. Rev.* **180**, 1583; **181**, 1984.
Gasiorowicz, S. (1966). "Elementary Particle Physics." Wiley, New York.
Gell-Mann, M. (1953). *Phys. Rev.* **92**, 833.
Gell-Mann, M. (1961). Cal. Tech. Rept. CTSL-20 (reprinted in Gell-Mann and Ne'eman, 1964).
Gell-Mann, M. (1962). *Phys. Rev.* **125**, 1067.
Gell-Mann, M. (1964). *Physics* **1**, 63 (reprinted in Gell-Mann and Ne'eman, 1964).
Gell-Mann, M. (1964a). *Phys. Letters* **8**, 214 (reprinted in Gell-Mann and Ne'eman, 1964).
Gell-Mann, M. and Ne'eman, Y. (1964). "The Eightfold Way." Benjamin, New York.
Glashow, S. L., and Rosenfeld, A. H. (1963), *Phys. Rev. Letters* **10**, 192 (reprinted in Gell-Mann and Ne'eman, 1964).
Glauber, R. J. (1959), *In* "Lectures in Theoretical Physics," Vol. 1 (W. E. Brittin *et al.* eds.), p. 315. Interscience, New York.
Goldberg, H., and Ne'eman, Y. (1963). *Nuovo Cimento* **27**, 1.
Gottfried, K. (1966). "Quantum Mechanics," Vol I. Benjamin, New York.
Greenberg, O. W. (1964). *Phys. Rev. Letters* **13**, 598.
Gürsey, F., and Radicati, L. A. (1964). *Phys. Rev. Letters* **13**, 173.
Gürsey, F., Lee, T. D., and Nauenberg, M. (1964). *Phys. Rev.* **135**, B467.
Hamermesh, G., (1962). "Group Theory." Addison-Wesley, Reading, Massachusetts.
Han, M. Y. and Biedenharn, L. C. (1970). *Phys. Rev. Letters* **24**, 118.
Han, M. Y. and Nambu, Y. (1965). *Phys. Rev.* **139**, B1006.
Herglotz, G. (1911). *Ann. Physik* **36**, 493.
Ikeda, M., Ogawa, S., and Ohnuki, Y. (1959), *Progr. Theoret. Phys.* **22**, 715.
Jacobi, C. G. J. (1884). "Vorlesungen Uber Dynamik," Werke, Supplementband. Reimer, Berlin.
Jones, L. W. (1969). "Intern. Conf. Symmetries and Quark Models." Wayne State University, Detroit.
Lee, T. D. and Wu, C. S. (1966). *Ann. Rev. Nucl. Sci.* **16**, 511.
Lichtenberg, D. B. (1965). "Meson and Baryon Spectroscopy." Springer, New York.
Lichtenberg, D. B., and Tassie, L. J. (1967). *Phys. Rev.* **155**, 1601.
Lichtenberg, D. B., Tassie, L. J., and Kelemen, P. J. (1968). *Phys. Rev.* **167**, 1535.
Lichtenberg, D. B. (1969). *Phys. Rev.* **178**, 2197.
Lie, S., and Scheffers, G. (1893). "Vorlesungen Uber Kontinuierliche Gruppen." Teubner, Leipzig.
Lipkin, H. (1965). "Lie Groups for Pedestrians." North-Holland, Amsterdam.
Lipkin, H., and Meshkov, S. (1965), *Phys. Rev. Letters* **14**, 670.
Lipkin, H., and Scheck, F. (1966). *Phys. Rev. Letters* **16**, 71.
McCusker, C. B. A. and Cairns, I. (1969), *Phys. Rev. Letters* **23**, 658.
McGlinn, W. D. (1964). *Phys. Rev. Letters* **12**, 467.
McNamee, P., and Chilton, F. (1964). *Rev. Mod. Phys.* **36**, 1005.
Meshkov, S., Levinson, C. A., and Lipkin, H. J. (1963). *Phys. Rev. Letters* **10**, 631 (reprinted in Gell-Mann and Ne'eman, 1964).
Messiah, A. M. L., and Greenberg, O. W. (1964). *Phys. Rev.* **136** B248.
Morpurgo, G. (1965). *Physics* **2**, 95.
Morpurgo, G. (1968). "14th Intern. Conf. High Energy Physics," Vienna, p. 225. CERN, Geneva.

NE'EMAN, Y. (1961). *Nucl. Phys.* **26** 222 (reprinted in Gell-Mann and Ne'eman 1964).
NE'EMAN, Y. (1964), "Proc. Intern. Conf. Nucleon Structure" (R. Hofstadter and L. I. Schiff, eds.), p. 172. Stanford Univ. Press, (reprinted in Gell-Mann and Ne'eman. 1964).
NE'EMAN, Y. (1967). "Algebraic Theory." Benjamin, New York.
NISHIJIMA, K., AND NAKANO, T. (1953), *Progr. Theoret. Phys.* **10**, 581.
OAKES, R. J. (1963). *Phys. Rev.* **131**, 2239 (reprinted in Gell-Mann and Ne'eman, 1964).
OKUBO, S. (1962). *Progr. Theoret. Phys.* **27**, 949 (reprinted in Gell-Mann and Ne'eman, 1964).
O'RAIFEARTAIGH, L. (1968). *In* "Lectures in Theoretical Physics" (A. O. Barut and W. E. Brittin, eds.), p. 527. Gordan and Breach, New York.
PAIS, A. (1966). *Rev. Mod. Phys.* **38**, 215.
PARTICLE DATA GROUP (1969). *Rev. Mod. Phys.* **41**, 109.
PONTRIAGIN, L. S. (1966). "Topological Groups." Gordon and Breach, New York.
RACAH, G. (1965). "Group Theory and Spectroscopy" (Springer Tracts in Modern Physics, **37**), p. 28. Springer, New York.
ROSE, M. E. (1957). "Elementary Theory of Angular Momentum." Wiley, New York.
RUBINSTEIN, H. R., SCHECK, F., AND SOCOLOW, R. H. (1967). *Phys. Rev.* **154**, 1608.
SAKATA, S. (1956). *Progr. Theoret. Phys.* **16**, 686.
SAKITA, B. (1964). *Phys. Rev.* **136**, B1756.
SCHIFF, L. I. (1968). "Quantum Mechanics." McGraw Hill, New York.
SCHÜTZ, J. R. (1897). *Gött. Nachr.* p. 110.
SCHWINGER, J. (1969). *Science* **165**, 757; **166**, 690.
STREATER, R. F., AND WIGHTMAN, A. S. (1964), "TCP, Spin and Statistics and All That." Benjamin, New York.
VAN DAM H., AND WIGNER, E. P. (1965). *Phys. Rev.* **138**, B1576.
WICK, G. C., WIGHTMAN, A. S., AND WIGNER, E. P. (1952). *Phys. Rev.* **88**, 101.
WIGNER, E. P. (1937). *Phys. Rev.* **51**, 106.
WIGNER, E. P. (1954). *Progr. Theoret. Phys.* **11**, 437.
WIGNER, E. P. (1959). "Group Theory." Academic Press, New York.
WIGNER, E. P. (1964). *Phys. Today* **17**, No. 3, 34.
WU, T. T. AND YANG, C. N. (1964). *Phys. Rev. Letters* **13**, 380.
ZWEIG, G. (1964), CERN preprint 8409/Th. 412, unpublished.

SUBJECT INDEX

A

B

C

D

E

F

G

H

S